Joining Processes

Joining Processes

Introduction to
brazing and diffusion bonding

M. G. Nicholas, BSc, PhD, FIM, C Eng

Some-time Group Leader,
Atomic Energy Research Establishment,
Harwell, UK
and
Visiting Scientist, Institute for Advanced Materials,
CEC Joint Research Centre Petten, The Netherlands

KLUWER ACADEMIC PUBLISHERS
DORDRECHT / BOSTON / LONDON

A C.I.P. Catalogue record for this book is available from the Library of Congress
A catalogue record for this book is available from the British Library

ISBN 0 412 79360 1

Published by Kluwer Academic Publishers,
P.O. Box 17, 3300 AA Dordrecht, The Netherlands.

Sold and distributed in North, Central and South America
by Kluwer Academic Publishers,
101 Philip Drive, Norwell, MA 02061, U.S.A.

In all other countries, sold and distributed
by Kluwer Academic Publishers,
P.O. Box 322, 3300 AH Dordrecht, The Netherlands.

Printed on acid-free paper

Typeset in 10/12 Times by AFS Image Setters Ltd, Glasgow, UK

Printed in Great Britain by TJ International Ltd, Padstow, Cornwall

For
Eleanor Lucy and Daniel James
Nicholas

Contents

Preface

This book provides an introduction to the joining of metals and ceramics by brazing, diffusion bonding and some related hybrid processes. Brazing is a process in which liquid metal flows between the components to be joined and there solidifies to produce a permanent bond, while diffusion bonding is a solid state process in which the components are joined by pressing together their mating surfaces at high temperatures.

The successful use of brazing, diffusion bonding and other joining processes is of crucial importance in countless applications of industrial technology. Joining is one of the most vital enabling technologies of modern industry, but this technology is not an important or even a significantly identifiable part of the formal training of many materials scientists or engineers. Their first encounter with joining problems often occurs after graduation, when it is assumed by employers that the scientist or engineer is fully qualified if not experienced.

This book is intended to substitute in part for the courses that might have been, providing an introduction for those who must for the first time consider the question of the suitability and applicability of brazing and diffusion bonding processes. The book will not make the reader an expert fabricator of joints but it should provide a foundation of knowledge about the processes and an appreciation of the usefulness and limitations of the processes sufficient to pose effective questions to experts when problems arise in their application.

In preparing the book, an attempt has been made to strike a balance between fundamental scientific aspects and consideration of practical problems in accord with the guidelines for qualification as a European welding engineer. It is assumed that readers have a scientific or technical background and knowledge equivalent to that of a final-year undergraduate studying materials science, engineering or other applied subjects. However, no knowledge or experience is assumed of either the science of capillary phenomena that underlie joining technology or of joining technology itself. This information is provided in the main body of the text, but an additional reading list is provided at the ends of some of the chapters and appendices as well as sources of specific data cited within the text when this is appropriate.

Acknowledgements

The errors of omission and commission in this book are mine and mine alone, but nevertheless the writing of the book has been greatly assisted both indirectly and directly by many colleagues within the technical community. The useful interactions with fellow scientists at the Harwell and Petten laboratories are too numerous to list, but it is a great pleasure to acknowledge the insight gained from working with Drs Julie Ambrose, Clive Knights, the late Barry McGurran, and Denise Mortimer at the Harwell Laboratory and with Drs Giacomo Ceccone and Stathis Peteves at the JRC Petten.

The work load involved in the actual writing of the book was eased by the kindness of Giacomo Ceccone and Stathis Peteves in reading draft versions. Similarly, the presentation of the book has been aided by the generosity of those who allowed access to their albums of photographs. Specifically, thanks are due to Julie Ambrose for Figures 7.1, 7.7 and 7.8, to Dr William Gale of the Materials Research and Education Center of Auburn University, Alabama, for Figure 7.10, to Dr Richard Grugel of the Universities Space Research Association, Huntsville, Alabama, for Figure 7.6, and to Dr Stathis Peteves for Figures 6.4, 7.2, 7.4, 7.5, 7.9 and 7.12.

Heartfelt thanks are due also to my wife, Judy, for her tolerance while I shirked many domestic tasks and absented myself from social occasions to keep writing. Exceptions to this industrious regime were made, however, to delight in the arrival of our two grandchildren.

List of symbols

A	Constant in equation {3.47}
A^*	Constant in equation {3.53}
A'	Constant in equation {7.3}
A_B	Half-width of the contact area between asperities
A_F	Fractional area
A_U	Proportion of joint area that is unbonded
B	The breadth of a plate forming one wall of a capillary gap
BS	Bend strength
B_J	The strength of a joint
B_0	Constant in equation {7.2}
C	Constant in equation {3.33}
D	Coefficient of diffusion
D'	Component diameter
D_{GB}	Coefficient of grain boundary diffusion
D_V	Coefficient of volume diffusion
D_S	Coefficient of surface diffusion
D_{vi}	Diagonal length of a Vickers hardness indentation
E	Modulus of elasticity, Young's modulus
E_H	Total enthalpy of a liquid
F_V	Force applied when making a Vickers hardness indentation
F_L	Bend test failure load
G	Constant in equation {6.6}
H	Constant in equation {6.6}
H_S	Constant in equation {3.34}
J	Mass flux
K	Equilibrium constant
K^*	Constant related to joint design
K_{Ic}	Plane strain fracture toughness
L	Joint length
L_G	Length of gauge section of tensile test piece
L_S	Length between shoulders of a tensile test piece

L_T Total length of a tensile test piece
M Molecular weight
N_A Atom fraction of constituent A
P Pressure applied to a liquid column
P_{eff} The effective pressure
P_{ext} An externally applied pressure
P_0 Constant in equation {3.49}
P_{vap} Vapour pressure
Q_{GB} Activation energy for grain boundary diffusion
Q_{RPL} Activation energy for the process controlling the growth of a reaction product layer
Q_S Activation energy for surface diffusion
Q_V Activation energy for volume diffusion
R The gas constant
R_S Shoulder radius of a tensile test piece
S Spacing between contacting asperities
T Temperature in degrees Kelvin
T_B Bonding temperature
T_E Eutectic temperature
T_L Liquidus temperature
T_M Melting temperature in degrees Kelvin
T_{opt} Optimum brazing temperature
U Constant in equation {6.7}
V Volume of a sessile drop
W Joint gap between two parallel plates
W_R Ratio of the actual to the nominal area of a rough surface
W^* The breadth of a butt joint
Y Correction constant used in equation {D.4}
a Notch plus crack depth of a bend test piece
a Chemical activity
b Breadth of a bend test piece
d_v Diagonal length of a Vickers microhardness indentation
e Distance between the points of load application and support of a bend test piece.
f Stoichiometry integer
f_k Force applied when making a Knoop microhardness indentation
f_v Force applied when making a Vickers microhardness indentation
g Acceleration due to gravity
h Height of capillary rise
h^* The thickness of a joint
h_v Height of a void
k Boltzmann's constant
m^* Evaporation rate
n Constant in equations {3.33} {3.47} and {3.53}

n^*	Creep power term
p	Penetration of a horizontal capillary
q	Stoichiometry integer
r	Radius of a sessile drop contact area
r_0	Radius of the contact area of a sessile drop at time $t = 0$
r_v	The radius of a void in the interface plane
t	Time
t^*	Component thickness
t_{opt}	Optimum brazing time
w	Width of an interlayer
w'	Thickness of a bend test piece
x	Distance from the centre of an interlayer
$\Delta\alpha$	Difference between two coefficients of thermal expansion
ΔT	Difference between two temperatures
ΔH_f	Enthalpy of fusion
ΔH_g	Enthalpy of evaporation
α^*	Slope of an asperity
α'	Inclination of a capillary gap to the horizontal
β	Activity coefficient
γ_L	Energy of a liquid surface
γ_S	Energy of a solid surface
γ_{SL}	Energy of a solid–liquid interface
δ	Width of an interface or grain boundary
δE	Energy change
ε_{SS}	Strain due to creep
η	Coefficient of viscosity
θ	Contact angle of a liquid front
θ_R	Contact angle on a rough surface
θ_a	Advancing contact angle
θ_r	Receding contact angle
θ_t	Contact angle at time t
θ_0	Contact angle at time $t = 0$
θ_∞	Contact angle after a long time
ρ_L	Density of a liquid
σ_J	Stress needed to deform a joint
σ_R	Residual stress
σ_S	Shear strength
σ_T	Tensile strength
σ_Y	Yield stress
τ	Characteristic time

Introduction

1

1.1 WHY JOIN?

The metal and ceramic materials available to the design engineer exhibit many useful characteristics and can be formed into a wide range of shapes but it is relatively rare for a product to be formed from a single piece of material and many products incorporate different types of materials. Thus while a single piece of metal may be used to form a simple spoon or ladle many more items of tableware or their kitchen cousins are produced using separate materials for the handles to improve their aesthetic appeal or performance. This use of different materials in a single product to achieve an enhanced performance is not new. An early example is provided by the improvement in cutting power produced by changing from hand-held flints, a natural ceramic, to an axe made by binding a flint to a wooden handle, a natural ceramic/natural polymer product. Similarly, the goose quill pen has nowadays been replaced by pens with separate nibs or disposable but complex ball or fibre points

Nowadays, the joining of dissimilar materials to make products best suited to particular applications is very common and this book pays considerable attention to the joining of ceramics to or with metals. The characteristics of these two types of material are quite distinctive with even laymen having little difficulty in deciding whether an object is made of a ceramic or metal. Thus ceramics in comparison to metals usually have low densities, are strong and hard but brittle, are poor electrical and thermal conductors, are refractory and expand little when heated. The data assembled in Table 1.1 provide quantitative values for some of these physical and mechanical characteristics, and will be referred to frequently throughout this book, but additional data for a wider range of both metals and ceramics are presented later in Chapter 4.

It is a common performance requirement that the joints between both similar and dissimilar materials used to manufacture products are both permanent and strong. Examples include the bonding of ceramic tips to

Table 1.1 Physical and mechanical properties of some metals and ceramics

Material	*Melting temp.*, °C	*Denisty*, Mg.m^{-3}	*Thermal expans.*, $\times 10^{-6}$ K^{-1}	*Strength**, MPa	*Elasticity*, GPa	*Thermal cond.* Wm^{-1} K^{-1}	*Elect. resis.* Ωm
Al	660	2.7	23.5	55	71	238	2.6×10^{-8}
Ag	962	10.5	19.1	172	83	128	1.5×10^{-8}
Cu	1083	9.0	17.0	216	130	397	1.6×10^{-8}
Ni	1455	8.9	13.3	310	199	89	6.2×10^{-8}
Fe	1535	7.9	12.1		211	78	9.8×10^{-8}
Ti	1677	4.5	8.9	241	120	22	3.9×10^{-7}
Nb	2467	8.6	7.2	240	105	54	1.3×10^{-7}
Mo	2615	10.2	5.1	435	325	137	5.3×10^{-8}
W	3387	19.3	4.5	550	411	174	5.3×10^{-8}
SiC	2700#	3.2	4.3	141	211	50	1.0×10^{-3}
TiC	3140	4.9	7.2	703	422	36	5.2×10^{-7}
WC	2777	15.8	5.2	598	704	84	
AlN	2197	3.3	4.6	400	310	165	1.0×10^{12}
Si_3N_4	1900#	3.2	3.2	750	280	35	6.0×10^{8}
Al_2C_3	2050	4.0	7.6	390	363	32	1.0×10^{14}
BeO	2530	3.1	7.4	246	401	210	9.1×10^{9}
MgO	2800#	3.6	11.6	281	394	62	1.0×10^{12}
ZrO_2	2690	5.8	6.5	176	140	2	3.3×10^{4}

* Four-point bend test data for ceramics, tensile data for metals.
# Does not melt but sublimes, decomposes or vapourizes.

metal drill shanks, application of gold decoration on glass or china ware, the fixing of copper bottoms to stainless-steel saucepans, the joining of copper piping in central heating systems, attachment of electronic components to circuit boards, the sealing of metal spark-plug electrodes to their ceramic insulators, and the joining of the glass envelopes to the metal bases of incandescent lamps.

Using joining processes as part of a manufacturing route can offer considerable technical and economic advantages to the designer and fabricator, provided careful and informed decisions are made about the processes to be applied, materials to be selected, joint configurations and the process parameters to be employed. Clearly some form of joining is essential when manufacturing structures requiring the use of dissimilar component materials but even single material structures are often produced more economically by joining geometrically simple sub-structures rather than machining or forming from a single piece of the material. This is particularly true when their design involves struts, box sections or internal channels. Similarly, the possibility of using hollow structures, with internal reinforcement where necessary, can help realize the frequent desire to make both single and multi-material products weigh as little as possible, as is evident by the many metal analogues of corrugated cardboard to be found in transport applications.

In principle, a very wide range of processes can be used to join both similar and dissimilar materials, but in practice the number that can be employed for a specific product application is often severely limited by economics as well as technical factors such as the characteristics of the particular materials that have to be joined and the service conditions that must be endured. Processes that form permanent joints either produce bonding directly between the component materials or make use of foreign materials introduced between mating surfaces. Great advantages can be derived by using a joining process but the decisions that have to be made when identifying which is most suitable and optimizing the application parameters are many and varied.

Figure 1.1 identifies five major processes for such permanent bonding that are available to the fabricator, identified as

- fusion welding, in which mating surface regions of components are melted and mixed before solidifying to form a permanent bond;
- diffusion bonding, in which solid mating surfaces are pressed together and heated to cause bonding by interdiffusion of the components;
- brazing, in which liquid metal flows into a narrow gap between the mating surfaces and solidifies to form a permanent bond;
- glazing or glass sealing, which uses a fluid glass to bond mating surfaces in processes analogous to brazing or fusion welding;
- adhesive bonding, in which component gaps are filled by fluid organic compounds that polymerize to form rigid bonding interlayers.

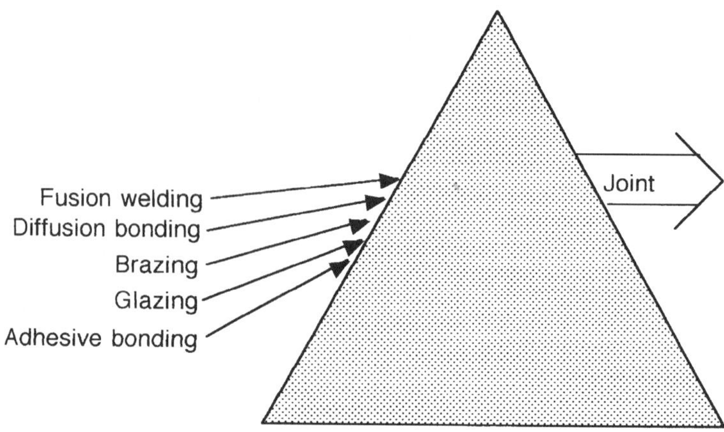

Figure 1.1 A spectrum of some major joining processes

Thorough discussion of this wide range of processes would require an encyclopaedia rather than a slim volume such as this, and therefore an overview is presented in Chapter 2 before the remainder of this present book concentrates on the two processes identified in its title, brazing and diffusion bonding.

Brazing is a process of considerable antiquity, as witnessed by metal artifacts from early Egyptian and Indus valley sites, and the form of diffusion bonding known as forge welding has long been practised by blacksmiths. While acknowledging the validity of an interest in the history of science and technology, this book is concerned primarily with the effects of the recent developments that have changed brazing and diffusion bonding into reliable engineering processes for joining both metals and ceramics. Some attention is paid, however, to hybrid processes such as transient liquid phase bonding.

1.2 WHAT DRIVES JOINING PROCESSES?

Brazing and diffusion bonding processes replace two mating surfaces by one interface and this change will occur readily if it results in a decrease in the energy of the system. The atoms within the bulk of a solid metal and ions in a solid ceramic have lattice configurations and spacings that minimize the total energy of the material. The network of bonding forces exerted by the surrounding neighbours creates an energy well around each atom or ion and imposes simple spatial arrangements such as a cubic or hexagonal structure that continues throughout the crystal. Even in liquid metals, and many liquid ceramics, there is a preference for short range ordering that corresponds to similarly close packed structures.

The interatomic forces acting within the bulk of a metal are nearly equal in all directions, but those at a surface are asymmetric and can be resolved as a net force pulling the atoms into the bulk. For ceramics also, the bonding forces are symmetrical within the bulk but asymmetric at the surface. Hence work must be done against these forces to move an additional atom or ion to enlarge the area of a surface and this work, measured as so many $J.m^{-2}$, is called the surface energy. Similarly, the force needed to stretch the surface so that it can accommodate another atom or ion is called the surface tension and when measured in $N.m^{-1}$ is numerically equal to the surface energy for liquids. Values for the surface energies of ceramics are rare but generally lie within the range of 0.3 to 1.3 $J.m^{-2}$, but surface energy data for metals are abundant and range from about 0.5 $J.m^{-2}$ to almost 3.5 $J.m^{-2}$, with the most refractory metals having the largest values.

Changes in lattice continuity also occur at interfaces and hence create asymmetric bonding forces that result in the existence of interfacial energies. The values for these are low if lattice continuity is high, that is if the lattice structure and orientation on either side of the interface is closely similar, but they can be very high if the lattices are very dissimilar as when a metal is joined to an ionic ceramic. Theoretical calculations of interfacial energies have been done for simple metal–metal systems but those for interfaces between alloys or for metal–ceramic systems are complex and methods are still being developed.

A frequently preferred and generally reliable method for deriving liquid surface energies and liquid–solid interfacial energies is the sessile drop technique, literally the sedentary or sitting drop technique. In this technique, a small drop of liquid is placed on a horizontal solid surface and its profile is viewed or recorded. The drop profile is determined by the interaction of the liquid surface energy, which will try to minimize the surface area by causing the drop to assume the form of a spherical cap, and gravity which will try to flatten the profile. Careful measurements of the deviation of a drop profile from that of a spherical cap enables a value to be derived for the surface energy of the liquid.

Another notable characteristic of such a drop is the contact angle identified as θ in Figure 1.2. If the edge of the drop has advanced over the solid surface the contact angle will be less than 90° as in the figure and the liquid is said to have wetted the solid. Whether the drop wets or does not depends on the relative sizes of the surface and interfacial energies as described by the Young equation

$$\gamma_S = \gamma_L \cos\theta + \gamma_{SL} \qquad \{1.1\}$$

where γ is an energy and the subscripts identify whether it is that of the solid surface, the liquid surface or the solid–liquid interface. The derivation of this and other quantitative predictive equations relevant to brazing

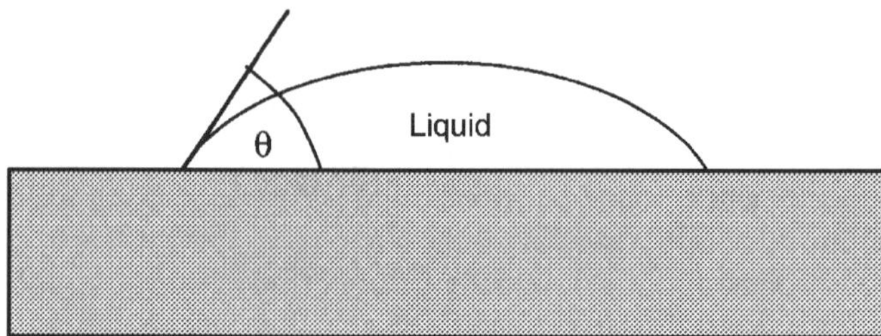

Figure 1.2 Outline of a sessile drop of liquid resting on a horizontal substrate. The contact angle is identified as θ

behaviour is discussed in Chapter 3, but for now we need concern ourselves only with noting that if the contact angle is less than 90°, then $\cos\theta$ is positive and γ_{SL} is smaller than γ_S. Thus simple measurement of an angle provides direct evidence about the relative sizes of two important energies. This angle also indicates whether or not the liquid metal can be used to braze the solid.

For brazing to be successful, the liquid metal must flow into the gap between components. If it does, two surfaces will be replaced by two liquid–solid interfaces and this will occur readily only if it results in an energy decrease, if $2\gamma_{SL} < 2\gamma_S$. The Young equation predicts that this will happen if θ is less than 90° so that $\gamma_L \cos\theta$ is positive. If the materials are different so that we have to consider different surface and interfacial energies for solid A and solid B then the condition for penetration is that $\gamma_{SLA} + \gamma_{SLB} < \gamma_{SA} + \gamma_{SB}$. The actual extent of penetration of gaps between components is greater for narrow gaps, but it should be noted also that impedance of liquid flow due to viscous drag will also increase as the gaps narrow.

Similarly addressing the question of what promotes the creation of an interface between solid mating surfaces during diffusion bonding leads to the conclusion that this will occur if the surfaces are replaced by an interface of lower energy. Thus it would appear that components of the same material will always experience a driving force for interface creation by diffusion bonding since $2\gamma_S$ will be invariably larger than that of any form of interface discontinuity. Only if the interface is between two dissimilar materials, such as a ceramic and a metal, is it possible that its energy could be larger than that of the surfaces it is to replace. However, this favourable energy situation does not necessarily make it easier to create interfaces by diffusion bonding than by brazing because the initial contact between surface asperities has to be converted to bonded interfaces and extended by slow solid state diffusion and creep processes.

1.3 APPLICATION OF JOINING PROCESSES

Safety is the first factor to be taken into consideration when deciding whether or not a joining process should be used and if so how it is to be used. The safety of the operators and the customers must take precedence over technical and economic factors. Most joining processes are not especially hazardous and joints are generally totally innocuous – unless they fail in service – but nevertheless care must be taken to minimize risk levels when selecting joining materials, equipment and process parameters. Some processes, however, involve clear risks and therefore require special care to be taken and sometimes special procedures to be followed. Thus braze alloys containing Cd need to be used with care in well ventilated areas or preferably dedicated booths as do braze fluxes containing, for example, fluorides. Similarly the gaseous and electrical heating sources used to achieve the joining temperatures can be lethally dangerous if used without care. The detailed and precise specification of safety procedures lies outside the scope of this introductory book but some guidance as to what may be involved is given in Appendix A and readers in any doubt should consult their national standards and technical organizations.

Effective application of brazing and diffusion bonding depends on careful selection of materials and fabrication parameters, a task represented in Figure 1.1 as a shaded triangle that can seem almost as mysterious as that off Bermuda. The purpose of this book, therefore, is to make understanding of joining processes less opaque by illuminating the factors that influence the acceptability and technical success of using brazing or diffusion bonding as part of a manufacturing schedule. These factors include economic aspects such as the availability and operational costs of the specialized equipment needed for some of the processes and the price of bonding materials, but in practice the selection of a joining process is often even more constrained by technical factors.

One of the most important technical factors that must be considered is the type of material that has to be joined. When using brazing or diffusion bonding as the joining process, the materials that can be joined are metals or ceramics, or composites or compacts of both types of material. When selecting component and joining materials for a particular application, attention must be given to the effects of their individual and separate characteristics on both the suitability of any proposed joining process as well as the probable ability of the bonded product to withstand the intended service conditions. Thus, bulk properties such as yield strength affect the ease of diffusion bonding, while surface characteristics affect both brazing and diffusion bonding. Similarly, the melting temperatures of brazes and other bonding materials must be reviewed in terms of what is achievable with the bonding equipment, the refractoriness of the component materials and the proposed maximum service temperatures. The detailed chemistry of the joining materials also needs consideration if unwanted diffusion of embrittling species into the components is to be avoided and for certain types of brazing it is also necessary to select a flux which can aid the joining process and provide a protective environment during its operation.

Having selected the materials and the type of joining process to be employed, decisions must be made about joint designs, which are usually examples of the butt, T, lap or sleeve configurations shown in Figure 1.3. Optimization of joint designs is discussed in Chapter 5 which describes how design details are adjusted not only to facilitate brazing and diffusion bonding mechanisms but also to permit easy assembly of the components before they are joined. Thus, many of the preferred designs are made self jigging by the use of burrs or crimped sections and other features that hold the components in place while they are heated to the joining temperature.

Before assembling the preferred joint designs it is necessary to prepare the component mating surfaces if successful brazing or diffusion bonding is to be achieved. Specification of this preparation procedure often involves decisions about the mechanical finish, or roughness, of the mating surfaces. Always decisions must be made as to how to ensure the surfaces are free from chemical blemishes that can impede joining, which can range from paint or rust layers to oil stains, finger prints and other virtually invisible mobile organic contaminants.

Brazing is a high-temperature process and in practice the temperatures used range from a minimum of 450 °C, the boundary between brazing and the related low-temperature process of soldering, to a maximum of about 1300 °C, a limit governed by the melting temperatures of the most refractory commercial brazes. Although there are no limits in principle to the temperatures that can be used for diffusion bonding, those employed in practice so far also normally lie within the range of 450–1300 °C.

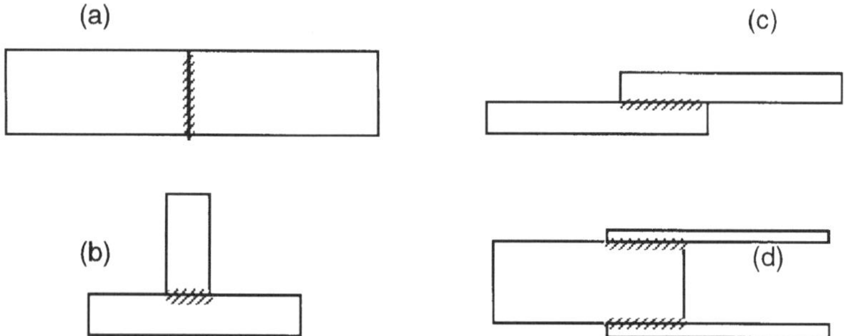

Figure 1.3 Cross-sections of configurations often used when brazing or diffusion bonding: (a) butt, (b) T, (c) lap and (d) sleeve joints. The bonded areas are shaded

Fortunately, this range is employed for many industrial processes so a wide variety of heat sources can be used when brazing or diffusion bonding. These include gas torches, heating elements, induction heaters, infrared lamps, microwave heaters and salt baths, as will be described in Chapters 2 and 6. Which heating method is chosen for a particular application is influenced by factors such as the size and number of components that have to be joined and the environment that is required. Some brazing and diffusion bonding processes can be conducted in air in the open workplace but others need fabrication chambers that can be evacuated or filled with inert gaseous atmospheres and some brazing techniques require the component mating surfaces to be covered with an inorganic flux that plays an active role in the joining process. Brazing is a relatively fast process with joint filling being achieved within a few minutes, but the heating and cooling stages can be slow. Similarly, the selection of diffusion bonding process and the equipment employed must take account of the fact that the cycle times can be hours long because joint formation depends on the operation of slow solid-state diffusion processes.

Many decisions have to be made before applying brazing or diffusion bonding as a joining process, but this book should provide a rational basis for taking action.

2 Joining techniques

This chapter provides a general introduction to the range of joining processes available for the fabrication of metal and ceramic products. Each of those described can be used by a skilled technician or engineer for the safe and economic fabrication of joints that satisfy demanding service specifications. Achievement of such success depends on an appreciation of the dangers, advantages and limitations of the process being applied. It is assumed throughout this book that consideration has been given to safety before initiating any action.

This chapter is divided into six parts for ease of presentation, five of which relate to the major joining processes for making permanent bonds referred to in Chapter 1 and shown as a spectrum in Figure 1.1. The spectrum is in fact virtually a continuum because of the numerous minor variants and hybrid processes that have been developed and the sixth small part of this chapter is concerned with hybrid brazing/diffusion bonding processes. In this chapter and later in the book, attention will be drawn to shared scientific foundations and to common as well as differentiating practical aspects of the various major and minor processes.

The number of process variations that can be used in principle to make a joint is very large, but those that can be applied in practice to satisfy a particular joining need are often very limited in number. This restriction arises because there can be incompatibility between the fabrication or joint service conditions and the characteristics of the product or joining material. Thus fusion welding is seldom suitable for making ceramic–metal joints and adhesive bonding is not suitable for high service temperature applications. Because of these and other difficulties, attention must be paid to the limitations of processes as well as their advantages.

2.1 FUSION WELDING

Fusion welding is the most important and widely used industrial technique for joining metals, but its use to join ceramics and ceramic–metal systems has yet to leave the laboratory. The joining of metal components by melting and mixing together their mating surfaces, fusion welding, has a long history. Swords and even a Chinese chariot were fabricated in the Bronze Age by pouring or casting molten bronze between the components to be joined. However, success was not easy to achieve and the technique remained a skilled craft rather than a production process. The majority of modern applications of fusion welding do not involve a casting technique but make use of a heat source that melts the component mating surfaces as it is traversed along a joint. Two types of effective heat source have become available and dominant within the last century: high-intensity combustion of gases and localized high-current electric discharges.

Gas welding with high-pressure oxy-acetylene gas torches is used to achieve fusion when small and moderately sized steel components are to be joined. Heat is provided by combustion of the gas mixture, which takes place in two regions as sketched in Figure 2.1: an inner flame in which CO and H_2 are produced, equation {2.1}, and an outer flame where these products are oxidized to CO_2 and H_2O, equation {2.2}.

$$C_2H_2 + O_2 \rightarrow 2CO + H_2 \qquad \{2.1\}$$

$$2CO + H_2 + 3/2O_2 \rightarrow 2CO_2 + H_2O \qquad \{2.2\}$$

Equation {2.1} is very exothermic and can generate temperatures of 3500 °C, sufficient to melt any metal. This inner flame can be best thought of as an explosion that is prevented from travelling back along the torch pipe by the pressure of the replacement gases flowing into the reaction zone. The outer flame is less intense but is useful for preheating the component mating surfaces ahead of the weld zone.

If the proportions of O_2 and C_2H_2 are 1:1 as they are in equation {2.1}, the flame is said to be neutral, but the flame can be oxidizing or reducing if the proportions are not 1:1. Oxy-acetylene flames are suitable for fusion welding steels and similar high-temperature metals but less exothermic oxy-hydrogen or oxy-propane flames are better for the less refractory metals if boiling and splattering is to be avoided.

The weld pool formed by impingement of the torch flame on the mating surfaces of the components is usually supplemented by additional metal,

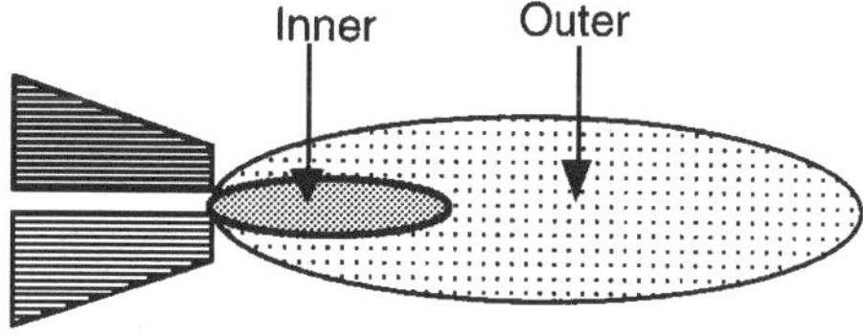

Figure 2.1 Schematic outline of a gas torch flame

with the same or closely similar composition to that of the components, that is melted from the end of a filler rod dipped in the pool. The weld pool is roughly hemispherical and should just penetrate through the joint section. If the heat is too intense, the pool will grow and the liquid metal will be blown away from the component, an effect that is exploited in the gas cutting process.

A wide range of metals can be welded using gas torches but fluxes are needed to ensure satisfactory results with some metals. The fluxes for steels and Cu alloys are generally borates while halides are used with Al alloy components and their function is to remove oxide films from the component surfaces and prevent excessive contamination of the weld pool by environmental gases such as O_2 and N_2.

Welding using high amperage electric arcs to provide the heat needed for fusion was first used at the end of the nineteenth century. This early practice employed a C electrode and while some initial success was achieved, the process was difficult to control and the welds produced were weak and brittle due to the introduction into the weld pool of excessive amounts of C. Modern practice is to use metal electrodes with compositions similar to those of the components. Further developments have led to the common practice of using DC rather than AC power since this avoids the need to re-ignite the arc 50 times a second and to encase the electrodes with specially formulated coatings that melt or decompose to enhance the ease of welding and the quality of the product. Steady development over several decades means that arc welding is now a reliable process that provides an intense heat source, that can generate a thermal flux in excess of $10^9\ \mathrm{Wm^{-2}}$ and temperatures of up to 20 000 °C within the arc.

Figure 2.2 illustrates the use of a coated electrode. The coating is crucial to the success of the process and fulfils at least three functions:

1. shielding of the arc from the workshop atmosphere;
2. promotion of arc stability;
3. coverage of the weld pool with a protective slag layer.

Not surprisingly, the formulations of the coatings are complex but they usually contain a mixture of minerals, metals, organic material and alkali silicate binders. They can be divided into three main types. Cellulosic coatings contain cellulose, $(C_6H_{10}O_5)_n$, and also TiO_2 and $MgSiO_3$, rutile coatings naturally contain TiO_2 and also some $CaCO_3$ and cellulose, and basic coatings are composed mainly of $CaCO_3$ and sometimes also a little CaF_2.

The compositional variations of the coatings reflect their differing modes of operation but all the ingredients of the electrode coatings assist in the optimization of performance. Thus the use of Na_2SiO_3 or K_2SiO_3 binders both assists application of the coatings and benefits arc stability since Na

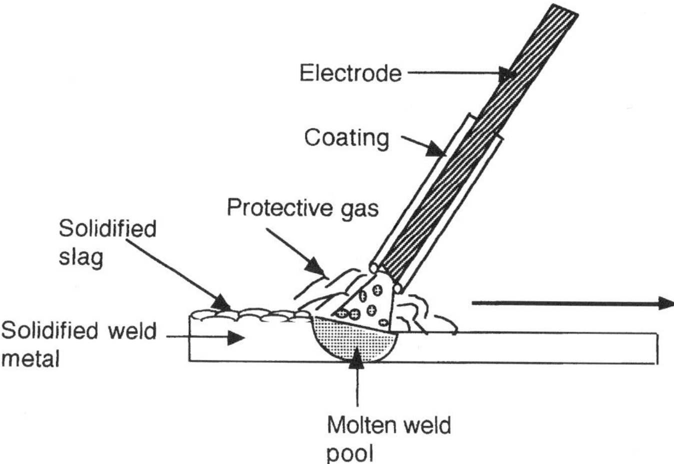

Figure 2.2 Representation of arc welding using a coated electrode

and K atoms have low ionization potentials. Again, the presence of cellulose beneficially affects the chemistry of the process by decomposing in the arc to produce reducing gases by the reaction

$$(C_6H_{10}O_5)_n + 0.5nO_2 \rightarrow 6nCO + 5nH_2 \qquad \{2.3\}$$

that helps prevent the weld metal from picking up O_2 or N_2. However, entrapment of H_2 can cause embrittlement of steels and hence it can be advantageous sometimes to use cellulose free basic electrodes whose $CaCO_3$ decomposes to yield CO/CO_2 mixtures. Similarly, the use of oxide and siliceous minerals promotes the formation of glassy slags that protect the surface of the weld pool behind the traversing arc and either confer an acceptable shiny appearance to the solidified weld metal or can be removed by chipping, while the incorporation of CaF_2 in basic coatings can enhance slag fluidity.

Many variations of arc-welding technique have been developed to meet specific needs. Thus the joining of particularly thick or thin components has been achieved by changing the process parameters such as the choice of AC or DC power sources, by using inert or reactive gas covers as well as electrode coatings or by striking the arc beneath a slag cover. The main varieties are:

- *manual arc welding* using coated electrode rods up to 500 mm in length to join a wide range of engineering components;
- *the continuous electrode process* using a flexible metal electrode drawn from a reel for the automatic, machine, welding of long joints;
- *tungsten inert gas, TIG,* process using a W electrode shrouded with Ar and a filler rod to supply additional metal for the weld pool;

- *metal inert gas, MIG,* process using a gas shielded continuous uncoated metal electrode;
- *submerged arc welding* strikes an arc with a continuous metal electrode beneath a layer of flux powder to join very thick sections.

Although many techniques of fusion welding have been referred to already, they are but a part of the wide range that is available. In some ways similar to arc welding is the use of resistance welding in which the contact resistance of two mating surfaces generates sufficient heat to produce fusion. This technique is used widely in automated processes for spot and seam welding of overlapping sheets and for joining studs to sheets. Another technique using ohmic effects is electroslag welding in which the heat is generated within a bath filled with liquid slag to melt the end of a continuous electrode brought into contact with the components.

Gas torches employ chemical reactions to generate the heat need for fusion, and this is true also for thermit welding of thick steel rails in which Fe_2O_3 and Al powder is mixed and reacts in accord with the equation

$$Fe_2O_3 + 2Al \rightarrow Al_2O_3 + 2Fe \qquad \{2.4\}$$

when ignited by a fuse to generate molten iron at temperatures of up to 2750 °C.

Other sources of heat used in fusion welding are electron beams and lasers. Electron beam welding is performed in a chamber evacuated to a pressure of about 10^{-4} mbar and uses the heat generated by impingement of 10–200 kV electron beams to produce deep but narrow weld pools. Careful control of the beam power and focusing is needed, but this necessity also enables thin sections and even foils to be welded and the process is used for the automated welding of, for example, high value nuclear or aerospace products. Laser welding using He shielding also produces deep weld pools but can be used with thin sections and its range of applications is now extending from the initial joining of micro-electronic products. Neither laser or electron beams, however, are suitable for the joining of Zn and other volatile materials because of the degradation of beam transmission caused by vapour clouds, and their use to join ceramics to metals presents major technical difficulties.

All of the welding processes referred to can be used to produce joints that satisfy very demanding service specifications by employing designs such as those sketched in Figure 2.3. These designs employ continuous weld seams but variants employing spot welds can be used when resistance welding. Effective though such designs might be, individual joints can be defective because of deficiencies such as inappropriate choice of a welding process, poor control of the process or metallurgical degradation. Thus electron beam and laser welding techniques are best suited for the joining

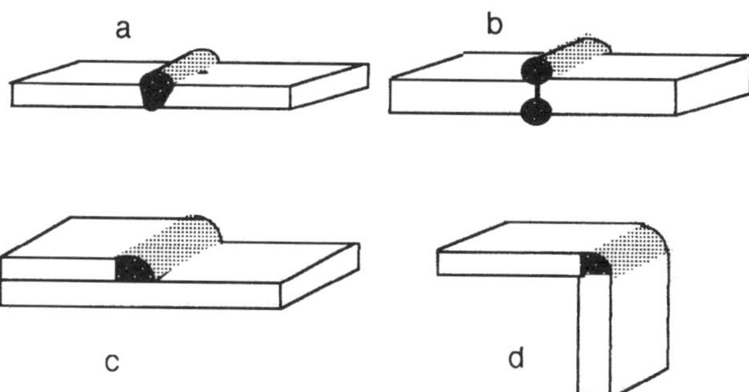

Figure 2.3 Some fusion welded joint configurations: (a) butt weld, (b) partial butt weld, (c) lap weld and (d) corner weld

of thin components and foils, while the electroslag and thermit processes were developed for the welding of sections up to 100 mm thick.

The choice of technique used affects not only the ability to form differently sized weld pools but also the thermal cycle experienced by the components. The traversing speed of the welding process and the cooling rate experienced by the material behind the weld front are particularly important characteristics. Thus rapid traversing and cooling can cause voids to form in the solidified weld seam, often as a line of porosity along the centre of the seam because premature partial solidification of the weld pool creates a skeletal structure that is too coarse to be filled by the residual liquid. However, slow traversing and cooling is not a universal panacea since it too can result in excessive void formation during the cooling of weld pool alloys with wide melting ranges. Slow cooling can also cause detrimental changes to the complex metallurgical structure of the solidified weld seam, coarsening the grain size and promoting grain boundary segregation of embrittling species such as S in steel and Ni alloys. The heat-affected zones in the components adjacent to the weld seam are subjected to a slower and less dramatic thermal cycle than the weld pool, but they too can experience excessive grain growth, segregation of harmful impurities to grain boundaries and coarsening of two phase structures.

The importance of these various structural effects lies in whether or not they can cause fracture or tearing of the welded components as they cool or in service. The presence and effect of such flaws can be assessed by either destructive or non-destructive techniques. Thus the usual prime requirement of a weld is that it should have a high mechanical integrity and information about the strength, ductility and toughness of the weld can be derived using samples cut from the welded components and machined into the standard configurations for tensile and other tests. Considerable insight can be gained also from examination of the metallurgical structures of the weld seams and heat-affected zones using both optical and scanning electron microscopy supplemented when necessary by micro-chemical techniques such electron probe micro-analysis. Non-destructive evaluation using ultrasonics and X-ray diffraction to detect voids, inclusions and tears is preferable to complete reliance on destructive assessments because it can be applied to the actual components and can monitor changes in service.

2.2 DIFFUSION BONDING

Diffusion bonding is a solid state process for the fabrication of metal–metal, ceramic–ceramic and ceramic–metal joints that is conceptually simple. In principle, the process requires no localized melting of components or introduction of foreign bonding materials but merely that mating surfaces are brought into intimate, atomic scale contact so that an interface can be formed by interdiffusion to create a structural continuum. Such interfaces, whether between metals, between ceramics or between a metal and a ceramic, can have good mechanical integrity even at high temperatures.

The introduction of diffusion bonding as an industrial technique has made great advances in recent years but has also resulted in the development of many varieties of the process, some of which exploit the transient presence of a liquid phase within the joint during the fabrication cycle. The simplest form of diffusion bonding involves the application of a low pressure at a high temperature, usually at least $0.75\,T_M$, where T_M is the melting temperature in degrees Kelvin, to achieve bonding of metal components that have smooth and well-matched mating surfaces. There need be no macroscopic deformation of the components but contacting asperities on the mating surfaces grow to create initially separate microscopic areas of contact. These are then transformed into interfaces by interdiffusion which removes barriers to structural continuum such as oxide films and enlarges the bonded area. This is in effect a sintering process that closes the small voids, no more than a few microns across, that are residues of the valleys and pits originally present on the mating surfaces.

If the mating surfaces are not flat but wavy, there can also be regions where the mating surfaces arch away from each other, as illustrated schematically in Figure 2.4, and these substantial voids can be slow to close. While very idealized, this simple picture reflects some of the motivation that has led to the development of differing diffusion bonding techniques. In the figure, the heavy marking of the mating surfaces is diminished as contact area grows as a schematic imitation of the dissolution of oxide and other films on the mating surfaces or the

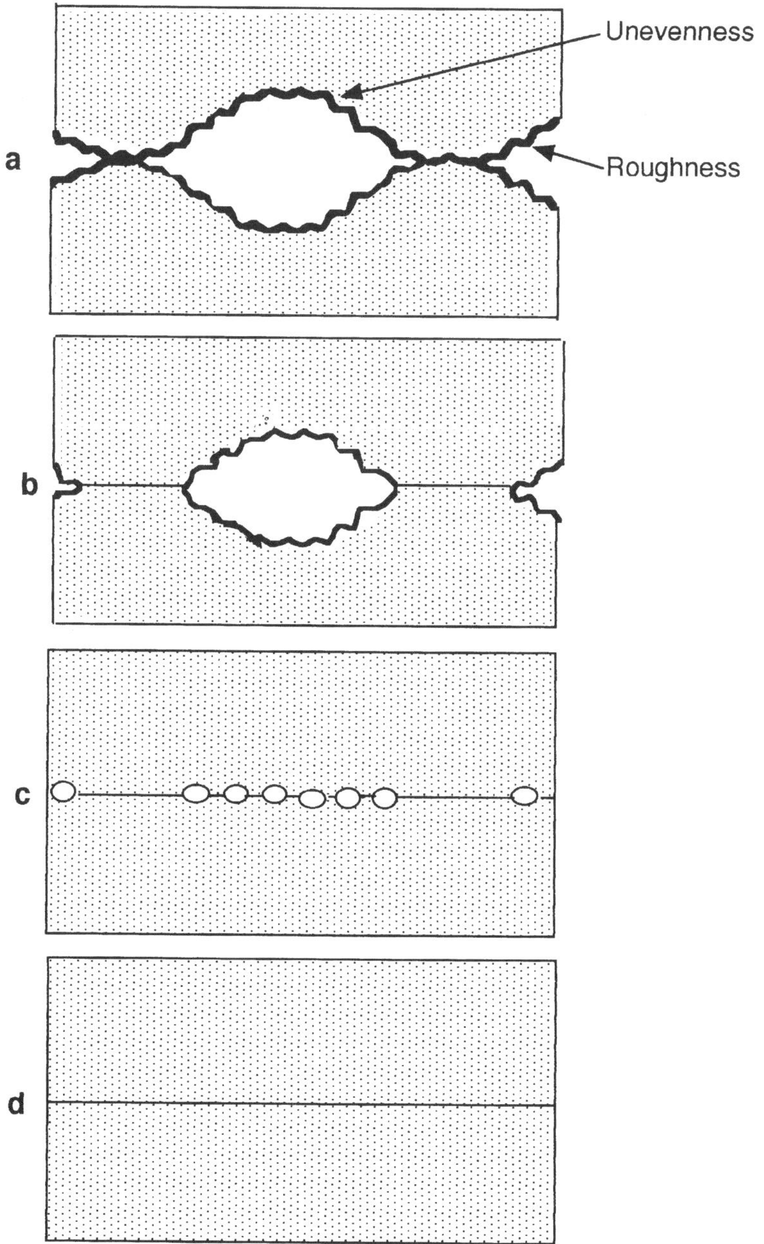

Figure 2.4 Schematic illustration of diffusion bonding induced growth of the contact area between two microscopically rough and wavy mating surfaces: (a) initial contact, (b), (c) and (d) progressive growth due to plastic flow and diffusion

desorbtion of segregated impurities and solutes that is an often necessary condition for successful diffusion bonding.

The process is best suited for the bonding of components made from the same ductile metal whose surfaces are oxide free. However, dissimilar metal combinations can be bonded provided they satisfy a number of additional conditions. The bonding temperature must not be so high as to cause melting of the less refractory component, but it is desirable that it should be high enough to permit beneficial diffusion within the refractory component to promote sealing of microscopic isolated porosity at the bond line and dissolution of soluble surface oxide films. If the oxide of the refractory component is not soluble, plastic flow of the asperities of the other component must be vigorous enough to scour its surface and remove the oxide barrier to bonding by mechanical disruption and dispersion. Once metal–metal contact has been achieved, beneficial interdiffusion will occur if the two metals are mutually soluble and the bonding temperature is high enough, but for some systems interdiffusion should be restricted because it will result in the growth of fragile layers of intermetallic compounds.

Uniaxial pressing has been used widely to produce butt joints by diffusion bonding metal components already machined to their final shape and dimensions and prepared with smooth and flat mating surfaces. There are a few examples of diffusion bonding being performed in an open workshop but the normal practice is to use a chamber that is evacuated and then sometimes back filled with inert gas before the components are raised to the proposed bonding temperature using heating elements or induction coils. The pressures used to bring the mating surfaces into contact are low to avoid macroscopic deformation and can be generated by applying a dead load, by using jigs made of low expansion materials that put the components under pressure when they are heated, or by using hydraulically operated rams that pass through the chamber walls.

Such processing has permitted successful diffusion bonding of a wide range of similar and dissimilar metals including Cu, Fe, Mo, Nb, Ta, Ti, W and some of their alloys. When wishing to bond Al, Be and other metals that form chemically stable oxides it can be advantageous to coat their mating surfaces with Ag or some other weak oxide former which will dissolve in the substrate during the diffusion bonding cycle. Coatings can also be useful when bonding some dissimilar metal combinations to avoid harmful effects of tenacious oxide films and the formation of fragile intermetallics, hence Al can be bonded to a wide range of other metals, such as Cu, Fe, Ti, Mo, and Nb. Similarly, Cu and Ni have been bonded successfully to steel and Mo, Nb, W even without the use of coatings by careful choice of the fabrication temperature.

Bonding of ceramics by uniaxial pressing has been achieved but it is not yet a significant industrial process. The plasticity of ceramics is generally so poor that deformation of asperities to obtain an initial contact and conformity of the mating surfaces is seldom possible and the refractoriness of ceramics, Table 1.1, means that the fabrication temperatures often can be unacceptably high for the equipment that is available. Nevertheless, some success has been achieved by using very smooth and very flat mating surfaces. Other examples of successful ceramic–ceramic joining has been achieved also by using component materials with glassy binder phases that flow to form what is really a glazed joint at the bonding temperature, and by the introduction between the components of a ceramic powder that is so fine that it creates a bond by sintering to their surfaces.

When ceramics have to be bonded to metals, it is a common practice to introduce a metal interlayer between the components, producing a joint structure similar to that created by brazing. Interlayers have also been used in laboratory studies of ceramic–ceramic bonding. For both requirements, the interlayer should be ductile so that it can deform readily to achieve intimate contact with both mating surfaces at pressures and temperatures that do not cause any macroscopic deformation of the components, that it should act as a stress relieving buffer layer if the thermal expansivities of the metal and ceramic components differ significantly, and of course that it should adhere strongly to both the metal and ceramic components. Many metals and alloys should be able to satisfy these requirements but those used in practice are usually Al, Au, Ni and Ni–Cr alloys. Adherence to the metal component can be relatively easy to achieve provided care has been taken to clean the mating surfaces of both the component and the interlayer, but adherence to the ceramic often depends on the formation of a reaction product layer that acts as a bridging compound and this reaction can proceed too far to produce thick and fragile reaction layer. The technique has been applied so far mainly for the joining of Al_2O_3 and SiO_2 but its utility has also been demonstrated in laboratories for the joining of Si_3N_4 and other new engineering ceramics. Optimization of the product designs and process parameters have allowed fabrication of very high-quality joints, one example of which is described in some detail as a case history in Chapter 8. In that example, as in others, the mechanical strength requirement is not high, but joints can be produced whose strengths approach that of the ceramic being bonded, even at temperatures above $0.5\,T_M$.

Despite these and other successes, the requirement of the uniaxial pressing method of diffusion bonding for a very high-quality finish to be imparted to the mating surfaces are significant technical and economic disadvantages that have led to the development of techniques to achieve conformity of poorly matched metal mating surfaces. The four most notable of these are pressure welding, roll bonding, superplastic forming and hot isostatic pressing.

Pressure welding uses a punch tool to achieve localized bonding of overlapping sheets of metal, and is therefore a solid state analogue of resistance spot welding. The mating surfaces are abraded and then degreased before the sheets are brought to the punch which applies a high pressure to produce severe and complex localized deformation of the metal sheets.

Punching often decreases the sheet thickness by more than 50% and this not only brings the mating surfaces in the bonding region into intimate contact but also disrupts and disperses surface oxides. The process is usually performed at room or moderate temperatures so that there is in fact little interdiffusion of the component materials, but it has been used successfully to join a wide range of ductile metals.

Roll bonding, in contrast, produces a uniform and unidirectional extension of metal sheets, whose mating surfaces have been abraded and degreased, by passing them through a rolling mill. Large reductions in the thickness of the sheets during a single pass through the mill are usually needed so that once more intimate contact is achieved and the oxide films on the original surfaces are disrupted and dispersed as the sheets are extended. For Al at room temperature the minimum single pass reduction needed to achieve bonding is 40% and for Fe it is 65%. Smaller reductions are needed to achieve bonding at higher temperatures, and of soft metals like Sn at room temperature. Dissimilar metal combinations can be joined but this is not the usual case. The technique found an early application in the production of Al refrigerator heat exchangers using sheets whose surfaces were first cleaned and then printed with a pattern of the fluid flow passages using an oxide powder, a ‘stop off’, to prevent bonding when they were passed through the rolling mill. Subsequently, hydraulic pressure was used to produce shallow tubular passages by injection of gas into the unbonded interfacial regions.

Superplastic forming/diffusion bonding is a new joining and fabricating process that depends on the fact that fine-grained materials such as Al and Ti alloys can experience very large extensions at moderate temperatures without fracturing. Sheet components whose mating surfaces have been cleaned and then printed with a stop off pattern are diffusion bonded and then the channels are massively inflated to produce a stiffened structure such as that shown schematically in Figure 2.5. When using Ti alloys, both diffusion bonding and superplastic forming can be accomplished at a temperature of about $0.6\,T_M$, 900 °C, because Ti readily dissolves its oxide at that temperature in inert environments. Very complex internal structures can be produced by careful design, as illustrated by the case history of a Tornado heat exchanger duct described in Chapter 8.

Hot isostatic pressing differs from the other diffusion bonding processes because of the use of a compressing fluid, often Ar. Components of the same or dissimilar materials are prepared for bonding by cleaning their

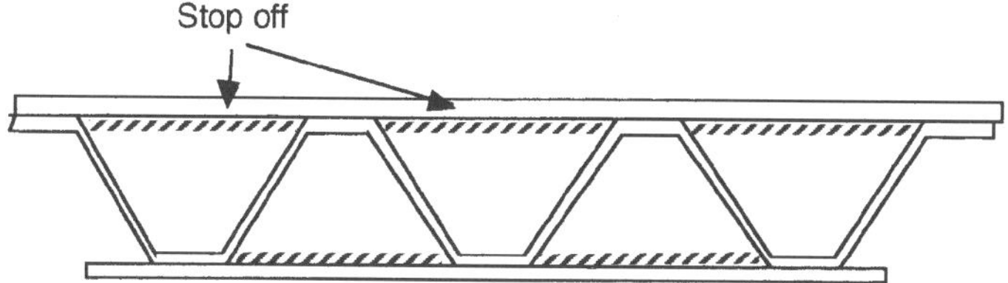

Figure 2.5 A simple sandwich structure produced by superplastic forming/diffusion bonding

mating surfaces and assembled into the form of the final product. This assembly is then encased in a tight-fitting metal can or glass flask and is placed in the hot isostatic press, HIP, unit and subjected to a diffusion bonding thermal and pressure cycle. A particular advantage of the process is that its isostatic nature means that higher pressures can be used when bonding to produce intimate contact of the mating surfaces without causing macroscopic deformation of the components. The capital costs of HIP units are high and the cycles times between loading and unloading can be long, but filling the working chamber of a HIP can enable many small products to be made at the same time and the technique is gaining use for the manufacture of both metal and ceramic products.

The varieties of diffusion bonding mentioned so far have all been solid state processes, but some joining methods often regarded as variants of diffusion bonding involve the transient existence of a liquid phase in the joint. An early example of this was the forge welding of wrought iron components in which the blacksmith tossed a handful of sand or borax into the joint that cleaned the mating surfaces by reacting with their iron oxide films to form a liquid that is driven out of the joint by hammering the components. More recently developed processes that can be viewed as diffusion bonding include friction welding in which similar or dissimilar components are held in a lathe while one is rapidly rotated and then rammed against its static partner for a brief time before a clutch disengages the rotation. The braking action of the static component in the early stages of contact generates heat which can melt the mating surfaces and the subsequent ramming drives out this liquid to leave clean surfaces that form the bond. Another related process is explosive welding in which slabs of the same or dissimilar metal are driven together by an advancing explosive front that causes the contacting mating surfaces to act like a pseudo-liquid and form a rippled interface.

Transient liquid phase bonding and partial transient liquid phase bonding are also sometimes regarded as variants of diffusion bonding but in this book are classified as hybrid brazing/diffusion bonding processes and are discussed in a separate section after the commentary on brazing processes.

2.3 BRAZING

2.3.1 General considerations

Brazing is an ancient craft that has become a widely used industrial process for fabricating products to meet service demands ranging from those of simple domestic utensils to complex structures for the aerospace and nuclear industries. No matter what the application, the requirements for successful fabrication of brazed joints are the same: the braze metal or alloy must melt and flow over the surfaces of the components to be joined, form a fillet or more usually fill a narrow gap between the components, and then form a permanent bond by remaining adherent while solidifying.

The formation of brazed joints depends on the flow of a liquid experiencing an attraction exerted by the component surfaces and therefore anything that affects this attraction is of crucial importance to the success of the brazing process. The surfaces of clean metal components should be readily wetted by liquid brazes in principle because similar material interfaces have low energies. However, difficulties can arise in practice due to surface contamination and the presence of chemically stable and physically tenacious oxide films. Thus cleaning procedures are of critical importance when defining a brazing procedure, but joint design is of almost equal importance because the extent and rate of penetration of a gap by a wetting liquid depends on its width. For example, Figure 2.6 shows the rise of a wetting liquid between two plates, the extent of which can be derived from the balance of the surface tension drawing the liquid up the gap and the gravitational drag so that

$$2.\gamma_L \cos\theta = h\rho_L gW \qquad \{2.5\}$$

where g is the acceleration due to gravity and ρ_L is the liquid density. A derivation of this relationship in terms of surface and interfacial energies is given in Chapter 3, but of more importance at present is the fact that equation {2.5} shows the liquid rise will be greatest in the narrowest gaps. The preferred gap for practical brazing of metal components is usually about 0.1 mm, and this particular dimension is some semantic interest because the phenomenon sketched in Figure 2.6 is called capillary rise and 0.1 mm is also the thickness of a human hair, or *capillaris* in Latin.

Braze alloys and procedures have been developed that promote the wetting of a very wide range of metallic components. Ideal flow of liquid into the capillary gaps of wettable components will be impeded only by viscous and gravitational drag and joints of a few millimetres in length should be filled in a fraction of a second. In practice, filling of joints can take a few minutes when the liquid braze experiences an additional impedance due to unwetted oxide films present on component surfaces. Nevertheless, joint formation times are usually only a small part of the total cycle time which includes heating and cooling stages and may also

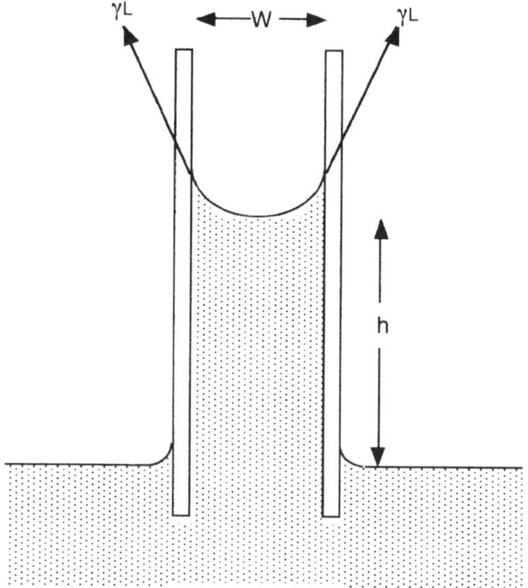

Figure 2.6 The rise of a liquid up a narrow capillary gap

accommodate holds of typically 10 to 30 minutes to achieve a uniform component temperature when heating or to encourage interdiffusion between the braze and the component material when cooling.

If the liquid braze does not wet, then the gap will not be entered and no joint will be formed. Thus it is technically important that few brazes wet ceramics and those few that do spread very slowly over the surfaces. This sluggishness not only affects the brazing cycle time but also the placing of the braze. While brazes for metal components can be placed at gap entrances and will flow to fill them, those for ceramics are normally placed within the gaps, the widths of which are then determined not by capillary factors but by the thickness of available braze sheet or thickness of the layer of braze powder that has been applied to the surface.

The design of brazed joints is discussed in Chapter 5 but it is not surprising that most preferred designs are lap or sleeve configurations, Figure 1.3, which provide capillary channels into which wetting brazes can be drawn. Like fusion welding or diffusion bonding, brazing can produce joints with very high structural and mechanical integrity, but it differs from these other processes in several important aspects. A brazed joint has a different composition to that of the components and this affects not only chemical characteristics such as corrosion resistance but also technically important physical properties. Thus the generally lower yield strengths and elastic moduli of braze materials can endow the joints with a desirable ductility, but also a regrettable weakness. Similarly, the low melting temperatures of brazes relative to the components means that brazed joints normally have a lower maximum service temperature than those produced by fusion welding or direct diffusion bonding of the components. There is no absolute need for a liquid braze to interdiffuse with metallic components even though this occurs quite often in practice, and for a few systems such interdiffusion can cause the melting temperature of the braze to increase so the joint is more refractory than might be expected. Further, like diffusion bonding but unlike fusion welding, brazing can be used to join ceramic as well as metal components although the two types of materials require quite different brazing practices. In particular, the fact that ceramics are wetted by very few braze metals or alloys makes it is necessary either to coat their surfaces with a metal before attempting brazing or else to use an 'active metal braze' alloy that reacts chemically with the ceramic to produce a wettable reaction product layer on its surface.

2.3.2 Materials

There are many varieties of brazing technique and which is most suitable for a particular application depends on many factors. Prominent among these are the melting temperature of the braze and

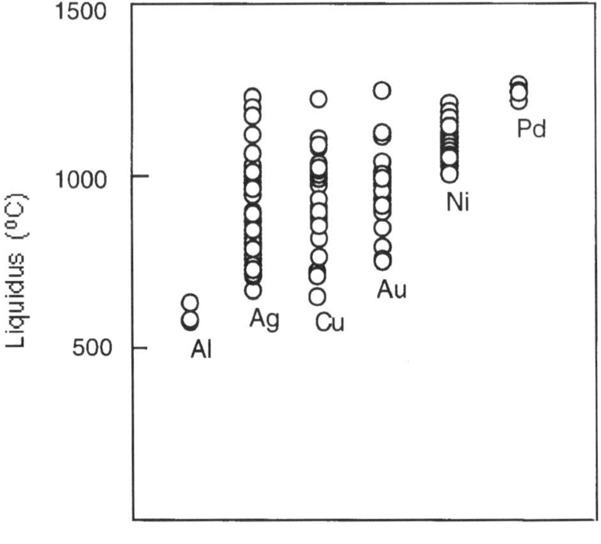

Figure 2.7 Melting temperatures of commercially available brazes for metal components

the chemistries of both the component and braze materials. There is an abundance of commercially available brazes based on alloys of Al, Ag, Cu, Au and Ni that can be used to join metal components and the reasons for these preferences are discussed in Chapters 3 and 4.2. While all the preferred braze alloys have good wetting characteristics, their flow and use temperatures vary widely. Thus, Figure 2.7 shows the liquidus temperatures, the temperatures at which the brazes become fully molten, range from 500 to 1250 °C and such a wide variation can have implications not only for what equipment is used but also for the brazing technique. It is clear from the figure that a multitude of brazes are available for joining metal components, but their effective use with oxide coated components can require surface treatments prior to brazing or the use of chemically aggressive fluxes. Further, the materials and techniques available for the brazing of ceramics is very restricted because so few liquid metals wet ceramics.

2.3.3 Techniques

The main techniques used to braze metal components involve the use of localized heating applied by gas torches, induction heaters or electrodes, or of generalized heating produced by dipping components in salt, flux or metal filled baths, or placing them in controlled atmosphere or vacuum furnaces.

Torch brazing

Torch brazing by hand is arguably the simplest and also possibly the most used brazing technique for joining metal components and this is particularly true for one-off and short run production applications. The practice is similar to that for fusion welding, Chapter 2.1, with the torch flames being directed close to but not at the joint regions but reflects the fact that it is not necessary to achieve such high flame temperatures. Thus the torch nozzles are usually larger when brazing so that the exothermic inner flame, Figure 2.1, is not so restricted in volume, gas pressures are generally lower than when fusion welding and sometimes only the outer cool flame is allowed to impinge on the component surface. A variety of gas mixtures are used ranging from O_2-C_2H_2, (oxy-acetylene), which produces very hot flames with an inner temperature of about 3500 °C, to natural gas–air mixtures with a maximum flame temperature of about 1850 °C.

The usual practice when manually using a torch is to supply liquid braze by melting thc tip of a brazc 'fillcr' rod that touches the component surface near the entrance to the joint. For flow of the liquid to form a joint, it is necessary that the component surface is clean and free of oxide films that inhibit wetting. This can be achieved by adjusting the gas mixture to produce a slightly reducing flame when joining Cu components with a Cu-8P alloy, in which the P not only depresses the melting temperature of the braze but also assists removal of oxide from the component surface by forming non-adherent phosphates. However, oxide removal by unassisted gaseous reduction is seldom sufficiently vigorous for effective brazing and in practice it is necessary to use a flux, generally a mixture of alkali borates and halides, that is applied to the component surface as a powder or paste or is present as a coating on the braze filler rod. During the brazing process, the flux melts at a lower temperature than the braze alloy and then spreads over the component surface and into gaps between components to dissolve or disrupt and disperse the oxide film before itself being displaced by the advancing liquid braze.

Torch brazing can be used also as part of an automated joining facility for high volume production runs. While the same gas mixtures can be used, it is usual practice to place both flux and braze alloy near the entrances to the component gaps and to preheat the component sufficiently to cause melting of the flux before impingement of flames hot enough to effect brazing.

Induction and resistance brazing

High-frequency induction brazing is a clean and rapid technique that lends itself to automated and semi-automated joining. As with any form of induction heating, power is transferred by a high-frequency electromagnetic field from a water-cooled induction coil to the component as a rapidly alternating eddy current. The power source can be used to effect fluxed or fluxless joining using preplaced braze and flux materials and can be conducted in the open workshop or in a controlled environment using a non-metallic chamber such as a glass tube with sealed ends.

In some respects, the resistance brazing technique is even simpler since it is merely a current to flow between electrodes through the joint region. A pressure is needed to achieve good electrical contact between the electrodes and the components, and once again the braze and any flux must be placed in the joint region prior to joining. The joint region normally has an initially high electrical resistance so that the resistance heating is very localized and often intense. Thus considerable skill may be required in adjusting the current and the design of electrodes to avoid overheating of the braze alloy or the component materials.

Dip brazing

Dip brazing involves the complete immersion of metal components in a bath that is filled with a chemically neutral salt that provides a substantial

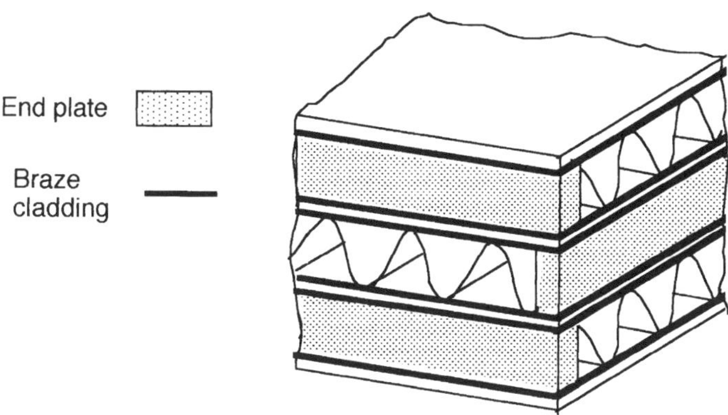

Figure 2.8 Corner of a heat exchanger made from braze clad Al alloy sheets, thin Al alloy corrugations and stout Al alloy end plates

and controlled heat source, as does an Aga cooker or a barbecue pit, or with a flux-covered liquid braze alloy or with only a brazing flux. The baths are maintained at a constant temperature by external gas torches or internal electrical heating elements. An advantage of this technique is that the components are less prone to distortion because heating is not localized. Uses of this technique are varied. Thus dipping into baths filled with liquid Ag or Cu based braze alloys covered by a flux layer of alkali borate and fluoride can be used to join small components such as wire connections and other simple designs. Similarly, immersion in a flux bath is a common method for joining of Al alloy components whose surfaces are clad with a braze alloy based on an Al-11 Si alloy composition that melts at 577 °C, or which have braze alloy circlips and other previously formed shapes placed near the joints.

The fluxes used when brazing Al are necessarily aggressive mixtures of chlorides and fluorides that strip the Al_2O_3 films from both component and braze surfaces, but the flux residues are also usually aggressive and must be removed from the component surfaces if continuing corrosion during service is to be avoided. This removal can be a difficult task when the component is, for example, a large heat exchanger with complex external fins or internal corrugated sections such as those sketched in Figure 2.8, but it can be accomplished by prolonged and agitated washing in acids such as HNO_3 or HF.

Furnace brazing

Furnace brazing is an industrial technique that has become progressively more important as performance specifications have increased for both mass production of components for the consumer industries and for very high-integrity components for the aerospace and nuclear industries. Brazing can be performed in continuous or batch furnace chambers that are evacuated or filled with a controlled atmosphere of, generally, an inert gas such as Ar. Continuous furnaces are particularly suitable for the mass production of small items such as automobile components. They use a conveyor belt, often of wire mesh, to move components into and out of a heated brazing zone and usually also into and out of controlled atmosphere regions. In contrast, batch furnaces find their main applications in the production of short runs, high added value and large components. The assemblies to be brazed are loaded into the furnace via a front opening door, a top lid or even from the bottom by lifting up a bell chamber and the chamber is then resealed. The closed chamber is evacuated or evacuated and back filled with a controlled atmosphere and the components are subjected to a temperature cycle.

When furnace brazing it is necessary to place the braze alloy in position before the assembly is in the furnace. The use of fluxing to promote braze flow is rare when using a batch furnace but is common when producing items by the conveyor belt technique, and such fluxed continuous brazing has recently been enhanced by the development of the Nocolok process, an Alcan Aluminium Ltd patented procedure that promotes the brazing of Al components by use of a K_2O-KF-Al_2O_3 flux that is almost insoluble in water and hence results in minimal service corrosion. Nevertheless, fluxless brazing in controlled atmospheres or vacuum has increased considerably in recent years despite the high capital costs of the equipment. A particular advantage of this practice is that there is no need to clean the product after brazing. Indeed, brazing in vacuum or reducing atmospheres can also enhance the appeal of the products by producing bright clean surfaces. It should be noted, however, that only certain grades of brazing alloy can be used in vacuum furnaces. Loss of volatile elements from unwisely selected braze alloys can degrade not only wetting and flow characteristics but also damage the furnace, condensing in cool regions and possibly causing breakdown of electrical insulation. In a few cases, such losses have to be accepted and the furnaces must be cleaned at regular intervals to prevent excessive build up of Mg and MgO.

Ceramic brazing

The brazing of ceramics presents particular difficulties because they are usually unwetted by the commercial alloys successfully developed for joining a wide range of metallic component materials. Special techniques have had to be developed to overcome this problem. Thus ceramic surfaces can be coated with pure metals or alloys by vapour or sputter deposition but the most developed technique is metallization by the 'moly-manganese' process which was originally devised for Al_2O_3 ceramics. A mixture of

Mo, Mn and a glass powder is applied to the ceramic surface and heated to about 1400 °C in relatively dry H_2. This causes the glass to become fluid and to wet and bond to both the Mo powder and the ceramic surface, with the bonding to the ceramic being promoted by segregation of MnO to the ceramic surface. This coating is sufficiently metallic to be electroplated with Ni and the metallized ceramic is then furnace brazed. This technique is very effective with Al_2O_3 ceramics but needs very precise optimization to avoid unacceptable reject rates.

Batch furnaces are very suitable equipments for brazing ceramic components. If the ceramic surfaces have been metallized by the moly-manganese process or by vapour deposition of pure metals, they can be treated as metallic components. However, batch furnaces also find use for the brazing of ceramic components that have not been metallized before being loaded. Thus ceramic component surfaces can be painted with TiH_2, which dissociates at 300–500 °C to deposit Ti during the subsequent vacuum brazing cycle and recent years have seen the development of 'active metal brazes' that can wet ceramic surfaces that have not been metallized or painted. These alloys are usually based on the Ag-28Cu composition which melts at 780 °C and to which has been added a small quantity of Ti that reacts with the ceramic surface during the brazing process to form a wettable surface layer, usually a metal rich form of TiO on oxides and of TiC or TiN on C, carbide or nitride ceramics. The reason why formation of these particular reaction products is beneficial will be discussed in Chapter 3 and an example of their use is described in one of the case histories given in Chapter 8, but two consequences are of relevance to this chapter. First, the high chemical reactivity of Ti that is exploited to achieve wetting also means that these braze alloys are almost always used in a furnace evacuated to a pressure of 10^{-4} mbar or less. In principle there is no reason why an inert atmosphere of Ar or He should not be used but it is difficult in practice to purify these gases to a sufficiently low partial pressure of O_2 to prevent formation of a flow restricting oxide skin on the liquid braze. Second, braze flow controlled by growth of a solid reaction product is very slow and hence active metal braze alloys are generally placed between ceramic components rather than being expected to flow into capillary gaps as when joining metal components.

2.4 HYBRID BONDING PROCESSES

The spectrum of joining processes contains many that could be classified hybrids of more than one variety. Thus what can be regarded as effectively brazed or fusion welded joints can be produced by the 'direct bonding' or 'eutectic bonding' techniques which were originally thought of as being derivatives of diffusion bonding. Direct bonding permits Cu and some other metals to be joined to Al_2O_3 by heating in a mildly oxidizing environment to form an oxide–metal composition that is liquid at a bonding temperature that is insufficient to melt the metal and wets the ceramic surface to produce a pseudo brazed joint. Similarly, eutectic bonding involves contact of dissimilar metal components followed by interdiffusion which for some combinations of component materials results in the formation of a eutectic alloy, a composition containing constituents from both components that has a singular melting temperature lower than that of either component material. This liquid then forms a pseudo welded joint by solidifying at the end of the thermal cycle to create a permanent bond.

In recent years the range of systems that can be joined by hybrid brazing/diffusion bonding processes has extended with the development of transient liquid phase bonding and partial transient liquid phase bonding. Transient liquid phase bonding is a process that consists essentially of placing a low melting temperature metallic interlayer between the mating surfaces of components and then heating the assembly to produce, ultimately, a bonded product with a near uniform composition. While it is simple to describe the end result of the process, the sequence of events leading to this are complex as illustrated schematically for an idealized system in Figure 2.9. Heating to the fabrication temperature causes the interlayer to become a liquid seam, either by simply melting or else by interacting with the component materials to produce an even lower melting temperature alloy. The assembly is held at that temperature for some time so there is diffusion of component materials into the liquid seam and of seam material into the components. The ingress of component material can cause an initial volume increase of the liquid seam but the longer-term effect is to decrease the volume of the liquid seam because increasing the concentration of the more refractory component material raises the melting temperature of the liquid and induces isothermal resolidification. Continued interdiffusion causes the liquid seam to resolidify completely and ultimately produces chemical homogenization of the bonded assembly with the final composition being similar to that of the original components.

Thus the technique can produce a structure resembling that of a diffusion bonded joint but it does not require the use of pressure or extremely careful surface preparation to achieve intimate contact of the mating surfaces. The equipment and control processes needed for application of this technique are essentially those required for furnace brazing. As with other joining processes, success depends on the correct choice of joining materials and fabrication conditions. Thus the interlayer should be similar in composition to the components, preferably differing only by containing small amounts of foreign species that will diffuse rapidly: for example the mobility of B in Ni

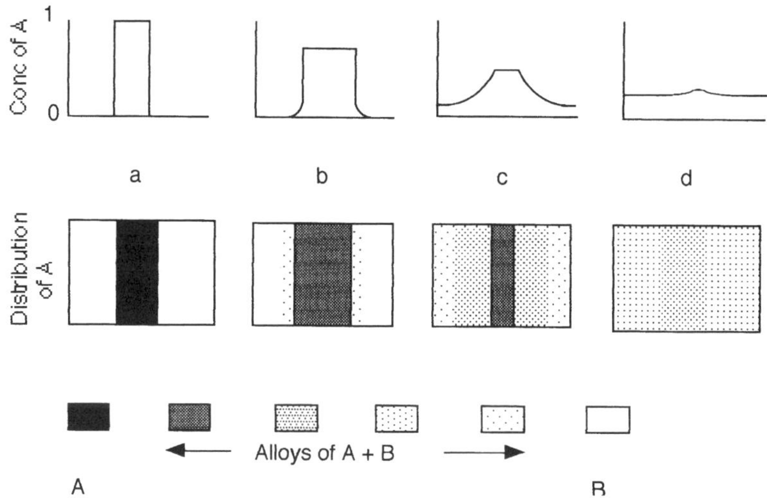

Figure 2.9 Stages in the transient liquid phase bonding of components of B using an interlayer of A: (a) formation of a liquid seam; (b) early thickening of the seam due to the ingress of B; (c) later thinning due to isothermal resolidification; (d) near homogenization of the completely solid system

alloys has been exploited successfully in the use of a Ni-15.5Cr-3.7B alloy interlayer to join M007, a Ni-10.Co-8Cr-6Mo-6Al-4Ta-1Hf-0.1C-0.02B super alloy. Similarly, decisions about processing conditions can be based on either the models described in Chapter 3 or practical experience, and preferably on a combination of modelling and experience.

The technique has been applied in the joining of ceramics but few systems are suitable because of the generally poor wetting of ceramics by liquid metals. Recognition of these difficulties has led to laboratory studies in which the derivative process, partial transient liquid phase bonding has been used successfully to join Al_2O_3 and Si_3N_4 ceramics. In this process, the interlayer consists of a ductile metal core that is coated with a lower melting temperature metal or alloy. The core or the coating alloy contains a reactive element that promotes wetting and bonding to the ceramic. Thus the coating melts on heating, often because of interdiffusion with the core to form a low melting temperature alloy, and reactive metal present within coating or derived from the core segregates to the ceramic surface to form wettable reaction product layers that also provide the chemical bridging needed for the bonding of metals to ceramics. Continued heating and interdiffusion causes the coating material to alloy with the core and produce a ductile interlayer with a composition little different to that of the original core. For example, Au coated Ni-20Cr has been used as the interlayer when joining Si_3N_4. The Au alloys with Ni to produce an alloy that melts at 950 °C and Cr is transported from the core to wet and bond to the ceramic, while continued interdiffusion ultimately results in the formation of a ductile Ni alloy slightly depleted in Cr and slightly enriched in Au and Si.

2.5 GLASS SEALING

The fabrication of many products requires that glass components be joined. Glass–glass joints are usually made by melting their mating surfaces and melting is also the most important technique for joining glass components to other materials. Similarly, thin glass interlayers can be melted to produce ceramic–ceramic and ceramic–metal joints. The craft of fusing glass to make joints is very old but the technology and science have advanced rapidly in recent years. This has permitted better control of the joining processes by close specification of design and fabrication parameters based on understanding of material characteristics.

2.5.1 The glassy state

Glasses are a special class of oxides that have unusual thermal characteristics. Most solid oxides and metals have regular lattice structures with fixed neighbour–neighbour distances, but these structures collapse when the materials are heated to their melting temperatures. The liquids produced display no rigid structural regularity, but the original solid lattice structures are regained when the liquids are cooled to their solidification temperatures. However, nucleation of the solid crystalline structure can be inhibited when a glass forming oxide such as SiO_2 is fast cooled from the liquid state. The resultant solid structure is not that of a regular lattice but rather that of a super cooled liquid, an irregular network rather than the tetrahedral ionic array of the crystalline form. Showing such irregularity for a three dimensional structure is difficult, but Figure 2.10 provides a schematic illustration of a two-dimensional analogue for an imaginary sesquioxide, Q_2O_3.

The irregular structure of glassy Q_2O_3 can be modified by introducing other cations which can either cause further deviation from the form of the crystalline lattice or can produce some recovery depending on the strengths of their interionic bonding. Thus a range of glass structures can be produced by additions to SiO_2 of materials such as B_2O_3, Al_2O_3, PbO and alkali oxides. The structural changes caused by these additions also affects properties such as the surface energy, viscosity and thermal expansivity.

The surface energies of oxide glasses are low compared to those of metals, thus the γ_L of SiO_2 is only 0.3 Jm^{-2} compared to the 1.65 $J.m^{-2}$ of

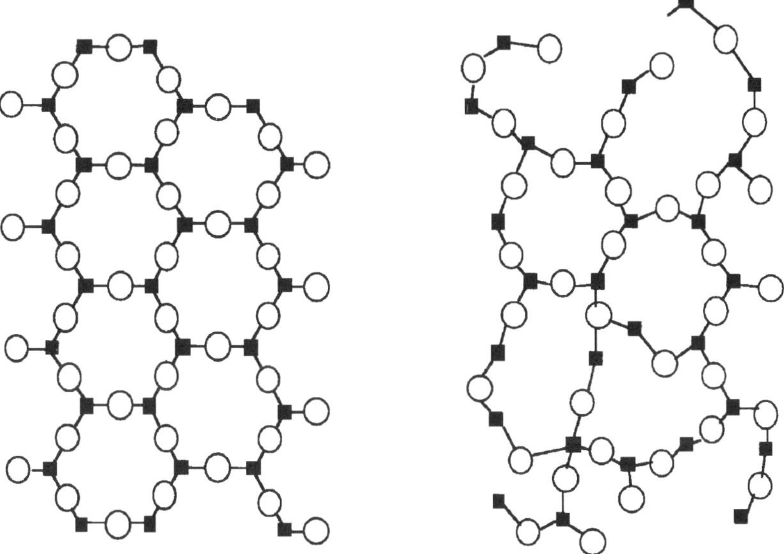

Figure 2.10 Two-dimensional crystalline, left, and glassy, right, forms of an imaginary oxide Q_2O_3 in which the O^{2-} ions are large circles and the Q^{3+} ions are small black squares

Ti which has a similar melting temperature. This difference will affect wetting and spreading behaviour, but of even more practical importance in glass sealing processes are the effects of such additives on viscosity and thermal expansion. Stressing a crystalline solid causes no permanent deformation until a well-defined yield stress is exceeded but the open and irregular structures of glasses permit some permanent deformation to be initiated by the application of any significant external stress – they are viscous. This viscosity is very large at low temperatures, window panes do not noticeably slump in their frames, but technically significant decreases in viscosity occur as the temperature is increased. Thus a glass can be regarded as rigid for practical purposes when its viscosity is at least $10^{13.5}$ Ns.m^{-2}, but it can be annealed to relieve residual stresses when the viscosity falls below $10^{12.5}$ Ns.m^{-2}, becomes soft when it decreases below $10^{6.5}$ Ns.m^{-2} and can be worked when the viscosity is less than 10^3 Ns.m^{-2}. (In comparison, the viscosity of glycerine is 1.5 Ns.m^{-2}, of olive oil is 8.4×10^{-2} Ns.m^{-2}, of water is 1×10^{-3} Ns.m^{-2} and of molten metals is a few 10^{-3} Ns.m^{-2}.) For SiO_2 glass, the working temperature is about 1700 °C but it can be decreased dramatically by introduction of other oxides, as shown in Table 2.1.

The coefficient of thermal expansion for SiO_2 glass is very low,

Table 2.1 Some glass compositions and thermal properties

Type	*Alumino-silicate*	*Boro-silicate*	*Soda lime*	*Lead*
		Composition weight %		
SiO_2	55.3	80.6	73	63
Al_2O_3	23	2.2	1	< 1
B_2O_3	7.5	12.6	< 1	–
CaO	4.7	< 1	5.2	< 1
Fe_2O_3	–	< 1	< 1	–
K_2O	< 1	–	16	6
MgO	8.5	–	3.6	< 1
Na_2O	< 1	4.2	7.6	–
PbO	–	–	–	21
		Technical temperatures, °C		
Strain	670	520	470	390
Annealing	720	565	510	430
Softening	940	820	700	620
Working	1220	1220	1100	970
		Coefficient of thermal expansion, K^{-1}, $\times 10^6$		
	3.3	3.2	9.2	9.1

$0.5 \times 10^{-6}\ K^{-1}$, but can be increased by the introduction of other oxides. Examination of the data in Table 2.1 shows that the coefficient are largest for the least refractory glasses, and statistical analysis of such data in terms of composition shows that alkali oxides have a particularly strong effect on expansion characteristics. Nevertheless, even these enhanced coefficients of expansion are low compared to those of most metals, as can be seen by comparing the data in Tables 1.1 and 2.1, and this difference has important implications for the design of glass–metal seals because of the brittleness of glasses.

Such is the importance of the expansion characteristics of glasses that recent decades have seen the development of glass ceramics, glassy materials to which have been added nucleating agents such as TiO_2, ZrO_2 or P_2O_5. The presence of these oxides causes crystals to form and grow when the glass is reheated and held at certain temperatures, producing an array of crystalline particles embedded in the glass matrix. This structural change affects the thermal expansion, and other material characteristics such as strength and toughness. Precise control of the heat treatment conditions allows structures to be produced with expansion characteristics that closely match those of metals, but few mass production applications have yet emerged.

2.5.2 Design of glass seals

Since glasses are inherently brittle materials, care must be taken when designing seals to minimize the strains endured by glass components and steps taken to minimize and accommodate the mismatched expansion characteristics when joining to other materials. Thus it can be advantageous to use a graduated seal in which a series of glasses with progressively different but intermediate coefficients of expansion are used to form a bridge between two different glass components. Particularly severe problems can be encountered when joining glass to metal components. For example, the embedding of metal electrodes in glass insulators when making feedthroughs presents challenging design problems because the greater shrinkage of the metal will impose tensile stresses in the insulation that can be markedly detrimental since glasses are 10 to 20 times weaker in tension than compression.

Glass seal designs should specify materials whenever possible with closely matching coefficients of thermal expansion and this desire has resulted in a preference for glass–metal components that use low expansion metals such as W or Mo for which matching glasses have been developed. Also widely used are Kovar, a Fe-29Ni-18Co alloy with an expansion coefficient of $4.8 \times 10^{-6}\,K^{-1}$ up to 400 °C, and Invar, Fe-36Ni alloy with an expansion coefficient of $2.3 \times 10^{-6}\,K^{-1}$ up to 200 °C, although the expansions of both alloys are substantially larger at higher temperatures. Commercial examples of this practice include the use of B37™,[1] a SiO_2-B_2O_3-Na_2O-Al_2O_3-K_2O glass with a coefficient of expansion of $3.75 \times 10^{-6}\,K^{-1}$ for joining to W, and B47™, a SiO_2-B_2O_3-K_2O-Al_2O_3 glass containing small additions of Li_2O, PbO and F with a coefficient of $4.7 \times 10^{-6}\,K^{-1}$ for joining to Kovar. Conventional high expansion metals can be used in glass–metal joints if the design is such that cooling from the fabrication temperature causes the glass to be subjected to a small compressive stress. A first idea of the magnitude of the stress, σ_R, caused by the mismatched coefficients of thermal expansion, $\Delta\alpha$, can be gained from the equation

$$\sigma_R = K^*E \int \Delta\alpha\Delta T \qquad \{2.6\}$$

where the limits of the integral are the service temperature and the setting temperature at which elastic stresses are generated, E is the elastic modulus of the glass, and K^* is a constant that reflects the design of the joint.

More accurate assessment of the stresses that may occur requires a detailed thermal history of the joint, including cooling rates, and the possibility of stress relief by plastic yielding of metal components.

[1] ™ Registered Trade Marks of Glass Bulbs Ltd, Doncaster, England

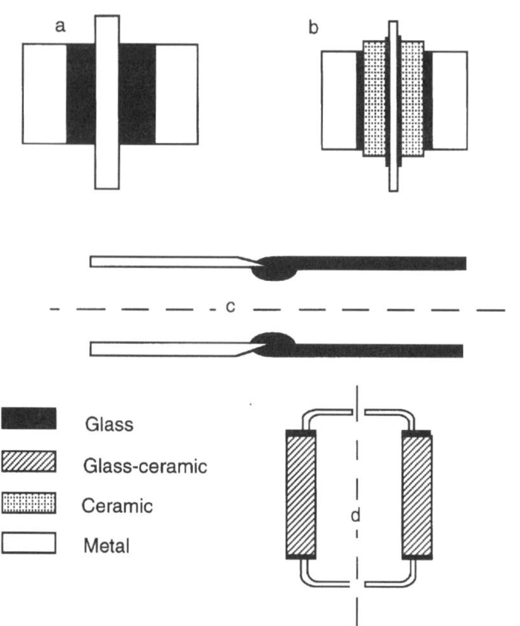

Figure 2.11 Some glass seal configurations used to fabricate, (a) and (b), electrical feed throughs, (c) a Housekeeper seal and (d) a vacuum interrupter

Nevertheless, the equation does provide some insight and in general it has been found that the maximum permissible difference in the coefficients of thermal expansion has a value of about $1 \times 10^{-6}\,K^{-1}$. The expansion of glass ceramics can be near linear functions of temperature, but that of glasses can show marked variability. Thus the maximum strain between a glass and other materials can occur at some intermediate temperature as the joint cools after fabrication rather than at room temperature and will be particularly detrimental if it is sufficient to cause cracking of the glass. This degradation can be mitigated by using a very thin glass interlayer in which cracks propagate with difficulty, but it is even more desirable to use joint designs that can accommodate the maximum strains thereby causing the values of K and σ_R to diminish.

Figure 2.11 shows four common forms of joint design which can overcome or accommodate mismatches in the coefficients of thermal expansion that would otherwise cause cracking of glass. Figure 2.11a shows a low expansion metal electrode encased in glass which is itself encased by a high expansion metal sheath which causes compressive stressing of the glass as the product cools. Figure 2.11b is a similar product that employs a ceramic insulator bonded to the electrode and the sheath by thin glass interlayers. Figure 2.11c is a Housekeeper seal produced by melting the end of a glass tube in contact with a thin and ductile metal tube with chamfered ends which deforms plastically to accommodate strains caused by the mismatched coefficients of expansion. It should be noted that the name Housekeeper refers to the American scientist W. G. Housekeeper, and is not a subtle reflection on some arcane aspect of domestic economy. Finally, Figure 2.11d is a prototype vacuum interrupter whose metal end caps are joined to a glass ceramic tube with a matching coefficient of expansion by thin glass interlayers.

In describing these various designs attention has been focused on ways of minimizing the stresses within the glass components or bonding layers. The implication has been that the glass–metal or glass–ceramic interfaces are strong and free of defects such as unbonded regions. This should be the case if good bonding practices are adopted but their specification requires understanding of the materials science that controls bond formation.

2.5.3 Bond formation

Adequate bond formation when glass sealing requires that the materials are in intimate, atomic, contact and that there is chemical bonding across their interface. High temperatures are used when sealing so that the glass is liquid and the achievement of intimate contact depends on its wetting behaviour, that is, on its ability to flow across surfaces and into gaps between components. As when brazing, this requirement can be specified quantitatively by the statement that γ_{SL} must be less than γ_S so that θ is less than 90°, and is preferably a lot less.

Experimental sessile drop studies to supplement observations made when glass sealing have shown that many glasses wet oxides very well, and that glasses based on naturally occurring binder phases wet non-oxide ceramics. The wetting of metals, however, can be difficult to achieve and this particular difficulty has been ascribed to the lack of atomic continuity or chemical bonding across glass–metal interfaces, as there is also across liquid metal–solid oxide interfaces. Sometimes, the difficulty with glass–metal interfaces can be overcome by oxidizing the surface of the metal before it is brought into contact with the liquid glass. This then permits an ionic continuity and bonding to be maintained across the interface as illustrated schematically in Figure 2.12 which shows SiO_2 glass in contact with a metal M whose surface is coated with an oxide, MO. The metal substrate is presumed to be saturated with O and hence in chemical equilibrium with the surface oxide. This in turn is in equilibrium with the glass into which some of it has dissolved to produce a local saturation with the metal oxide, represented by including some M ions in the Si–O sequence. This continuity is maintained even if only one mono-layer of the

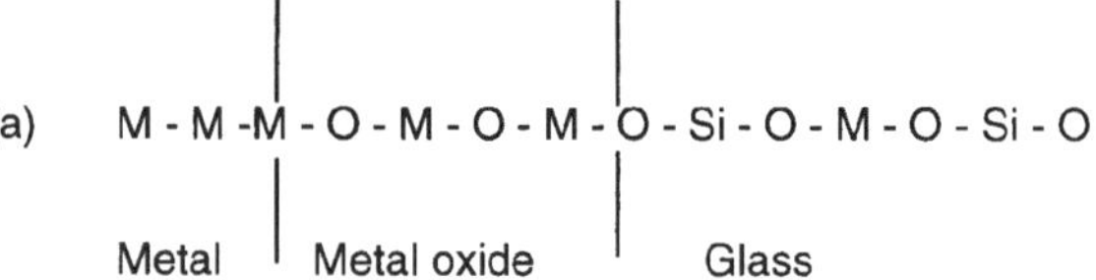

Figure 2.12 Schematic representation of the interfaces formed by glass with (a) a metal coated with an oxide; (b) a metal covered by a monolayer of oxide; and (c) a bare metal surface

metal oxide remains, Figure 2.12b, but is lost if all the oxide is dissolved and dispersed or if the metal surface had never been coated with oxide. In these circumstances, only weak van der Waals forces can operate across the metal–glass interface, Figure 2.12c.

Research has shown that stable chemical bonding of metals to glasses is associated with the presence of low valent oxides such as Cu_2O or FeO. Surface films of higher valent oxides produced by prior oxidation can be wetted by glasses, but this can be accompanied by structural failure within the oxide as a layer of low valent oxide is produced at the interface with the glass. It is also essential that the surface oxide is adherent to its metal substrate and this can be achieved in practice usually by forming relatively thin films at modest temperatures. Thus the 1–2 μm thick oxide films formed on Kovar by heating in air for about 20 minutes at 800 °C are strongly adherent, while the 5 μm thick films formed by similarly heating at 900 °C are not.

Some research workers have suggested chemical reactions promote the wetting in glass–metal systems because the accompanying energy release effectively decreases the value of γ_{SL} and hence of θ. Whatever the details, it is clear that reactions can be beneficial. Thus it is a long-established practice to include some CoO in glasses used to enamel steel surfaces. This additive is reduced by the Fe which has a higher oxidation potential, resulting in the formation of FeO and promoting interface formation even with unoxidized steel sheet. The Co released alloys with the Fe to form small crystals of $FeCo_X$ that provide mechanical keying to supplement the chemical bonding.

2.6 ADHESIVE BONDING

This is almost certainly the most widely used joining process since there can be hardly a man, woman or child who has not glued together some domestic item, has not sealed an envelope or pasted paper. Adhesive bonding is a capillary joining process that uses the flow of organic filler materials to form joints between components. The fillers, either simple liquids or else solid particles carried in a liquid, must wet if intimate contact is to be achieved with the component mating surfaces and subsequently must harden to produce a permanent bond.

Adhesive bonding is a room or low-temperature process that can employ any one of a wide range of natural and synthetic adhesives, such as those summarized in Table 2.2. Selection of an adhesive for a particular application must take into consideration its ability to wet the component materials, whether it may attack or corrode the component, its ease of hardening and its toughness and elasticity after hardening.

A frequent choice for technological applications is one of the thermoplastic or thermosetting synthetic adhesives. Both groups consist of covalent bonded C chain molecules that interact with each other through weak van der Waals forces. Thus the surface energies of the adhesives are low, typically

Table 2.2 Main groups of generally available organic adhesives

Protein derived materials such as glues made from hides and bones

Starches and vegetable gums made from gum arabic

Cellulose adhesives such as methylcellulose and ethylcellulose

Thermoplastic synthetic resins such as polythene, polyvinyl chloride and polytetra fluoroethene that soften when heated

Thermosetting synthetic resins such as melamines and phenol- and urea-formaldehyde resins that remain rigid when heated because of cross-linking of the polymer molecules

Natural resins and bitumens such as shellac and rosin

Rubber adhesives such as latex that are used to bond flexible materials such as cellophane tapes

$0.05\,Jm^{-2}$ or less, and they will be strongly attracted by and therefore wet the bonding surfaces of clean metal and ceramic components that have energies ranging from about 0.3 to $3.5\,Jm^{-2}$. It is good practice, therefore, to remove low surface energy organic contaminants and moisture from metal and ceramic bonding surfaces, and to bond in dry environments. If the component materials themselves are low surface energy materials such as polyvinylchloride, PVC, it is also usually necessary to activate, generally to oxidize, their surfaces by chemical etching or exposure to radio frequency plasmas or the corona produced by high-voltage electrical discharges. For both high- and low-surface energy components, it is good practice to roughen mating surfaces before bonding.

Having achieved intimate contact, permanent bonding must be produced by hardening the adhesive. This may involve simple drying of carrier liquids, coalescence of emulsions, cooling of thermoplastic adhesives or by cross-linking, thermosetting, reactions as when Bakelite is formed by the interaction of phenol and formaldehyde or a diamine is used to form the epoxy resin from diglycidyl ether, a reaction that can be written as the equation

$$2(\underbrace{CH_2-CH}_{O}-R'-\underbrace{CH-CH_2}_{O}) + H_2N-R-NH_2 \longrightarrow \begin{matrix} R'-\underset{}{\overset{OH}{\overset{|}{C}H}}-CH_2 & & & & CH_2-\overset{OH}{\overset{|}{C}H}-R' \\ & \diagdown & & \diagup & \\ & & N-R-N & & \\ & \diagup & & \diagdown & \\ R'-\underset{OH}{\underset{|}{C}H}-CH_2 & & & & CH_2-\underset{OH}{\underset{|}{C}H}-R' \end{matrix}$$

and typically takes 15–30 minutes to complete at room temperature. One notable exception to the need for hardening additives is 'superglue', a cyanoacrylate adhesive that can cross link completely within a few seconds at room temperature in the presence of hydroxyl groups, such as those provided by the moisture adsorbed on even a nominally dry metal surface. The cross linking referred to so far produces hardening of the adhesive, but similar interactions can also promote strong and permanent adhesion to metals or ceramics by forming covalent bonds with their mating surfaces.

As with brazed or glass-sealed joints, lap and sleeve configurations are usually preferred for adhesive bonding applications, but the overlap lengths are usually substantially longer because of the relative weakness of the organic filler materials. Whatever the design, the introduction of adhesively bonded joints can and does suffer resistance because of two important *perceived* limitations. The first of these is a suspected poor resistance to water ingress along the bond line or through the adhesive. This was a genuine problem once but care in design and specification of appropriate filler materials and fabrication process parameters now permits moisture resistant joints to be fabricated. The second is a low maximum service temperature and this remains a restriction for the use of adhesively bonded joints because the available filler materials cannot withstand temperatures of more than a few hundred degrees Celsius. However most applications of manufactured goods are at room or near room temperature and automated application of adhesives is being increasingly used in industry to replace mechanical attachment of components. Adhesive bonding is also impinging on some of the traditional applications of other joining techniques, such as the spot welding of automobile body structures.

FURTHER READING

Diffusion Bonding, 2, (1991) Proceedings of the 2nd International Conference on Diffusion Bonding (ed. D. J. Stephenson), Elsevier Applied Science, London.

Kinloch, A. J. (1987) *Adhesion and Adhesives: Science and Technology*, Chapman & Hall, London.

Lancaster, J. F. (1993) *Metallurgy of Welding*, 5th edn, Chapman & Hall, London.

Milner, D. R. and Apps, R. L. (1968) *Introduction to Welding and Brazing*, Pergamon Press, Oxford.

Rodriguez, F. (1989) *Principles of Polymer Systems*, 3rd edition, Hemisphere Publishing Corporation, New York.

Thwaites, C. J. (1992) *Capillary Joining – Brazing and Soldering*, Research Studies Press, Chichester.

Tomsia, A. P. (1990) Glass-metal seals. In M. G. Nicholas (ed.) *Joining of Ceramics*, Chapman & Hall, London.

3 Scientific aspects of joint formation

This chapter treats the scientific aspects of joint formation for each process separately. The order adopted – first brazing, then diffusion bonding and finally transient liquid phase bonding – reflects the completeness and acceptance of current scientific understanding of the controlling mechanisms and the present industrial importance of the joining processes. Also included in this chapter are some material property data such as surface energy, viscosity, solubility and diffusivity values needed to quantify model predictions and correlations of joining behaviour with material characteristics.

Joining processes are important and sometimes crucially important enabling technologies, but no technology is firmly based unless it is built on proven scientific foundations. Technology without this underpinning is not engineering but merely witchcraft or alchemy. Thus it is essential to have an awareness and understanding of the scientific principles involved in the fabrication of joints by brazing, diffusion bonding and transient liquid phase bonding if they are to be specified and used with confidence. Without such understanding, an attempt to overcome production difficulties can degenerate rapidly from empirical adjustment to mere random tinkering with process parameters. While awareness of the scientific foundations of a joining process will not necessarily provide an immediate answer to a production problem, it will at least provide the map that will help locate possible causes. Thus understanding of the mechanisms responsible for the creation of interfaces should enable rapid identification of whether a high reject rate is due to incorrect selection of materials, of surface preparation procedures or of processing conditions.

The most fundamental scientific description of the processes of brazing, diffusion bonding and transient liquid phase bonding is that they change the configuration of the component/joining material system replacing pairs of mating surfaces by interfaces and thereby causing the total energy of the system to decrease. Thus the change will be permanent unless there is outside intervention to replace this energy – the processes create permanent bonds and work must be done to break them and dismantle the product. While this is a characteristic common to all three joining processes, the details of the scientific analysis and description of each individual process reflects its distinctive characteristics. Thus all scientific descriptions of brazing must be based on the fact that flow of a liquid metal or alloy over a solid surface or into a capillary gap between components to create a brazed joint is always impeded by viscous drag within the liquid. No such impedance operates when diffusion bonding but the scientific descriptions must pay close attention to which of many solid state diffusion and plastic flow mechanisms are controlling the extent and rate of interface formation, while descriptions of the successful application of the transient liquid phase bonding technique are even more complex because its progress depends on the sequential operation of solid, liquid and again solid state mechanisms.

One of the most obvious results of the operation of these different controlling mechanisms is the difference in the kinetics of joint formation. Brazing is controlled by inherently fast liquid flow mechanisms, diffusion bonding depends on slow solid state mechanisms and transient liquid phase bonding may be controlled by either liquid or solid state mechanisms. Thus insight into the scientific foundations of a joining process can be gained by consideration of the kinetics of joint formation as well as the final equilibrium configurations of the component/joining material system.

3.1 BRAZING

Successful brazing depends on the ability of a liquid metal or alloy to wet the component being joined so that it flows readily over their surfaces and fills gaps to form joints. It is important that the flow of liquid should fill the joint gaps completely and not result in voids attached to the component surfaces that degrade mechanical properties. Thus, every part of the component surface should be in intimate contact with the liquid, wetted, even though it is impossible to ensure that everywhere on the solid surface has identical characteristics. It is necessary, therefore, for the wetting of a component by a liquid braze to be good enough for small degradations of surface characteristics to be accommodated, and to be confident of this requires some means of measuring the extent of wetting.

3.1.1 Wettability and the Young equation

The most widely accepted and best established scientific measure of wetting behaviour is the contact angle. This was identified earlier as the

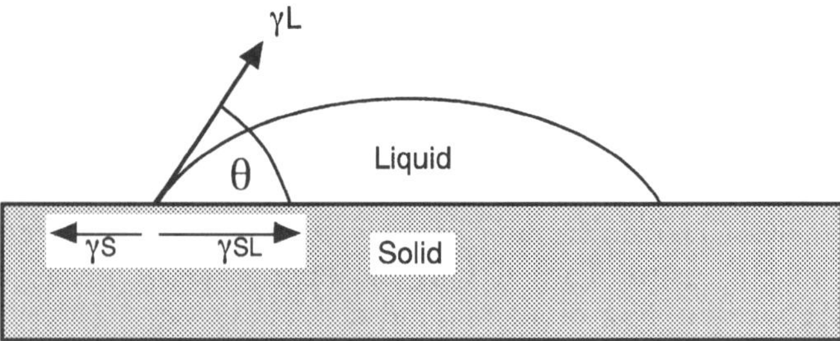

Figure 3.1 Identification of the surface and interfacial tensions acting at the periphery of a sessile drop resting on a horizontal solid substrate

angle at a liquid front subtended by the liquid surface and the solid–liquid interface, Figure 1.2, and thus measurements of drop profiles can be used to derive contact angle values. The technical literature contains many contact angle data derived from the profiles of sessile drops with volumes of typically 0.01 ml, and also from other techniques such as the heights of capillary rise. Contact angles derived from careful experimentation are reproducible and independent of the particular technique employed and close examination of a variety of such experimental data led Thomas Young to conclude as long ago as 1805 that 'for each combination of a solid and a fluid there is an appropriate angle of contact', that is, that the contact angle is a unique materials characteristic. Young concluded also that the contact angle depended solely on the 'surface tensions' of the materials and that its value was governed by the relationship

$$\gamma_S = \gamma_L \cos\theta + \gamma_{SL} \qquad \{1.1\}$$

where γ is the surface tension and the suffixes S and L refer to the solid and liquid surfaces and SL refers to the solid–liquid interface identified in Figure 3.1. This particular relationship is referred to frequently in discussions of brazing and wetting and is generally called the Young equation.

As commented earlier, liquid surface tensions measured in $N.m^{-1}$ have the same numerical values as surface energies measured in $J.m^{-2}$, and treating the γ terms identified in Figure 3.1 as energies permits a derivation of the Young equation, {1.1}. Thus, consider the energy changes that will occur when the edge of a sessile drop advances over a smooth horizontal solid substrate to approach an equilibrium configuration. In a short time during this approach, the contact area radius will increase slightly from r to $r + \delta r$ and the contact angle will decrease from $\theta + \delta\theta$ to θ, as illustrated in Figure 3.2. Thus the area of the solid surface will decrease by $\pi[(r + \delta r)^2 - r^2]$ and that of the interface will increase by a similar amount. The radial change from r to $(r + \delta r)$ will also increase the area of the liquid

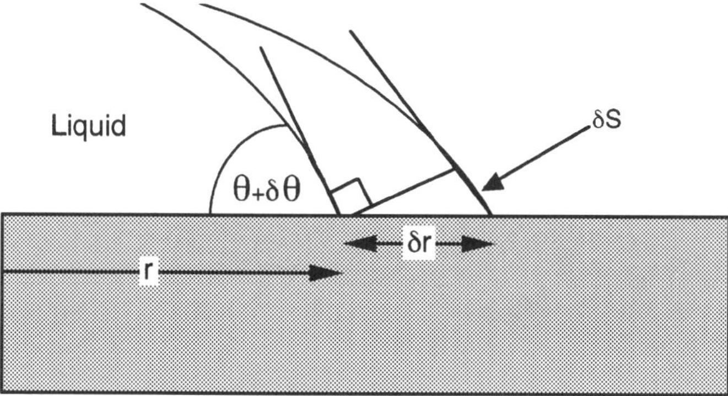

Figure 3.2 Schematic illustration of the advance of a liquid front over a smooth solid substrate

surface by $2\pi r.\delta r \cos\theta$ so that the energy change, δE, can be described by the expression

$$\delta E = \pi[(r + \delta r)^2 - r^2]\gamma_{SL} + 2\pi r.\cos\theta.\gamma_L - \pi[(r + \delta r)^2 - r^2]\gamma_S \qquad \{3.1\}$$

which can be rearranged as

$$\delta E = \pi[(r^2 + 2r\delta r + \delta r^2) - r^2]\gamma_{SL} + 2\pi r.\delta r \cos\theta.\gamma_L - \pi[(r^2 + 2r\delta r + \delta r^2) - r^2]\gamma_S \qquad \{3.2\}$$

$$\delta E = \pi[2r\delta r + \delta r^2]\gamma_{SL} + 2\pi r.\delta r \cos\theta.\gamma_L - \pi[2r\delta r + \delta r^2]\gamma_S \qquad \{3.3\}$$

$$\delta E/\delta r = \pi[2r + \delta r]\gamma_{SL} + 2\pi r \cos\theta.\gamma_L - \pi[2r + \delta r]\gamma_S \qquad \{3.4\}$$

No further advance will occur or be energetically favourable when the drop attains an equilibrium configuration so that both δr and $\delta E/\delta r$ will be zero

$$0 = \pi 2r\gamma_{SL} + 2\pi r \cos\theta.\gamma_L - \pi 2r\gamma_S \qquad \{3.5\}$$

which may be rewritten as the Young equation by rearranging the terms and dividing throughout by $2\pi r$

$$\gamma_S = \gamma_L \cos\theta + \gamma_{SL} \qquad \{1.1\}$$

The Young equation can be derived even more readily also by regarding the γ terms as tensions and resolving those shown in Figure 3.1 to identify the horizontal forces acting on the drop periphery. When the drop has assumed an equilibrium configuration, the resultant horizontal force must be zero, so

$$0 = \gamma_L \cos\theta + \gamma_{SL} - \gamma_S \qquad \{3.6\}$$

where forces acting to the right are regarded as positive, and transferring the negative γ_S force to the left hand side of this expression again yields the Young equation. However, this treatment also yields a vertical force of

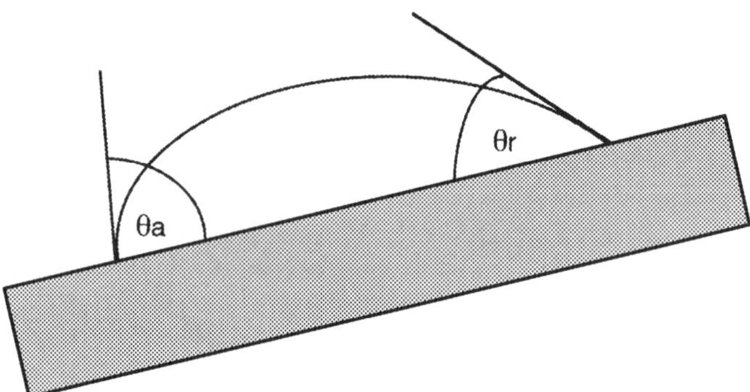

Figure 3.3 A sessile drop of liquid resting on an inclined solid substrate. The advancing and receding contact angles are identified as θa and θr

$\gamma_L . \sin\theta$ acting at the drop periphery without any apparent balancing force. Some workers have suggested that this force will cause the solid at the edge of the drop to be lifted, either elastically or plastically but there is as yet no body of evidence to support this suggestion. Indeed, analyses based on tensions may be questionable because while it is easy to conceive that the creation of surfaces and interfaces requires work it is not so obvious that a solid surface or a solid–liquid interface will experience a tension that is numerically equal to that energy.

While the Young equation has both theoretical and experimental bases, the uniqueness of the contact angle as a materials characteristic cannot be maintained. Thus the angle is affected by inclination of the substrate on which a sessile drop is resting. Tilting the substrate does not cause immediate movement of the drop down the slope and this pinning or hysteresis produces an asymmetrical profile as illustrated in Figure 3.3. The contact angle of the advancing, downhill, front is larger and that of the receding, uphill, front is smaller than that assumed by a drop with an equilibrium configuration. The values of the advancing and receding angles cannot be predicted readily since they depend not only on the system materials but also on causes of the hysteresis. One commonly suspected cause is heterogeneity of substrate surface chemistry, but the extent and effects of this are not easy to characterize or model. Another frequent cause of apparently aberrant contact angle values, and the display of hysteresis effects by advancing and receding liquid fronts, is lack of surface smoothness, and this has been analysed and modelled as described below.

3.1.2 Wetting of idealized rough surfaces

The derivation of the Young equation given above assumed that the substrate surface was perfectly smooth. This circumstance will occur rarely in practice and it has been known for many decades that the wetting of a surface is affected by roughening. However, early scientific studies of the effect of substrate roughening on wetting behaviour were limited in scope and predictions based on theoretical models were in conflict.

Early experimental work by Wenzel with chemically inert systems showed that the contact angles of non-wetting drops were increased by roughening their substrates. At that time, it was not possible to characterize the topographic changes produced by roughening so it was assumed implicitly that the surface features were insignificantly small compared to the dimensions of a sessile drop and that their geometry was of no consequence except in so far as it increased the surface area of the substrate. Increases in the area of the substrate directly increases the contributions of the solid surface and solid–liquid interface to the total energy of the system, so that the derivation of the Young equation needs to be modified to

$$0 = \pi 2\mathrm{r}\gamma_{\mathrm{SL}}.\mathrm{W}_{\mathrm{R}} + 2\pi\mathrm{r}\cos\theta_{\mathrm{R}}.\gamma_{\mathrm{L}} - \pi 2\mathrm{r}\gamma_{\mathrm{S}}.\mathrm{W}_{\mathrm{R}} \quad \{3.7\}$$

$$\gamma_{\mathrm{S}}.\mathrm{W}_{\mathrm{R}} = \gamma_{\mathrm{SL}}.\mathrm{W}_{\mathrm{R}} + \cos\theta_{\mathrm{R}}.\gamma_{\mathrm{L}} \quad \{3.8\}$$

$$(\gamma_{\mathrm{S}} - \gamma_{\mathrm{SL}}).\mathrm{W}_{\mathrm{R}} = \cos\theta_{\mathrm{R}}.\gamma_{\mathrm{L}} \quad \{3.8\}$$

$$\mathrm{W}_{\mathrm{R}}.\cos\theta = \cos\theta_{\mathrm{R}} \quad \{3.9\}$$

where W_R is the ratio of the actual to the nominal area of the solid surface and the solid–liquid interface and θ_R is the contact angle assumed on a roughened substrate. This analysis correctly predicts that roughening a substrate will increase the contact angles assumed by non-wetting drops, and also predicts that the angles assumed by wetting drops will decrease.

In contrast, later experiments by Shuttleworth and Bailey (1948) of both wetting and non-wetting inert organic systems using coarsely scribed substrates showed that asperities on rough surfaces could pose significant barriers to the flow of liquids attempting to assume equilibrium configurations. They found that advancing liquid fronts came to rest on down grades as shown in Figure 3.4 and that the receding liquid fronts rested on up grades. Further they observed that the actual contact angle between the substrate surface and the liquid surface was unaffected by the asperity gradient, and this angle was assumed to be θ, the equilibrium angle, rather than some variable parameter such as θ_a or θ_r. Thus their observations can be described by the simple expression

$$\theta_{\mathrm{R}} = \theta + \alpha* \quad \{3.10\}$$

where $\alpha*$ is the gradient of the asperity where the liquid front has come to rest. Hence their work leads to the prediction that all advancing angles will be *increased* by an amount $\alpha*$ and that all receding angles will be *decreased* by $\alpha*$ so that the system will exhibit a hysteresis of $2\alpha*$.

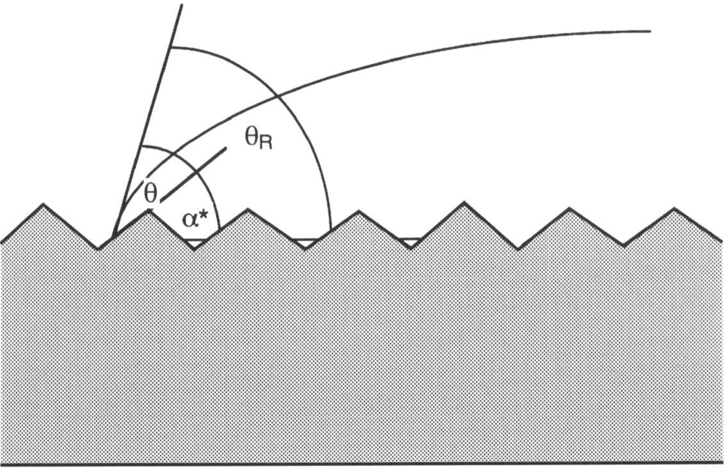

Figure 3.4 Sketch illustrating the Shuttleworth and Bailey description of the effects of substrate roughness on the apparent contact angle assumed by a sessile drop

Thus the analyses of both Wenzel and of Shuttleworth and Bailey predict that the contact angles of non-wetting liquids will be increased by roughening of the substrate, but Wenzel predicts that the contact angles of wetting liquids will be decreased while Shuttleworth and Bailey again predict that they will be increased. Further, the Shuttleworth and Bailey interpretation predicts that roughening causes hysteresis while that of Wenzel does not.

This stark conflict between the predictions for wetting liquids and other lesser disagreements was resolved by the more recent theoretical analyses. Thus Johnson and Dettre (1963) considered idealized substrates covered by mathematically tractable sinusoidal features which they regarded as being not only physical obstacles but also a series of energy barriers that the liquid could surmount as it advanced from one metastable configuration to another if it had sufficient vibrational energy. This vibrational energy was not quantified and indeed was treated as an adjustable parameter, but they did relate the substrate energy barriers to the heights of surface asperities. It was then shown by detailed calculations that the contact angle observed for a sessile drop resting on one of their idealized substrates would approach that predicted by the Wenzel model if the vibrational energy of the liquid was large but would approximate to the predictions of the Shuttleworth and Bailey model if the vibrational energy was small. Thus as illustrated schematically in Figure 3.5, the effect of substrate roughening was to increase contact angle values for wetting as well as non-wetting liquids with low vibrational energies, but this enhancing effect of roughening diminished progressively as the vibrational energy of the liquids was increased.

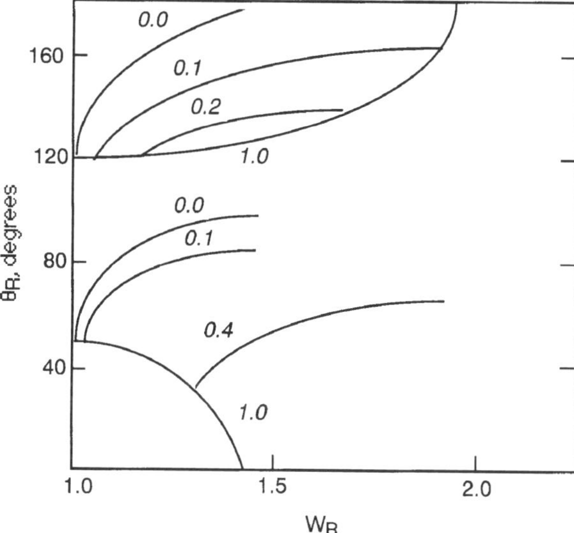

Figure 3.5 Predicted effects of vibrational energy, shown as italicized labelling of curves, on the contact angles assumed by wetting and non-wetting sessile drops on roughened substrates. After Johnson and Dettre (1963)

Application of the Johnson and Dettre model to an energetic liquid also confirmed the Wenzel prediction that a liquid will behave as if it had a zero contact angle, will wick, when the substrate roughness is such that

$$W_R = (1/\cos\theta) \tag{3.11}$$

The detailed calculations using the energy barrier model also confirm the early predictions of Cassie and Baxter that the steep valleys on the substrate surface will be incompletely filled if the liquid is very non-wetting, as illustrated schematically in Figure 3.6, causing the interface formed by a substrate in nominal contact with the liquid to be a composite of solid–liquid and solid–gas regions.

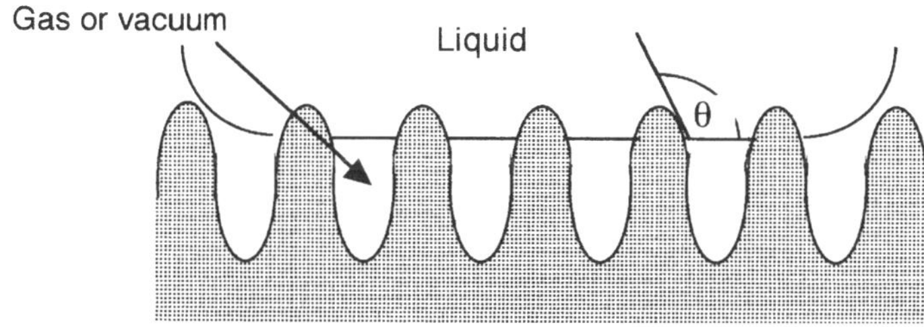

Figure 3.6 Idealized substrate surface in contact with a sessile drop that has produced a composite, solid–liquid and solid–gas, interface

3.1.3 Idealized filling of capillary gaps

While analysis of the configuration of sessile drops gives insight into wetting processes, the configurations of more usual importance in brazing involve flow of the liquid metal into vertical or horizontal capillary gaps.

An expression for the rise of a wetting liquid in a vertical capillary between two plates was derived earlier as equation {2.5}

$$2\gamma_L \cos\theta = h\rho_L gW \qquad \{2.5\}$$

by considering the action of a liquid surface tension illustrated in Figure 2.6. However, this equation can also be derived using surface and interfacial energies. Thus a height increase of δh while the liquid rises to its equilibrium configuration between two plates of breadth B will *increase* the total interfacial energy of the system by $2(W+B)\delta h.\gamma_{SL}$ and the potential energy of the liquid by $BWh\delta h.\rho_L g$ and will *decrease* the total solid surface energy by $2(W+B)\delta h.\gamma_S$. The rise does not involve any extension, or decrease, in the area of the curved liquid meniscus and hence will have no affect on the energy of the liquid surface. Thus the total energy change δE caused by the rise of δh, can be described by the expressions

$$\delta E = \delta h.\gamma_{SL}.2(W+B) + h\delta h.\rho_L gWB - \delta h.\gamma_S.2(W+B) \qquad \{3.12\}$$

$$\delta E/\delta h = \gamma_{SL}.2(W+B) + h\rho_L gWB - \gamma_S.2(W+B) \qquad \{3.13\}$$

No further rise will be energetically favourable when the liquid column attains its equilibrium height so that $\delta E/\delta h$ will be zero and the expression can be rewritten as

$$(\gamma_S - \gamma_{SL}).2(W+B) = h\rho_L gWB \qquad \{3.14\}$$

and by substitution in the Young equation as

$$2\gamma_L \cos\theta(W+B) = h\rho_L gWB \qquad \{3.15\}$$

which becomes equation {2.5} if the capillary gap is so narrow that $(W+B)$ can be approximated to B and both sides are then divided by B. (Capillary gap dimensions relevant to brazing are a width of typically 0.1 mm and a breadth of perhaps 10 mm.)

The equilibrium height of a column of wetting liquid between plates that are inclined to the horizontal at an angle α' is also given by equation {3.15}, but it should be realized that the length of capillary penetrated will be greater and equal to $h/\sin\alpha'$. Thus penetration of a horizontal capillary, $\alpha' = 0$, by a wetting liquid should be infinitely deep.

Energy analyses of the equilibrium configurations of a wetting liquid in other shapes of capillary gap yield similar results. Thus the equilibrium height in a circular tube of radius R' is

$$2\gamma_L \cos\theta = R'h\rho_L g \qquad \{3.16\}$$

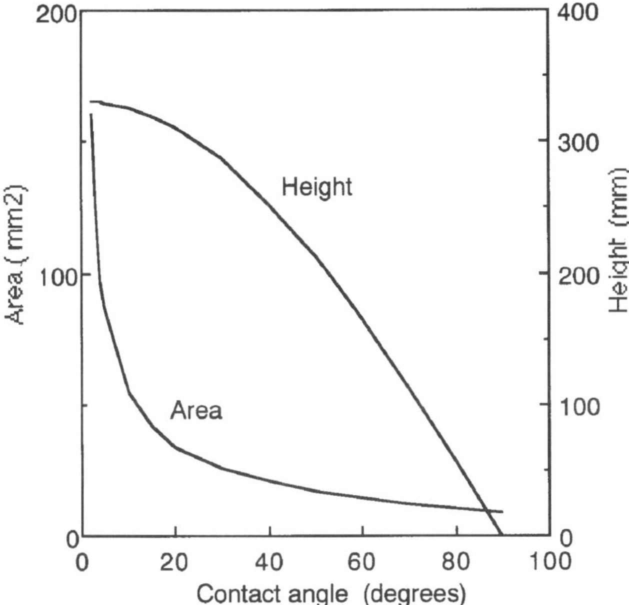

Figure 3.7 The effect of contact angles on the contact area of a sessile drop and the rise of a liquid up a vertical capillary. The drop volume was taken as 0.01 ml, the capillary gap as 0.1 mm and the liquid was assumed to be Cu with a surface energy of 1.3 $J.m^{-2}$ and density of 8000 $kg.m^{-3}$

Equations {1.1} and {2.5} describing the equilibrium configurations of sessile drops and liquid columns demonstrate that these depend on the contact angle. Thus no capillary penetration will occur unless the angle is less than 90° so that $\cos\theta$ is positive, and the actual extent of penetration and spread depends on the specific value of the contact angle as illustrated in Figure 3.7. It can be seen from this that changes in contact angle values have profound but different effects on both the spreading of sessile drops and the penetration of capillary gaps. The spreading of a sessile drop is increased by any fall in the value of the contact angle but the major increase in the contact area occurs only after the angle falls below about 20°. In contrast, the effect on penetration of a vertical capillary is most marked for falls in value of the contact angle from 90° but continuing falls below about 20° cause relatively little additional effect.

This dependence of capillary rise heights on wettability means that height measurements can be used to derive values for the contact angle of the system if data are available for the density and surface energy of the liquid, and the effects of capillary rise on the buoyancy of partially immersed vertical plates is frequently used to assess the wetting behaviour of low-temperature solder systems.

3.1.4 Wetting kinetics of ideal systems

Achievement of equilibrium capillary and sessile drop configurations is not instantaneous since there is a velocity gradient within a liquid flowing in a capillary gap or over a substrate surface and this causes it to experience a viscous drag. The effect of this on flow within channels is well understood and for flow in a horizontal channel between parallel plates depends on the velocity, viscosity and density of the liquid and the width of the channel. This dependence is normally expressed in terms of the Reynolds number,($vW\rho_L/\eta$), where v is the velocity and η is the coefficient of viscosity which is usually measured and reported in terms of $Ns.m^{-2}$ or the equivalent Pa.s. Flow into a channel is initially rapid and turbulent but slows as penetration continues and becomes laminar or streamlined. The change to laminar flow occurs when the dimensionless Reynolds number falls below about 10^3, and this condition is usually satisfied when brazing except for a short distance of perhaps a millimetre after the entrance of a typically 0.1 mm wide capillary gap in which the velocity is greater than about $10\,m.s^{-1}$. When flow is laminar, there is no movement of the liquid in immediate contact with the channel walls and advance of the liquid front along the walls is by movement of liquid from within the bulk.

Liquids advance into channels because they are responding to a pressure, which can be applied mechanically or is a reflection of an imbalance of the solid surface energy and the solid–liquid interfacial energy. When the advance is by means of laminar flow, the rate is directly proportional to the pressure experienced and the area of the channel but is inversely proportional to the viscosity of the liquid and the length, p, of the liquid column within the channel. For a horizontal channel formed by parallel plates, this interdependence can be described by the expression

$$p^2 = (PW^2/6\eta)t \qquad \{3.17\}$$

where P is the pressure. The value of P can be derived for unforced flow of a wetting liquid since its continued advance requires work, force × distance, to be done that is equivalent to the change in the solid surface and interfacial energies of the system. Thus for a small advance of δp

$$PWB\delta p = (\gamma_S - \gamma_{SL}).2B\delta p \qquad \{3.18\}$$

$$P = 2(\gamma_S - \gamma_{SL})/W \qquad \{3.19\}$$

$$p^2 = 2W^2(\gamma_S - \gamma_{SL})t/6\eta W \qquad \{3.20\}$$

$$p^2 = (W\gamma_L \cos\theta/3\eta)t \qquad \{3.21\}$$

It should be noted that this analysis introduces the contact angle only when substituting for $(\gamma_S - \gamma_{SL})$ and, therefore, θ is indeed the equilibrium contact angle assumed by a static liquid front rather than any dynamic value associated with a moving front.

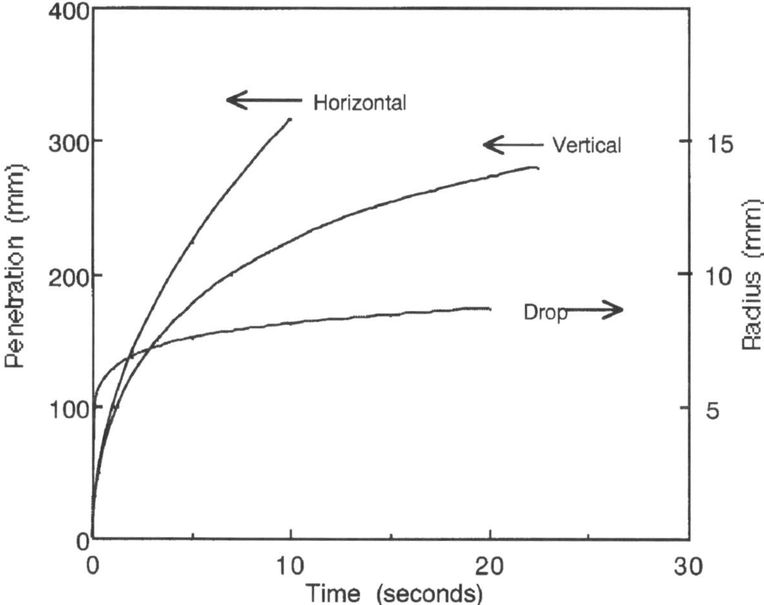

Figure 3.8 Predicted penetration and spreading kinetics for horizontal and vertical parallel plate capillaries and for a sessile drop. The values were derived for liquid Cu that was assumed to have a zero contact angle with the capillary walls and the substrate. The capillary width was taken as 0.1 mm and the sessile drop volume as 0.01 ml while the value of r_0 was taken as zero. The surface energy, viscosity and density values for liquid Cu were taken as $1.3\,J.m^{-2}$, $4.34 \times 10^{-3}\,Ns.m^{-2}$ and $8000\,kg.m^{-3}$ respectively

A similar albeit more complex analysis can be performed for a liquid penetration of a vertical capillary and that takes into account the additional impedance offered by gravity. This analysis results in an expression

$$t = (12\eta/\rho_L^2 g^2 W^3)\{-\rho_L ghW - 2\gamma_L \cos\theta . \ln_e[1 - \rho_L gWh/2\gamma_L \cos\theta]\} \qquad \{3.22\}$$

Thus as shown in Figure 3.8, penetration of a horizontal capillary gap should continue indefinitely albeit at a steadily decreasing rate but the initially fast flow of liquid into a vertical gap decreases rapidly as the level asymptotically approaches the maximum, equilibrium, height.

Predictions of the penetration kinetics of capillary gaps derived by substitution into equations {3.21} and {3.22} have been in existence for many years but predictions of the spreading of sessile drops to assume their equilibrium configurations have been sparse until relatively recently when, for example, de Gennes produced a rigorous analysis of the behaviour of a perfectly wetting drop. This took notice of the fact that the dynamic contact angle of a wetting liquid flowing into a capillary gap is larger than the equilibrium value and can be related to the velocity of the advancing front. De Gennes (1985) assumed a similar variation of dynamic contact angle would be exhibited by the advancing periphery of a sessile drop and this led to a derivation of the expression

$$r^{10} - r_0^{10} = 0.2(4/\pi)^3(\gamma_L/\eta)V^3t \qquad \{3.23\}$$

where r_0 is the radius of the contact area of a sessile drop at time $t = 0$ and V is its volume. As can be seen from Figure 3.8, this expression also predicts a fast initial spreading rate that slows even more dramatically than does penetration into vertical capillaries.

3.1.5 Physical properties of liquid metals

The various models of idealized behaviour described so far relate the extent and rapidity of interface formation to the surface energy, viscosity and density of liquids that wet and flow over solid substrates. Quantitative predictions of the behaviour of actual brazes, therefore, depend on the availability of physical property data for liquid metals and alloys.

Table 3.1 summarizes physical property data for a number of metals that are brazes or braze constituents. Most of the values are well established with a variability of only a few percent, but the reported viscosity data for Al are rather scattered as to a lesser extent are those for Ni.

Table 3.1 Physical properties of some liquids

Liquid	*Melting temp.*, °C	*Surface energy* $J.m^{-2}$	*Viscosity* $mNsm^{-2}$	*Density* $Mg.m^{-3}$
In	156	0.56	1.80	7.03
Sn	232	0.56	1.81	6.98
Cd	321	0.57	2.40	8.01
Pb	327	0.46	2.61	10.67
Zn	420	0.78	3.50	6.58
Mg	649	0.56	1.25	1.59
Al	660	0.91	2.71*	2.38
Ag	961	0.97	4.28	9.33
Au	1063	1.17	5.38	17.40
Cu	1083	1.30	4.34	8.00
Ni	1453	1.78	5.45**	7.90
Pd	1552	1.50	–	10.49
Ti	1667	1.65	5.20	4.13
Hg	−39	0.50	2.04	13.69
H_2O	0	0.07	1.00	1.00

* ±50%
** ±15%

The values quoted were obtained by a variety of experimental techniques. Those for liquid surface energies come principally from measurements of the profiles of sessile drops or pendant drops hanging from the tips of rods or from precise measurements of the pressure needed to blow a bubble in a bath of liquid metal. Similarly, the viscosity values were derived from techniques such as measurement of the drag exerted on a rotating cylinder immersed in a bath of liquid metal or the time taken for a liquid to be driven through a capillary by a known pressure. Finally, the density data were calculated from the results of Archimedean displacement experiments, from measurement of the mass of liquid needed to fill a vessel of known volume or from profile measurements of a sessile drop of known mass.

Also included in the table for the sake of comparison are data for Hg and H_2O. The various braze metals and components are ranked according to their melting temperatures which cover a fourfold range when expressed in degrees Kelvin, and it can be seen that the surface energy and viscosity values correlate with these temperatures, being larger for the more refractory metals. This correlation is physically meaningful because all three parameters reflect the strength of atom–atom bonding. No such correlation is seen or is to be expected when densities are compared to melting temperatures because the volume of the atoms is the dominant factor, as shown by the systematic variations revealed by plots of density against atomic number.

The surface energy, viscosity and density values all diminish as the temperature is raised and atom–atom distances increase. The rates of these decreases are quite small but not uniform, thus for Cu the value of $(d\gamma_L/dT)/\gamma_L$ is $-1.8 \times 10^{-4}\,K^{-1}$, of $(d\rho_L/dT)\rho_L$ is $-10^{-3}\,K^{-1}$ and of $(d\eta/dT)/\eta$ is $-6.3 \times 10^{-4}\,K^{-1}$, and these differences are significant because their use to adjust the values of data substituted into equations {3.21} and {3.22} lead to the prediction that the penetration rates of both horizontal and vertical capillaries will increase as the temperature rises.

Much larger variations in the physical properties of liquid metals can be produced by alloying. The presence of solutes can have particularly strong effects on melting characteristics, with the liquidus temperature at which an alloy composition is fully molten often, but not always, becoming progressively lower than the melting temperature, T_M, of the solvent metal as the solute concentration increases. This effect is due to the fact that a solid pure metal is in equilibrium with its liquid at the melting temperature so their chemical activities as exemplified by their vapour pressures are equal, but the introduction of a solute dilutes the solvent concentration and decreases its chemical activity and vapour pressure. Hence alloying a solvent liquid will generally cause it to be in equilibrium with a solid that has a lower vapour pressure, that is at a lower temperature. The resultant liquidus temperature can be calculated from the Clausius–Clapeyron equation if a regular solution is formed in which there is no preferential attraction or repulsion of solute atoms. This can be written as

$$\Delta H_f(1/T_M - 1/T_L) = -R.\ln_e N_A \quad \{3.24\}$$

where T_L is the alloy liquidus temperature, ΔH_f is the enthalpy of fusion of the pure solvent metal, T_M is its melting temperature and N_A is its atom fraction.

The depression of the liquidus temperature will be even greater than predicted by the Clausius–Clapeyron equation if there is a strong solvent–solute attraction since this will result in the formation of solvent–solute atom clusters and hence the decrease in the chemical activity and vapour pressure of the solvent will be even greater than with a regular solution, and such a strong attraction can also increase the viscosity of the liquid due to the diminished manoeuvrability of bulky clusters. Calculation of liquidus temperatures can be even more complicated if there is a strong repulsion because this could lead to the formation of immiscible liquids.

Increasing the concentration of a solute above 50% causes it to become the solvent and then the melting temperature of the former solute can be depressed by the presence of what was the solvent. Thus plots of the liquidus temperature as a function of composition can exhibit a low-temperature cusp called a eutectic. The liquid will be in equilibrium with both the solid solvents and further cooling will transform the liquid directly into a mixture of the solids without it needing to pass through a two phase solid–liquid regime. If a strong solvent–solute interaction excessively depresses the liquidus it can result in a 'deep eutectic' with an unusually low liquid–solid transformation temperature. This type of information is presented in the form of phase diagrams which show the stability regimes for various phases as a function of composition and temperature and such a diagram is presented in Figure 3.9 for the technically important Ag-Cu system. This shows that cooling a Ag-28 weight% Cu alloy causes it to transform at 780 °C directly from a liquid into two solid phases. (Readers who are unfamiliar with phase diagrams can find a brief account of their derivation, significance and use in Appendix B.)

Figure 3.9 shows that alloying also affects the nature of the stable solid phases. Thus the solid pure Ag has incorporated some Cu to form a solid solution phase, and Cu has formed a similar solid solution phase by incorporating some Ag. There is usually a limit to how much foreign material can be absorbed without causing the lattice structure to disintegrate, and Figure 3.9 shows this to increase to about 8% for both Cu in Ag and Ag in Cu as the temperature falls from 1063 to 780 °C and then to decrease to nearly zero at room temperature. Thus cooling a liquid alloy of Ag-28Cu to 780 °C will cause it to transform directly from the

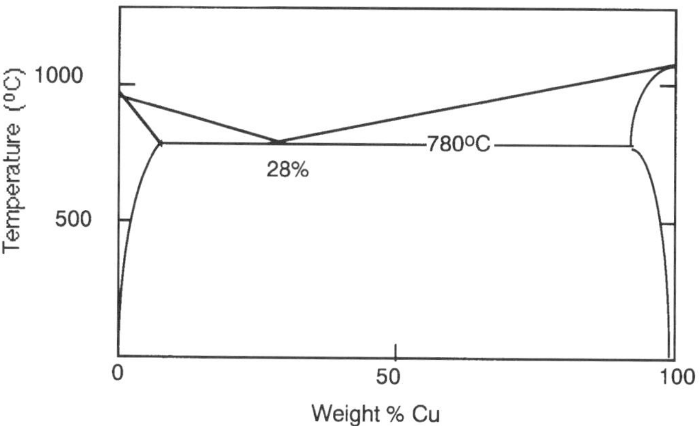

Figure 3.9 The Ag-Cu phase diagram

liquid to a mixture of solid Ag containing 8Cu and solid Cu containing 8Ag. The Ag-Cu system has a simple phase diagram but those for many others are made more complex by the formation of new phases at intermediate compositions. These new structures can be either solid solutions that are stable over a wide compositional range or intermetallic compounds with relatively precise compositions. Reference will be made later to some of these systems that are of importance as brazes.

Eutectic alloys are attractive as braze alloys and the majority of commercially developed braze alloys are based on such compositions. This is because the lowered melting temperatures of eutectic compositions minimize the possibility of macroscopic distortion and microstructural degradation of components while being heated to the brazing temperature, and also saves on energy and capital equipment costs. Further, the direct transition from solid to liquid when heating eliminates the possibility of liquation, that is of the premature flow of liquid while the alloy is in the two phase solid–liquid region that can result in incomplete joint filling.

The range of braze alloys that are used is very large and their compositions are seldom those of the simple binary alloy because melting behaviour is not the only important characteristic of a braze. Thus additions and modifications are made to improve the wetting behaviour, mechanical properties and corrosion resistance of joints and even to alter the colour of joints. Nevertheless, most brazes are designed to have narrow two phase solid–liquid temperature ranges, with those intended for moderate temperature applications being based on the Ag-28Cu and Al-13Si eutectic compositions, while those for higher temperature applications are often based on the Ni-11Si and Ni-11P eutectic compositions even though they contain brittle intermetallic phases that can complicate subsequent service performance.

Table 3.2 Physical properties of some eutectic alloys

Alloy weight %	*Eutectic temp.* °C	T_E/T_M†	*Surface energy* $J.m^{-2}$	*Viscosity* $mNs.m^{-2}$	*Density* $Mg.m^{-3}$
Ag-28Cu	780	0.85	0.97	1.44	(8.9)
Al-13Si	555	0.91	(0.9)	1.31	(2.4)
Ni-11P	880	0.67	(1.3)	15.00	(5.9)
Ni-11Si	1143	0.82	(1.3)	–	(6.3)
Sn-38Pb	183	0.90	0.45	2.70	(8.1)

† Ratio of the melting temperatures of the eutectic and the solvent.
Values in brackets are estimates, those for surface energies were derived by comparison with analogous systems and those for densities by interpolation.

Physical property data for actual brazes are rarities and even those for simple eutectic compositions are not as readily available as for pure metals, as shown by that assembled in Table 3.2, which also includes for comparison data for the Sn-38Pb alloy that is the basis of most low temperature solders. Thus all the liquid density data are estimates, although the values are unlikely to be in error by more then a few percent. The viscosity values reported for Ag-28Cu and Al-13Si are low compared to those of the solvent metals, but this may be merely a reflection of the generally low viscosities of low melting temperature metals. In contrast, the viscosity of Ni-11P is high but the low value of (T_E/T_M), the ratio of the melting temperature of the eutectic compared to that of the solvent, shows it to be a deep eutectic and hence its flow may be rendered sluggish by the presence of solvent–solute clusters. Ni-11Si is not a deep eutectic and hence its unknown viscosity can be expected to lie within the normal range of 1 to 3 $mNs.m^{-2}$. The data in Table 3.2 refer to temperatures close to those at which the eutectic compositions melt and actual use temperatures will be somewhat higher, but this difference should not have a significant effect on the property values.

3.1.6 Brazing of metal components

Wetting behaviour for simple systems is determined by the relative size of their surface and interfacial energies. Values are readily available for the surface energies of liquid metals as described earlier and those for solid metals can be derived experimentally for solid metals and a number of alloys by methods such as the zero creep technique. This uses the metal in the form of a thin foil or wire whose shrinkage at high temperatures to minimize its surface area and hence total surface energy is prevented by application of a small tensile load – so there is zero creep – and the solid surface energy can be derived by equating the gravitational effects of the

Table 3.3 Interfacial and surface energies of metal–metal systems, $J.m^{-2}$

	γ_{SL}								
	Al	Ag	Au	Cu	Ni	Fe	Ti	W	γ_L
Al	0.15	0.07	*	*	*	*	*	*	0.91 Al
Ag		0.18	0.09	0.24	0.60	0.84	*	1.18	0.97 Ag
Au			0.20	0.08	0.44	0.42	*	0.98	1.17 Au
Cu				0.26	0.48	0.67	*	0.96	1.30 Cu
Ni					0.36	0.33	*	0.35	1.78 Ni
γ_S	0.98 Al	1.10 Ag	1.40 Au	1.78 Cu	2.28 Ni	2.10 Fe	1.96 Ti	2.80 W	

balancing load with the solid surface energy times the peripheral length of the foil or wire. The actual solid surface energy values obtained by use of such techniques are typically about 15% larger than those for the liquids, and hence intuition suggests that uncontaminated metal substrates should be wetted, and well wetted, by liquid metals. Experiment shows this to be the case, indicating that the solid–liquid interfacial energies of metal–metal systems are generally significantly smaller than the surface energies of the solid metal substrates.

Few values for the solid–liquid interfacial energies of metal–metal systems have been reported even though they can be derived readily from observations of the behaviour of sessile drops if the relevant solid and liquid surface energies are known. However, theoretical values have been calculated for these energies, as exemplified by the model of Miedema and den Broeder (1979) which takes account of the physical changes experienced by the surface layers atoms of the solid A and liquid B, γ_{SL}^{A} and γ_{SL}^{B}, as well as dissimilar atom–atom interactions, γ_{SL}^{A-B}. Values of these parameters have been calculated from the enthalpies of fusion and solution and tabulated results for individual metals and combinations permit easy calculation for systems of interest and Table 3.3 shows that these predicted interfacial energies are invariably smaller than those for the relevant solid surface energies, so all the listed metal–metal systems should wet.

The interfacial energy values predicted for solid Al-liquid Al and other similar metal systems fall within the narrow range of 0.17 to 0.20 of their liquid surface energy while the ratios for dissimilar metal systems range from 0.07 to 1.22. The values for similar metal systems are in reasonable accord with experimentally measured values derived from undercooling experiments as shown in Figure 3.10. This accord generates some confidence about the predictions for most dissimilar systems from which it can be concluded that uncontaminated metal–metal systems without any

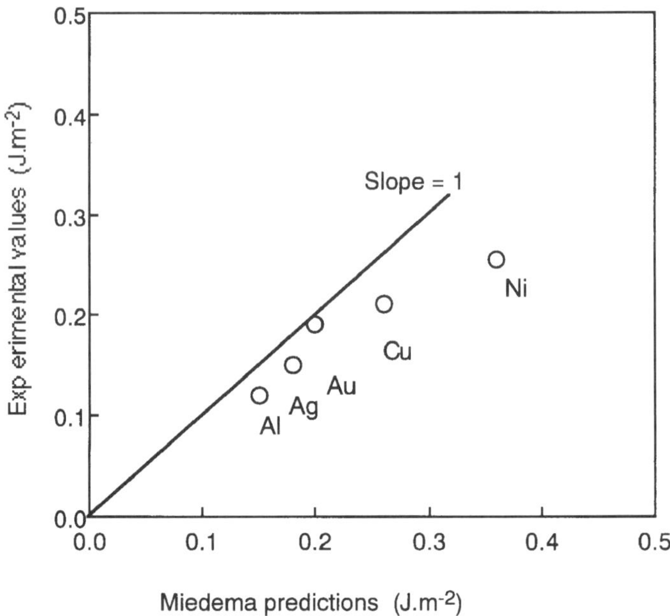

Figure 3.10 Comparison of the theoretical predictions and experimental measurements of the solid–liquid interfacial energies of similar metal systems

mutual solubility should wet well with their contact angles approaching or equalling zero. This conclusion is largely borne out by experience with real metal–metal systems which are seldom without some mutual solubility and often interact so strongly that they form intermetallic compounds. This behaviour is typified by the extensive and rapid spreading of the fluxed solder used to make electrical connections or join Cu pipework, and the liquid metals used as brazes also spread well on uncontaminated substrates.

There are few reported experimental measurements for the contact angles or interfacial energies with which to compare theoretical expectations, and most of these are for low melting temperature metals of little relevance to brazing. However, the interfacial energy values of $0.43\,J.m^{-2}$ reported for Cu-Fe, $0.98\,J.m^{-2}$ for Cu-W and $1.37\,J.m^{-2}$ for Ag-Fe are in fair accord with predictions taking into account the simplifying assumptions of the Miedema model. Substitution of these values into the Young equation predicts that the Cu-Fe and Cu-W systems will have zero contact angles. The similarly derived prediction of 41°, $\cos\theta = 0.75$, for the Ag-Fe system has been broadly confirmed by sessile drop experiments that yield contact angle values ranging from 40 to 16°, $\cos\theta = 0.77$ to 0.96.

The ability of many metal–metal systems to form intermetallic compounds invalidates calculation of interfacial energies by the Miedema model. However, semi-empirical analyses of the wetting behaviour show that systems exhibiting extensive mutual solubility generally wet well and those that form intermetallic compounds often wet particularly well. The beneficial effect of mutual solubility was advanced also to explain enhancement of the wetting behaviour of the Ag-Fe system produced by alloying either the liquid or the solid with Pd, which is completely soluble in Fe and has some solubility in liquid Ag at 1000 °C so that it can be taken into solution by Fe from the liquid and by Ag from the solid.

While the behaviour of pure liquid metals is important, that of liquid eutectic alloys is often of even more importance when brazing metal components. Similar theoretical and semi-empirical analyses have not been developed to describe the behaviour of interfaces formed by liquid eutectic alloys, but it seems reasonable to assume that their energies will again be small. In practice, therefore, the capillary behaviour of both liquid metal–solid metal and liquid eutectic–solid metal systems should be predictable by substituting liquid physical property data into the equations describing the equilibrium and dynamic behaviour of ideal systems and assuming that $\cos\theta$ is unity. The liquid metal or eutectic alloy should flow to fill even infinitely long horizontal capillaries because $\cos\theta$ is positive. Flow into vertical capillaries will be finite because of the effects of gravity but nevertheless the heights achieved should be substantial and far larger than the typical 10 mm maximum required for most brazing applications, as shown by the values in Table 3.4 for a 0.1 mm wide parallel-sided gap that were derived by substitution in to equation {2.5}. Equations {3.21} and {3.22} that describe the kinetics of flow into horizontal and vertical

Table 3.4 Predicted capillary behaviour of some liquid metals and eutectic alloys

Liquid	*Capillary rise**, mm	γ/η m.s^{-1}	*Time in seconds*		
			penetrate 10 mm* *vertically*	*horizontally*	*spread to* 4 mm *radius***
Ag	210	226	0.0141	0.0132	0.1050
Al	780	337	0.0090	0.0074	0.0705
Au	140	217	0.0145	0.0138	0.1095
Cu	330	299	0.0102	0.0100	0.0795
Ni	460	326	0.0093	0.0092	0.0792
Ag-Cu	220	673	0.0456	0.0445	0.0353
Al-Si	760	687	0.0044	0.0043	0.0346
Ni-P	440	87	0.0351	0.0346	0.2732
Pb-Sn	110	167	0.0191	0.0180	0.1423

* For a perfectly wetted 0.1 mm wide parallel sided gap
** For a perfectly wetting sessile drop with a volume of 0.01 ml

capillaries relate these to the physical properties of the liquids and specifically to the γ/η ratios of the liquid and to their densities. The ratios and the densities for the metal and eutectic alloys listed in Table 3.4 vary by an order of magnitude, but substitution into equations {3.21} and {3.22} yield expected times to achieve a 10 mm penetration of 0.1 mm wide gaps of only a few milliseconds or tens of milliseconds. Further, it is noteworthy that the values for penetration of 10 mm into a vertical gap are only slightly longer than those for horizontal capillaries. Similar expectations of rapid flow are generated by substitution into equation {3.23} which describes the spreading of a sessile drop to assume its equilibrium configuration.

The application of relationships developed for idealized liquid–solid systems leads us to expect that liquid metals should flow rapidly and extensively to fill capillary gaps and cover substrate surfaces. These predictions assume that metal–metal contact is achieved at the liquid–solid interface and that the physical properties of the liquid and solid are not changed by such contact. In reality, neither assumption is true for most braze/component systems. Both solid and liquid metal surfaces are usually coated by unwettable oxide films that are effective barriers against metal–metal contact and many liquid metal–solid metal systems change their physical properties because of interdiffusion. Perhaps because of these recognized discrepancies, few attempts have been made to use even amenable real systems to test the quantitative relevance of predictions based on analyses of idealized systems. A notable exception to that neglect is the work reported by Latin (1946), who found good agreement between theoretical and experimental values for the penetration behaviour of the virtually inert system of Sn rich Sn-Pb solders flowing into fluxed parallel-sided channels of tinned Cu, the experimental rates being on average 84% of those predicted by equation {3.21} for a perfectly wetting system. However, the penetration rates were slower for a reactive system in which interdiffusion and intermetallic compound formation could occur, the flow of Sn rich Sn-Pb solders into fluxed parallel-sided channels of Cu that had not been tinned being only 41% of that predicted by equation {3.21}.

The amenability of the system used by Latin was due to the removal of oxide barrier films from the metal surfaces by the flux and the similar compositions of the liquid and solid so that chemical interaction was comparatively minor. Fluxes are widely used when brazing as well as soldering and must not only remove oxide films but also

- be fluid at a temperature somewhat below that at which the component is to be brazed;
- cover the component surface to prevent reoxidation before braze alloy flow has converted it to an interface;
- must wet the component surface well but not as well as the braze alloy so that this can displace the flux to produce a joint.

Few scientific descriptions of the performance or characteristics of brazing fluxes have been reported despite, or perhaps because of, their considerable commercial importance. Effective fluxes have been developed for a wide range of braze and component materials using empirical approaches occasionally leavened by serendipity. Thus the development of the $Na_2B_4O_7/KBF_4$ mixtures successfully used for the brazing of Al components with Al-Si alloys was based on a perceived need to maximize the solubility and speed of dissolution of the Al_2O_3 films present on both the solid and liquid surfaces. However, later scientific studies showed used fluxes to contain flakes of undissolved Al_2O_3 and that the success of the flux was an example of serendipity because oxide removal was achieved by flow of the flux down cracks in the oxide, produced by tearing as the component or braze was heated, to create electrolytic cells at the oxide–metal interface that undermined the oxide films.

The fluxes used with brazes based on the Ag-Cu eutectic are also fluoride/borate mixtures and some measurements have been made of the effect of composition on their physical properties. These show the fluxes to have a glassy structure based on a B_2O_3 network that is broken by increasing the temperature above the normal use range of 550–800 °C or increasing the F content of the mixture. The effect of this chain breaking is to decrease the viscosity of the flux, and raising the temperature from the lower to the upper working limit typically causes a decrease from $10^2\,Ns.m^{-2}$ to $10^{-2}Ns.m^{-2}$. Similarly, electrochemical measurements show the fluxes to provide effective protection against reoxidation of component surfaces, with the O diffusion coefficients in the fluxes of 2×10^{-9} to $2 \times 10^{-11}\,m^2.s^{-1}$ being closely comparable to those for diffusion in **solid** Ag at similar temperatures.

Fluxing, however, is not the only way of removing oxide films from their metal substrates and is sometimes not even desirable. Other ways include

- dissociation in low O environments;
- dissolution in the substrate;
- disruption and dispersal of the resultant debris.

Dissociation of an oxide M_xO_y will occur when the O_2 pressure of the environment is so low that the reaction

$$M_xO_y \Longleftrightarrow xM + 0.5y\,O_2 \qquad \{3.25\}$$

progresses to the right. The pressure of O_2 needed for equilibrium with neither dissociation or formation of M_xO_y can be calculated from the free energy of formation of the oxide, $\Delta G_{M_xO_y}$, by substitution into the expression relating it to the equilibrium constant, K, the ratio of the product and reactant activities, a, and hence directly to the O_2 pressure since the activities of solid M and M_xO_y can be taken as unity

$$\Delta G_{M_xO_y} = RTln_eK \qquad \{3.26\}$$

$$= RTln_e([aM]^x.[aO_2]^{0.5y}.[aM_xO_y]^{-1}) \qquad \{3.27\}$$

$$= RTln_e[pO_2]^{0.5y} \qquad \{3.28\}$$

$$= 0.5y.RTln_e[pO_2] \qquad \{3.29\}$$

Free energy data are readily available for a wide range of oxides of relevance to brazing, and their use to predict the O_2 pressures below which dissociation will occur over a range of temperatures leads to estimates that vary by many orders of magnitude as can be seen by examination of Figure 3.11. Omissions from the figure include Ag_2O for which the dissociation pressures are too high for convenient inclusion and Au which does not form a stable oxide, so the surfaces of both it and Ag should be oxide free. It should be noted that the values of the free energy of formation per gram mole of O_2 are equal to $(\Delta G_{M_xO_y}/0.5y)$.

Figure 3.11 is particularly relevant to vacuum brazing, during which the total pressure is typically 10^{-4} to 10^{-6} mbar and the partial pressure of O_2 remaining in the furnace is perhaps 10^{-6} to 10^{-8} mbar. While low, the figure shows these typical O_2 pressures are not low enough to cause dissociation of many of the oxides present on metal surfaces with the possible important exceptions of Cu and Ni. The dissociation pressure of

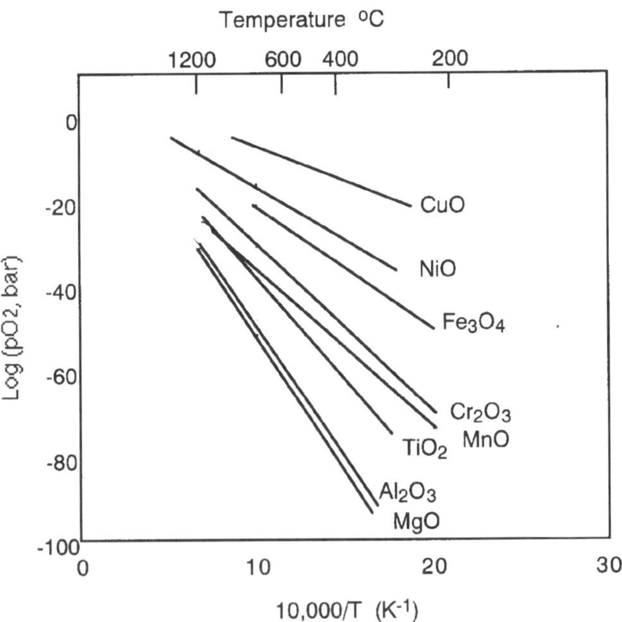

Figure 3.11 Dissociation pressures of some oxides

CuO is particularly high and the maximum rate of its dissociation in a perfect vacuum can be estimated as a few $\mu m.s^{-1}$ at brazing temperatures by substitution into the Langmuir evaporation equation

$$m^* = (3P_{vap}[M/T]1.2 \times 10^{-4}) \qquad \{3.30\}$$

where m^* is the dissociation rate in $kg.m^{-2}.s^{-1}$, P_{vap} is the equilibrium pressure expressed in Pascals and M is the molecular weight of the evaporating species. In contrast, dissociation of NiO at about 1200 °C, however, proceeds much more slowly at approximately $10^{-6}\,\mu m.s^{-1}$.

While seldom causing dissociation, high-quality vacua do promote the brazing of a wide range of metals that form chemically stable oxides. This is because the rate at which O_2 arrives at the metal surface is less in high vacua, in accord with the Langmuir equation and hence the oxide films will grow less quickly while the metal is being heated to the brazing temperature. The resultant effect, therefore, is to diminish the task that must be achieved by the oxide removal process.

Partial pressures of O_2 low enough to cause dissociation can be achieved by means other than creating an ultra high vacuum. Thus the O_2 partial pressure in a liquid metal can be exceptionally low and some workers have observed the oxide film on a Ni alloy to dissociate where it lay underneath a Ni braze. Similarly, a low O_2 pressure can be achieved in the gas phase by very careful purification of inert gases or by using reducing environments such as H_2 or CO that cause dissociation of surface oxides by the reactions

$$M_xO_y + yH_2 \rightarrow yH_2O + xM \qquad \{3.31\}$$

$$M_xO_y + yCO \rightarrow yCO_2 + xM \qquad \{3.32\}$$

for which the free energies are $(-\Delta G_{M_xO_y} + y\Delta G_{H_2O})$ and $(-\Delta G_{M_xO_y} + y\Delta G_{CO_2} - y\Delta G_{CO})$ respectively. These reactions will progress and render the metal surfaces oxide free if the H_2O/H_2 and CO_2/CO ratios of the gaseous environments are low enough and it is worthwhile noting that the H_2O/H_2 ratio of a hydrogenous environment can be regarded also as its dryness and this is the basis of the technical use of 'Dew Points' to characterize the suitability of H_2 environments as elaborated in Chapter 6.

Dissolution of oxide films in their metal substrates clearly depends on the solubility of O in the metal and the relative thicknesses of the oxide film and the substrate. Reported solubility values are generally low with the notable exceptions of Ag and, particularly, Ti, as shown by the data assembled in Table 3.5. Thus even though oxide films are usually thin, typically 10 to 100 nm and rarely more than 1000 nm or 1 μm, quite substantial depths of substrate are needed to accommodate the oxygen – several cm of Cu would be needed to dissolve even a 10 nm thick oxide film.

Table 3.5 Oxygen solubility in some metals

Metal	*Solubility at* 1000 °C, *weight %*
Ag	0.29†
Al	Very, very low
Cr	0.02
Cu	5×10^{-3}
Fe	< 0.009
Ni	0.014
Ti	1 in βTi, 14 in αTi
W	Very low

† 0.04 in solid Ag at 931 °C

It would seem, therefore that dissociation and dissolution processes are often or even generally unable to remove the oxide films that prevent metal–metal contact between mating surfaces. However, contact does occur and therefore it can be concluded that other mechanisms of disruption and dispersion play important roles in many brazing processes. Several suggestions have been made as to the identity of these other mechanisms for specific systems. Thus the high coefficient of thermal expansion coefficient of solid Al, $26 \times 10^{-6}\ K^{-1}$, and the further volume increase experienced when melting can cause tensile tearing of the surface films of Al_2O_3 to permit localized metal–metal contact, but how this then results in dispersion of the oxide film is not clear. Similarly, Al-Si alloys were modified for use as unfluxed vacuum brazing alloys by the introduction of reactive species such as Mg which were thought to reduce the surface oxide. However, later studies using hot stage microscopes for the direct observation of the surfaces of Al-Si-Mg alloys during vacuum brazing have shown that disruption of the surface is in fact caused by the violent evaporation of Mg that occurs when the 555 °C eutectic temperature of the ternary alloy is exceeded and that thereafter flakes of the oxide float on the metal surface. This disruption process exposes clean metal but nevertheless the resultant rate of spreading is relatively slow, with it taking two to three minutes to form a fillet a mere 0.2 mm deep when brazing Al alloy heat exchangers.

Other less violent mechanisms have been advanced to explain the practical success of Ni alloy brazes used to join high alloy steel and Ni alloy components. Thus attention has been drawn to the possible effect of the C content of Fe-18Cr-10Ni-1Nb substrates in reducing the surface oxides when being vacuum brazed by a Ni-14Cr-10P alloy at 950 °C, a suggestion qualitatively supported by the observation that such vacuum treatment caused the substrate material to become shiny. Other studies have assigned important roles to the fortuitous presence of cracks in the surface oxide which permit localized direct metal–metal contact. In the case of Ni-11P brazes on the surfaces of Fe-Cr and Ni-Cr substrates, detailed real-time observation using both optical and scanning electron microscopes led to the conclusion that this metal–metal contact led to dissolution of the substrate and hence undermining of the oxide film which then floated away on the liquid surface. Not surprisingly, variations in the composition of the substrates affected the extent of spread of sessile drops of the braze alloy, with this being progressively restricted as the Cr contents of the substrates increased due to the formation of chemical stable and physically tenacious Cr_2O_3 in the oxide films.

Spreading by disruption and dispersal of the substrate was found to continue for long times before ceasing due to isothermal resolidification of the sessile drop caused by ingress of high melting temperature constituents from the dissolving substrate. For most of its duration, spreading of the drops could be described by the expression

$$r^n = Ct + r_0^n \qquad \{3.33\}$$

where C and n are constants, r is the radius at time t and r_0 at the onset of timing. Measurements made by Ambrose *et al.* (1993) during a series of experiments with Fe-Cr substrates and sessile drops produced by melting 0.038 mm thick discs of braze alloy with a radius, r_0, of 0.5 mm showed the values of n to decrease from 10.4 to 8.2 as the Cr contents of the substrates rose from 2 to 20%. This experimental equation is similar in form to that derived for the idealized spreading of a sessile drop by de Gennes

$$r^{10} - r_0^{10} = 0.2(4/\pi)^3(\gamma_L/\eta)V^3t \qquad \{3.23\}$$

Assuming that the experimental values of n are actually approximations to the power of 10 in equation {3.23}, calculated values of C can be compared with the term $0.2(4/\pi)^3(\gamma_L/\eta)V^3$. The experimentally derived values of C decreased, from 21×10^{-34} to $4.7 \times 10^{-34}\,m^{10}.s^{-1}$ as the Cr contents of the substrates were raised from 2 to 20%, while the value of $0.2(4/\pi)^3(\gamma_L/\eta)V^3$ was an unchanging $1.45 \times 10^{-30}\,m^{10}.s^{-1}$. Thus, the spreading kinetics of this complex reactive metal–metal system are qualitatively similar to those of an ideal system. However, control by a substrate dissolution process rather than a balance between a surface energy driving force and a viscous impedance causes the spreading rates to be far slower, with it taking from 50 to 225 seconds for the sessile drop radii to increase from 0.5 to 0.8 mm, as compared to the 0.07 second predicted by the de Gennes equation.

The final aspect of the behaviour of real metal systems with which comparison can be made with that of idealized materials is the effect of roughening of the substrate. It is generally recognized that roughening does have an effect, with enhanced liquid flow along the grain of anisotropically roughened surfaces. The model studies predict changes in contact angles produced by isotopic roughening, but experimental quantitative measurements of changes in contact angle values produced by these structures are rare. However, some measurements have been reported by Nicholas and Crispin (1986). Thus work with the Cu-Ni system using an experimental temperature of 1100 °C and a vacuum furnace show that abrasion to produce isotopic roughening caused the advancing contact angle to increase linearly with the asperity slope, from 19 to 21.5° as the slope, α^*, increased by 13°. This linear increase is in accord with those of the geometric, Shuttleworth and Bailey, model but the actual increase is less than that predicted.

Similar shortfalls in the linear increase of contact angles caused by roughening metal substrates have also been measured for Cu on mild steel and stainless steel substrates, but roughening caused a decrease from 10 to about 0° in the contact angles assumed by a Ag-Cu braze alloy. Isotropic roughening also caused contact angle increases with the Sn-Ni and Hg-Ni systems at lower temperatures, and these increases more nearly approached the predictions of the geometric model.

Thus roughening can enhance the behaviour of a well wetting system but generally causes a degradation which becomes progressively less marked as the experimental temperature is raised, in accord with the Johnson and Dettre energy barrier model. More detailed analysis of the experimental data reveals that the shortfalls depend on the identity of the liquid metal as well as the temperature and that the unifying parameter is the total enthalpy of the liquid, which can be regarded as a measure of the vibrational energy cited in the Johnson and Dettre model, as illustrated in Figure 3.12. The data points in this figure reveal a simple relationship describing the behaviour of a

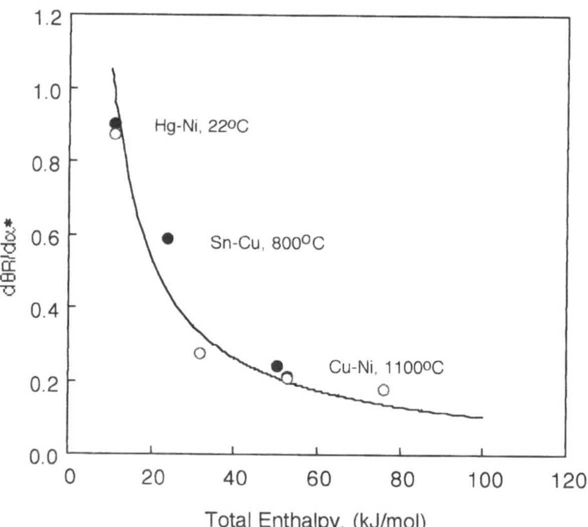

Figure 3.12 Influence of liquid enthalpy on the effect of isotropic roughening on contact angle values of a few metal–metal systems. The hollow symbols refer to liquid metal–solid ceramic systems that are discussed later

range of liquid–solid, metal–metal and metal–ceramic, combinations that can be expressed as

$$\Delta\theta_R/\Delta\alpha^* = H_S/E_H \qquad \{3.34\}$$

where E_H is the total enthalpy of the liquid metal and H_S is a constant with a value of 10.5 kJ.mol^{-1}.

3.1.7 Brazing of ceramic components

The prime requirement for successful brazing is that the liquid metal wets the solid component, but ceramics are generally not wetted or at best poorly wetted by liquid metals and alloys. Fortunately there are a few exceptions to this generalization, as there usually are to any generalization, and the main objective of scientific studies of the brazing of ceramics has been to explain these exceptions.

Contact angle data are available for numerous liquid metal–solid ceramic combinations. However, there are few systematic studies of the behaviour of extensive families of ceramics and therefore analysis is often based on compilations such as that of Naidich (1981), from which the values presented in Table 3.6 are taken, for smooth stoichiometric ceramic substrates heated in evacuated furnaces to temperatures just above those at which the metals melt.

Table 3.6 Contact angle data for some ceramic–metal systems, taken largely from Naidich (1981)

	Metal					
	Ag	Al	Au	Cu	Ni	Fe
Ceramic	*Contact angle, degrees**					
Al_2O_3	144	128	138	138	128	141
BeO	137	–	–	140	152	147
MgO	136	–	–	128	152	130
SiO_2	139	140	–	134	125	–
TiO_2	–	–	–	141	104	–
UO_2	122	145	–	113	65	–
C	136	157	–	140	45	37
B_4C	137	–	–	136	87	36
SiC	128	127	138	130	65	–
TiC	153	118	–	128	32	–
WC	–	–	–	30	0	0
AlN	–	155	–	135	–	–
BN	140	154	145	137	75	–
Si_3N_4	138	157	–	135	90	–
TiN	141	–	113	112	–	–

* Data obtained from studies conducted in evacuated furnaces at temperatures close to those at which the metal melts

Table 3.7 Pauling ionicities for a number of element pairs

Element	O	N	C
Zr	67	47	26
Ti	63	43	22
Al	63	43	22
Cr	59	39	19
W	55	–	15
Ni	51	–	–
Si	51	30	12
Cu	47	–	–
B	43	22	6

The basic reason for the poor wetting of ceramics by liquid metals is thought to be the differing natures of their binding forces. Thus a metal lattice can be viewed as an array of positive ions located in a 'gas' of delocalized electrons, the movement of which confers the characteristics of generally high electrical and thermal conductivity. In contrast, ceramic lattices are held together by the sharing of specific pairs of electrons to form a covalent bond or else by the transfer of electrons from one type of atom to another to create arrays of electrically charged ions that are held in place by electrostatic forces. Thus there is a lack of the delocalized electrons that characterize metal lattices and a metal–ceramic interface represents a major electronic discontinuity that requires a considerable energy to form. In practice, the energies of metal–ceramic interfaces are often so high that they exceed those of the free surface of the solid ceramic, $\gamma_{SL} > \gamma_S$, and hence the contact angles assumed by the liquid metal are larger than 90°.

In practice, ceramics are neither perfectly covalent or ionic and Pauling (1960) has related the degree of ionicity of bonding to differences in the electronegativity of the constituent elements. Pauling ionicity values for a number of M–O, M–N and M–C pairs are quoted in Table 3.7, and these show that the most chemically stable oxides are also the most ionic materials in the table, and that ionicity decreases in the sequence oxide→nitride→carbide. These changes in bonding are reflected in characteristics such as electrical conductivity and surface energy. Thus the data assembled in Table 3.8 show that while the γ_S values for oxides are generally less than $1.0\,J.m^{-2}$, those for carbides can be much higher. Similarly, while very few of the systems cited in Table 3.6 have contact angles less than 90°, wetting was achieved when using WC substrates and when liquid Ni and Fe were brought into contact with non-oxide ceramics. Semi-empirical analyses have shown that the variations in reported contact angle data can be related to the ionicity of the ceramic and to their thermodynamic stability, as illustrated in Figure 3.13 which suggests that systems most prone to metal–ceramic reactions are likely to be best wetted.

Table 3.8 Surface energies of some solid ceramics, J.m^{-2}

Al_2O_3	0.90	Cr_2C_3	1.05
BeO	1.00	HfC	0.55
SiO_2	0.30	NbC	2.35
UO_2	0.85	TiC	1.05
ZrO_2	0.60	VC	1.35
		WC	2.80
		ZrC	0.90

The data in Figure 3.13 are somewhat scattered, as might be expected since they were generated by many separate studies made using possibly significantly different experimental conditions. In particular, there may have been variations in vacuum quality or substrate roughness, but some general conclusions can be drawn about the possible importance of such variations.

Thin oxide films may have been present on a number of the metals for which data were complied by Naidich but it is only for Al that there is clear evidence of a marked increase in the measured contact angle being

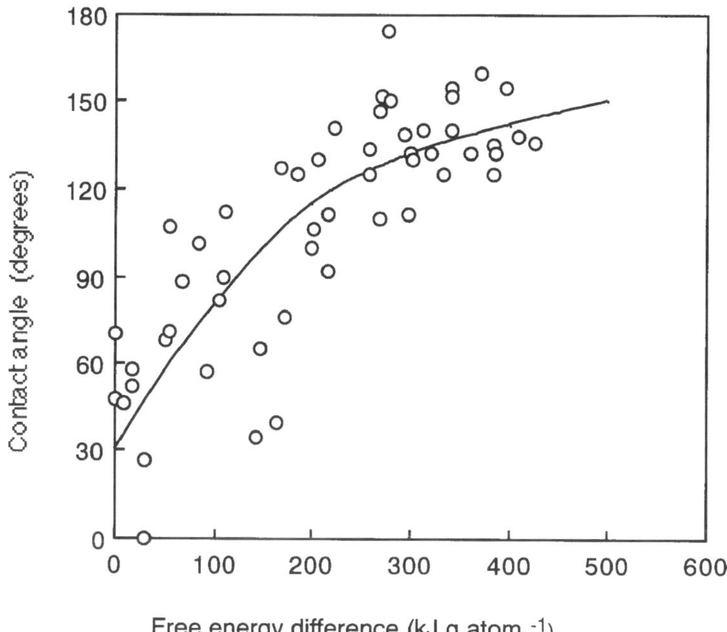

Figure 3.13 Comparison of reported wettabilties and the free energies of reaction of oxide–metal systems. After Naidich (1981)

caused by the restricting encapsulation in an oxide film. The scientific basis of ways of eliminating this restriction when brazing metal components were discussed in section 3.1(3), but fluxing is not used when brazing ceramics, dependence being placed instead on the use of a high quality vacuum and a high temperature so that the reaction

$$Al_2O_3 \Longleftrightarrow Al_2O + O_2 \qquad \{3.35\}$$

progresses to the right to form volatile Al_2O. Thus, Al has been found to wet a range of ceramics at temperatures above about 1000 °C in technical quality vacua of 10^{-4} to 10^{-5} mbar, Table 3.9, and the temperature for this change can be decreased to about 800 °C in ultra-high vacuum equipment. The data for liquid Al plotted in Figure 3.13 were measured at temperatures of 700 °C or less and hence are of questionable validity because they do not relate to metal–ceramic contact. However, the data points for Al represent such a small proportion of the total number plotted that their adjustment or elimination will have a negligible effect on the apparent broad correlation between chemical reactivity.

A much wider range of systems was affected by the second source of uncertainty, the roughness of the substrate, which was often described merely qualitatively as 'polished', 'smooth' or even 'as received'. However, quantitative measurements of contact angle changes caused by roughening have been made for a number of metal–ceramic systems and the revised version of Figure 3.12 shows that they can be described in the same way as for metal–metal systems. Since brazes for ceramics are used at high temperatures and therefore will have high enthalpies it is unlikely that large variations in contact angle values will have been caused by the use of modestly rough substrates.

Thus neither possibly marked adjustment of the few data plotted in Figure 3.13 for Al-ceramic systems or minor adjustments for the many other systems to take account of substrate roughness effects should change the conclusion that the most reactive metal–ceramic systems generally display the best wetting behaviour. Detailed chemical, optical and scanning electron microscope studies of wetting metal–ceramic systems show that many of them have formed reaction product layers at the liquid–solid

Table 3.9 Effect of temperature on the wetting of ceramics by liquid Al

	Contact angle, degrees			
	Ceramic			
Temperature °C	Al_2O_3	AlN	BN	Si_3N_4
700	135	154	153	152
1000	80	47	41	70

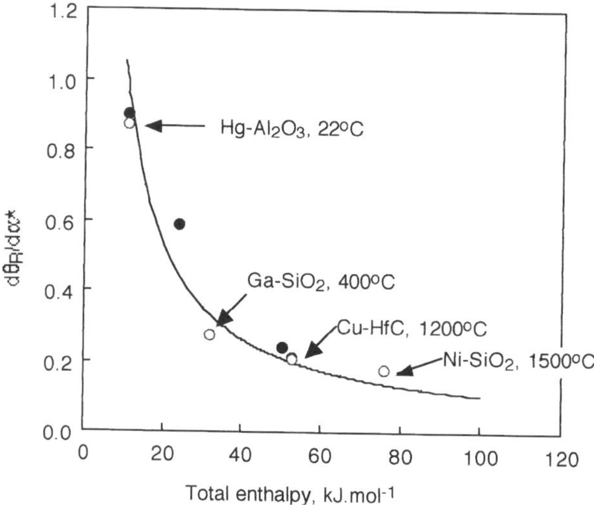

Figure 3.12 Revised. Influence of liquid enthalpy on the effect of isotropic roughening of ceramic substrates on contact angle values. The full symbols refer to liquid metal–solid metal systems discussed earlier

interfaces and it is the replacement of the original substrate–metal interface with a reaction product–metal interface that is responsible for the better wetting behaviour of reactive systems, as will be elaborated in Chapter 7. Regardless of the specific mechanism responsible for enhanced wetting behaviour, the broad correlations shown by data plots such as Figure 3.13 have led to attention being focused on the behaviour of Al, Cr and Ti and other reactive metals during attempts to develop brazes to join ceramic components. Many development programmes have evaluated the wetting of ceramics by Al and its alloys, of the Ni-Cr-Si alloys developed to braze high temperature metal–metal joints and, particularly, the effects of additions of Ti to Cu and Ag-28Cu eutectic alloys.

A notable characteristic of Ti is its ability to form ceramic compounds with wide compositional ranges. Thus TiC is stable at compositions ranging from $TiC_{0.60}$ to $TiC_{0.94}$ at 1000 °C and the effects of such a shortfall in the C, or N or O, content of the ceramic is that not all the available Ti electrons are required to form covalent or ionic bonds. The presence of these uncommitted electrons give the ceramic somewhat metallic characteristics such as an enhanced electrical conductivity and wettability. Figure 3.14 illustrates this effect for wetting by Cu but similar effects are displayed by other braze metals. These marked changes in the wettability of Ti ceramics caused by variations in stoichiometry provide an explanation for the success of the commercial Ag-Cu-Ti developed as active metal brazes for joining a wide range of ceramics.

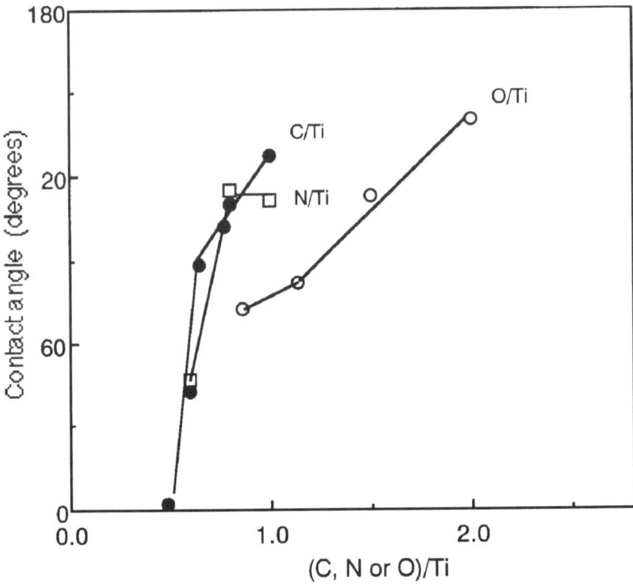

Figure 3.14 The effect of ceramic substrate stoichiometry on wetting by Cu

Wetting of many ceramics could be promoted by coating their surfaces with layers of hypostoichiometric TiC, TiN or TiO but this would be a complex process and active metal brazes work by reacting with the ceramic substrates to produce the wettable compounds during the brazing cycle. Stoichiometric TiX compounds can be produced on the surfaces of ceramics by reactions such as

$$Al_2O_3 + 3\underline{Ti} \rightarrow 3TiO + 2\underline{Al} \qquad \{3.36\}$$

$$C + \underline{Ti} \rightarrow TiC \qquad \{3.37\}$$

$$Si_3N_4 + 4\underline{Ti} \rightarrow 4TiN + 3\underline{Si} \qquad \{3.38\}$$

where solute elements are underlined. The energy changes of these reactions, derived from the free energies of formation of reactants and products, are related to the chemical potential, μ, ΔG and to the activity ratios of the products and reactants. If the activities of the solid phases are taken as 1.0, the activity ratios of the reactant and product solutes for the first reaction can be calculated from the expressions

$$3\Delta G_{TiO} - \Delta G_{Al_2O_3} = \mu_{TiO} + \mu_{\underline{Al}} - \mu_{\underline{Ti}} - \mu_{Al_2O_3} \qquad \{3.39\}$$

$$= -RT\{\ln_e(a^3 TiO) - \ln_e(a^2 \underline{Al}) + \ln_e(a\underline{Ti}^3) + \ln_e(aAl_2O_3)\} \qquad \{3.40\}$$

$$= -RT\ln_e(a\underline{Al}^2/a\underline{Ti}^3) \qquad \{3.41\}$$

and similarly for the other reactions

$$\Delta G_{TiC} = -RT\ln_e(aTiC/a\underline{Ti}.aC) = -RT\ln_e(1/a\underline{Ti}) \quad \{3.42\}$$

$$4\Delta G_{TiN} - \Delta G_{Si_3N_4} = -RT\ln_e(a\underline{Si}^3.aTiN^4/a\underline{Ti}^4.aSi_3N_4) \quad \{3.43\}$$

$$= -RT\ln_e(a\underline{Si}^3/a\underline{Ti}^4) \quad \{3.44\}$$

Substitution of thermodynamic data predict that TiO will be formed by reaction with Al_2O_3 if $a\underline{Ti}$ is at least 10^{-4} at 1100 °C and will continue to do so until the ratio ($a\underline{Al}^2/a\underline{Ti}^3$) reaches the equilibrium value of 0.57. Alloys with $a\underline{Ti}$ values of less than 10^{-4} can also react but unwettable Ti oxides such as Ti_2O_3 and TiO_2 will form.

The nature of the liquid solvent does not change the $a\underline{Ti}$ value needed to form TiO but it does affect the concentrations through its influence on activity coefficient, $\beta_A = aA/N_A$ which tends to be high for systems in which the solute is relatively insoluble. Thus manipulation of such data predicts that formation of TiO at 1100 °C will be initiated by about 0.3 atom% Ti in Cu and 4 atom% in Au, while 8 atom% are needed with a Ni solvent at 1500 °C, in accord with experimental observations that 2 atom% Ti in Cu form TiO but 2 atom% in Au or Ni produce Ti_2O_3.

Similar semi-quantitative agreement between prediction and experience has been found for other oxide and non-oxide ceramics, but the Ti concentrations needed for normal brazing practice are higher to ensure that wettable hypostoichiometric reaction products are formed rapidly. However, enriched alloys such as Cu-5Ti or Cu-10Ti have found few applications because they lack the ductility desirable in a solid braze and the substantial Ti concentrations can result in the formation of thick and fragile reaction product layers. More desirable compositions are those in which the Ti activity is high but the concentration is low and these requirements can be satisfied by using solvents in which Ti is relatively insoluble. Thus attention has been paid to the usefulness of Ag in which the Ti solubility is only 7 atom% at a brazing temperature of 1000 °C as compared to 68 atom% in Cu at its brazing temperature of 1100 °C, and in particular to Ag–Cu alloys in which the solubility of Ti can be even lower than that in Ag. Commercial development has focused particularly on the usefulness of the Ag–28Cu eutectic alloy in which the Ti solubility is less than 2 atom%. The activity of Ti in this alloy is about 0.15 at a brazing temperature of 900 °C, far higher than the 0.0001 needed to form TiO by reaction with Al_2O_3, and alloys based on Ag-28Cu-2Ti have been found to wet a wide range of ceramics more readily than Cu-Ti alloys. Thus the data plotted in Figure 3.15 show that a Ag-28Cu-2Ti alloy wets Si_3N_4 better than does a Cu–5Ti alloy and matches the wetting behaviour of a Cu–10Ti alloy at about 1000 °C. Further, the Ag-28Cu-2Ti alloy wets the ceramic well at temperatures lower than those needed to melt the Cu-Ti alloys and does not suffer from being excessively reactive and stiff. Some recent developments have focused on further reduction of the Ti concentration by dilution of the Cu content of such brazes by introducing

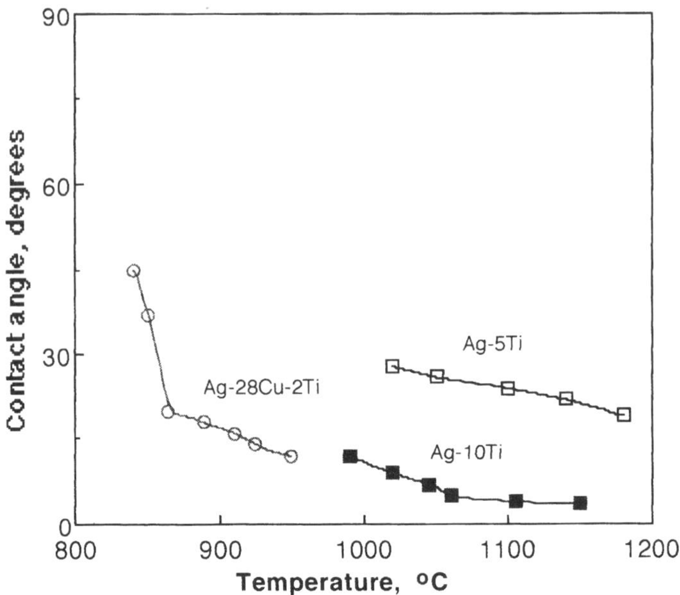

Figure 3.15 The wetting of Si_3N_4 by sessile drops of active metal braze alloys

In as in brazes such as Ag-27Cu-9.5In-1.25Ti and it is significant, therefore, that the solubility of Ti in In is only 3 atom% at 900 °C.

Prolonging the duration of contact between a liquid active metal braze and a ceramic causes the reaction product layer formed at their interface to thicken, and also to grow laterally. If the layer is of a wettable compound, this lateral growth permits the liquid front to advance, as in the spreading of a sessile drop illustrated schematically in Figure 3.16. The growth of such wettable layers is controlled by solid-state diffusion processes and, therefore, the resultant advance of the sessile drop periphery and decrease in the contact angle will be slow compared to the rates predicted by de Gennes for idealized wetting systems.

zExperimental measurements of the variation of contact angles with time have been reported for a wide range of reactive metal–ceramic systems and Figure 3.17 presents some data characterizing the wetting behaviour of liquid Al as well as active metal brazes that contain Ti. Mathematical analysis of such data show the decrease in contact angle values to be an exponential function of time that can be described by an expression

$$\theta_t - \theta_\infty = (\theta_0 - \theta_\infty)e^{-t/\tau} \quad \{3.45\}$$

where θ_t, θ_0 are the contact angles at time t and t = 0, θ_∞ is the stable contact angle assumed after a long time, and τ is a characteristic time that

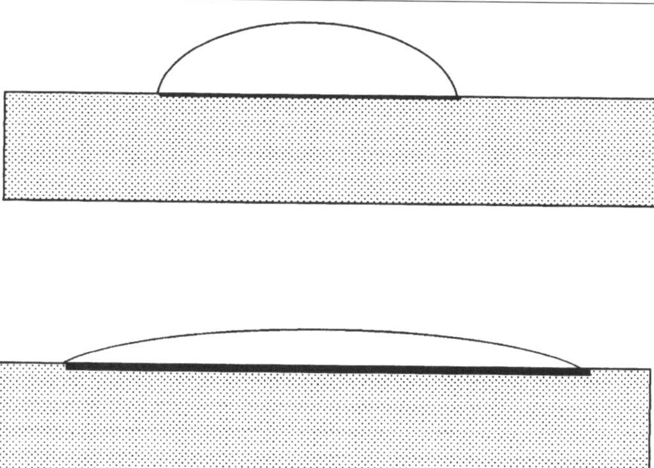

Figure 3.16 The effect of the thickening and lateral growth of a wettable reaction product layer on the contact angle assumed by a sessile drop in a reactive metal/ceramic system

has values of 16 and 0.6 minutes for the Ag-27Cu-9.5In-1.25 alloy on Si_3N_4 and the Cu-7.5Ti alloy on Al_2O_3. These angular changes are brought about by increases in the drop radii that again can be described by the relationship

$$r_n = Ct + r_0^n \qquad \{3.33\}$$

where n now has a value of 0.5 to 1.0, as compared to the 10 found for the spreading of reactive brazes on metal substrates.

These kinetic effects have been confirmed by several studies, but a generally accepted explanation has yet to be advanced. However, it is important to realize that the contact angles measured in such sessile drop experiments are not equilibrium values and that they cannot be used to predict behaviour by substitution into expressions developed for idealized systems. For example, the decrease from 30 to 5° shown to occur in Figure 3.17 a Ag-27Cu-9.5In-1.25Ti in contact with Si_3N_4 for 30 minutes at 900 °C does not imply that the liquid will rise immediately up a 0.1 mm gap between Si_3N_4 plates to attain the height of 220 mm consistent with $\theta = 30°$ or that it will then rise further during the following 30 minutes further to a final height of 254 mm as suggested by equation {2.5}. In fact the liquid should rise a distance similar to the radial increase of the sessile drop, a fraction of a millimetre, and this accounts for the practice of placing active metal braze alloys within a joint gap rather than expecting them to penetrate and fill the gap after melting.

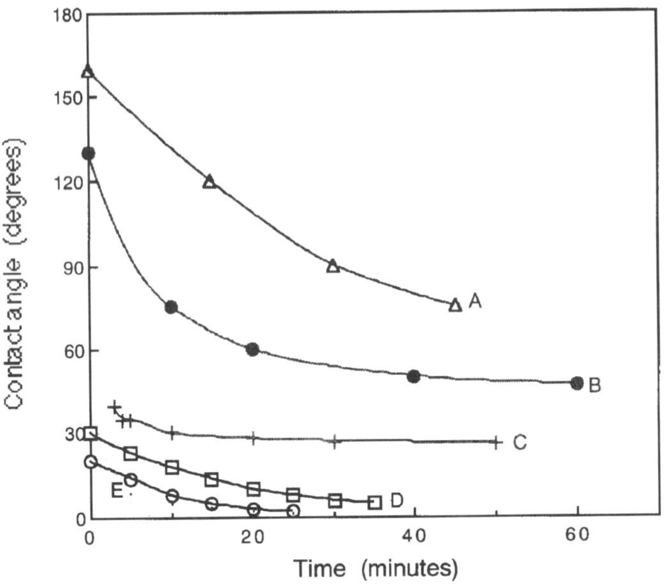

Figure 3.17 The effect of time at temperature on the contact angles of several reactive liquid metal/solid ceramic systems:

A – Al/SiC 700 °C, B – Al/Si_3N_4 1100 °C,
C – Cu-7.3Ti/Al_2O_3 1100 °C D – Ag-27Cu-9.5In-Si_3N_4 900 °C,
E – Ag-28Cu-1.5Ti/Si_3N_4 900 °C

3.2 DIFFUSION BONDING

For the present purposes, diffusion bonding can be defined as the solid state joining of similar or dissimilar material components at high temperatures by the application of external pressures insufficient to produce permanent deformation of the components. The components may be joined directly to each other as illustrated in Figure 2.5 or by means of an interlayer inserted between their mating surfaces but the temperatures employed are always below those needed to melt the component or interlayer materials. The component materials may be metals or ceramics but the interlayers are almost always metals. Particularly in technical literature, the term 'diffusion bonding' is also used sometimes to describe processes in which joining depends on interdiffusion of component or component and interlayer materials to produce transiently liquid phases that facilitate joint formation. These variants are excluded from this discussion, but some will be addressed in the next sub-section concerned with hybrid processes.

The process of diffusion bonding can be viewed as consisting of four steps.

1. achievement of contact between mating surfaces;
2. interface formation;
3. interdiffusion of component materials;
4. for some systems, grain growth to eliminate the original interface.

The first two steps are concerned with the formation of the joint and their scientific aspects will be discussed in this chapter. The last two steps relate to the optimization of its mechanical and other properties of a joint rather than its formation, and they will be considered in some detail in Chapter 6 which is concerned with process optimization.

3.2.1 Modelling joint formation by idealized systems

Models of ideal systems assume that there is no barrier to the creation of interfaces between contacting surfaces so that steps (1) and (2) can be compounded and joint formation merely requires contact between component mating surfaces. Achieving contact between mating surfaces that are not completely smooth requires mechanical work but the subsequent creation of an interface is invariably energetically favourable for diffusion bonding components of the same material. Thus such an interface is comparable to the natural boundaries between the grains of polycrystalline material, the energies of which for metals are typically $0.3\,\gamma_S$ and for ceramics usually lie in the range $0.3\text{–}0.5\,\gamma_S$ so there is a gain by creating a diffusion bonded interface between identical materials of at least $1.5\,\gamma_S$.

Because there is no inherent obstacle to conversion of contact area between the mating surfaces of the same material into interfaces, models of diffusion bonding encompass both the initial formation of contact areas and the subsequent joint formation by the growth of bonded interfaces. Bonding is usually assumed to be between two blocks of identically pure ductile material whose flat mating surfaces are clean and uncontaminated but covered by a regular array of identical small topographic features such as pyramids or domes whose peaks make the initial contact. This contact flattens the peaks, creating bonded neck areas with sharply curved surfaces and then isolated voids, Figure 3.18.

The driving forces for the subsequent growth of the bonded neck areas and shrinkage of the isolated voids when diffusion bonding identical material systems are the accommodation of the externally applied pressure and the reduction in the total surface energy of the system caused by the interface formation. The pressure causes the asperities to creep, the continuous slow deformation at high temperatures that can be caused by a small constant stress, and also enhances material transport by decreasing the chemical potential of vacancies near the bonded interface. This potential is decreased also by the curvature of bonded neck and void

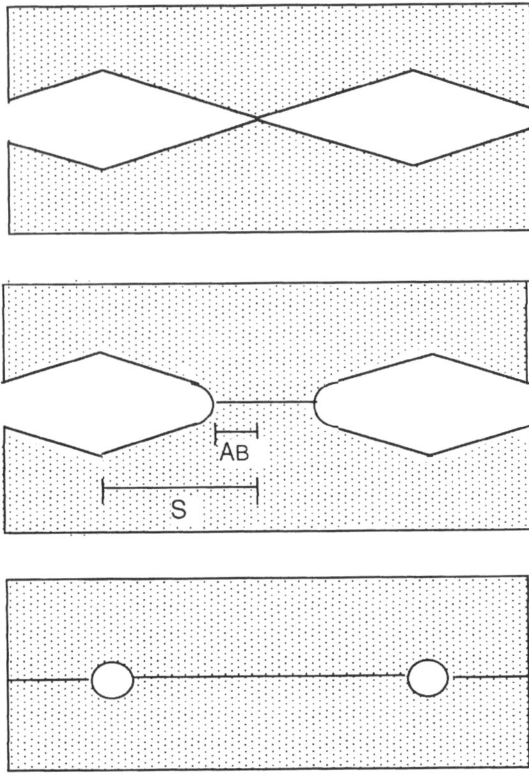

Figure 3.18 Stages in the growth of the contact area between asperities, showing the formation of a bonded neck that then grows to create isolated spherical voids.

surfaces, the Gibbs-Thompson effect, so there are gradients within the system down which material will diffuse to the bonded necks and voids.

This creation of bonded necks and their subsequent growth requires the operation of a family of processes. The initial contact is achieved by a near instantaneous plastic deformation of the peaks of contacting surface features which will create a contact/interface area of width A_B such that

$$A_B = P_{ext}.S/Constant.\sigma_Y \qquad \{3.46\}$$

where P_{ext} is the externally applied pressure, S is the spacing between the ridges, σ_Y is the yield stress of the material below which no permanent plastic deformation can occur. The constant reflects a geometrical constraint on yielding, and for a triangular ridge geometry has a value of about 2.8.

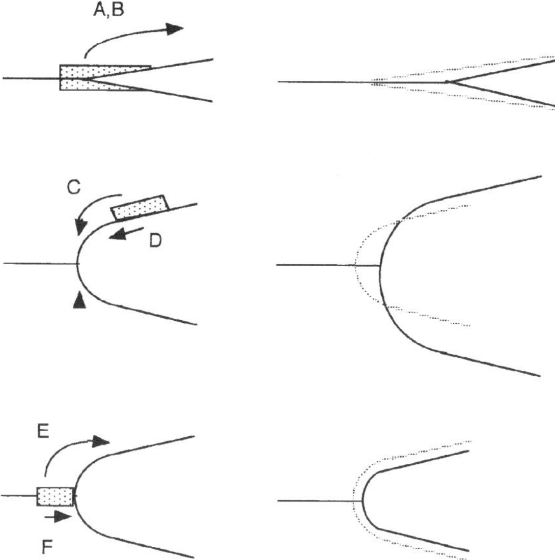

Figure 3.19 Local material transport mechanisms involved in the growth of a bonded neck: A – instantaneous plastic deformation, B – creep, C – volume diffusion, D – surface diffusion, E – volume diffusion from an interface, F – diffusion along an interface

The subsequent growth of bonded necks and the closure of voids is achieved by a variety of processes that can be classified in terms of their use of different sources from which material is transported, Figure 3.19. As indicated by the figure, creep closes the voids by flattening, while vapour phase transport and some diffusion processes cause them to spheroidize and others subsequently cause them to shrink. Even for the simplest of systems, the rates at which these processes operate depend on a variety of material characteristics and temperature. Thus the steady state stage of creep which usually lasts longer than any other has a linear deformation rate of $d\varepsilon_{SS}/dt$ that can be derived from the expression

$$d\varepsilon_{SS}/dt = A(\sigma/E)^n \exp[-Q_V/RT] \qquad \{3.47\}$$

where σ is the stress, E is the modulus of elasticity, n is a power term with a value of typically between 4 and 5, Q_V is the activation energy for volume diffusion and A is a constant related to fundamental deformation characteristics of the material such as the Burgers vector. Similarly, the maximum rate of vapour phase transport, dm^*/dt, is given by the Langmuir evaporation equation

$$dm^*/dt = (3P_{vap}.\rho_G)^{1/2} \qquad \{3.48\}$$

where ρ_G is density of the transported gaseous material and P_{vap} is its equilibrium vapour pressure, which can in turn be related to the temperature by the expression

$$P_{vap} = P_0.\exp(-\Delta H_g/RT) \quad \{3.49\}$$

where P_0 is a constant and ΔH_g is the enthalpy of evaporation.

Describing the operation of diffusion processes for even the simplest of systems is more difficult because several can operate to transport material to the voids as indicated in Figure 3.19. However, the driving force for all of them is an activity gradient in accord with Fick's first law,

$$J = -D(da/dx)_t \quad \{3.50\}$$

where J is the material flux, D is the diffusion coefficient and (da/dx) is the gradient which can change with time. Three different diffusion mechanisms can operate simultaneously when diffusion bonding – transport from surface sources being by surface and volume diffusion and from interfacial sources being by interface and, again, volume diffusion as indicated in Figure 3.19. Thus any description of material transport by diffusion requires knowledge of the coefficient for volume diffusion, D_V, for surface diffusion, D_S, and for interface diffusion, D_{int}. In practice, the coefficient for interface diffusion can be equated to that for grain boundary diffusion, D_{GB}, when the components being bonded are of the same material. Values for these coefficients are available for some metals and few ceramics, but predictions of the material transport due to surface or interfacial diffusion processes is complicated by the fact that these can operate only within a very thin surface or interface region which can approach atomic dimensions.

Modelling joint formation during diffusion bonding requires a substantial body of material property data, which can change with temperature. However, the temperatures of interest when bonding are generally high, such as the $0.9\,T_M$ for which some material transport characteristics are presented for a number of metals in Table 3.10. While

Table 3.10 Material transport characteristics for some metals at a temperature of $0.9\,T_M$

Metal	T_M, K	D_V, m^2s^{-1}	D_{GB}, m^2s^{-1}	D_S, m^2s^{-1}	P_{vap}, Pa
Al	933	2.4×10^{-13}			4.7×10^{-9}
Ag	1234	8.5×10^{-14}	7.0×10^{-10}		1.7×10^{-2}
Au	1338	2.5×10^{-13}	1.3×10^{-10}		2.2×10^{-5}
Cu	1356	7.3×10^{-14}		3.4×10^{-9}	1.4×10^{-3}
Fe	1808	3.8×10^{-14}	1.2×10^{-9}	5.8×10^{-7}	1.5×10^{-1}
Ni	1726	4.5×10^{-14}	1.9×10^{-10}		1.5×10^{-2}
Ti	1940	$5.0 \times 10^{-1?}$ [illegible]			$2.7 \times 10^{-?}$ [illegible]
W	3660	1.8×10^{-14}	5.2×10^{-10}		1.9×10^{-1}

the vapour pressure data in the table are very varied, the diffusion coefficient data are closely grouped. Thus most of the D_V values lie within the narrow range of 10^{-13} to $10^{-14}\,m^2s^{-1}$ while the fewer values quoted for D_{GB} and D_S lie within the ranges of 10^{-9} to $10^{-10}\,m^2s^{-1}$ and 10^{-7} to $10^{-9}\,m^2s^{-1}$ respectively. These relatively narrow ranges reflect the fact that both diffusion and melting are bond-breaking processes and hence the energies required are related to each other and to the melting temperature, and for many metals both Q_V and Q_S are about $0.15\,T_M$ kJ.mol^{-1} while Q_{GB} is about $0.1\,T_M$ kJ.mol^{-1}. For ceramics, the presence of two constituents produces complications and the activation energies for diffusion tend to be higher than for comparable metals, thus values for Al^{3+} in Al_2O_3 are about $0.22\,T_M$ for Q_V and Q_S and $0.18\,T_M$ for Q_{GB}.

Predicting interface formation rates using these and other data is complicated because all of the closure processes may operate simultaneously and a shape change caused by one process may affect the transport of material by another. Nevertheless calculations have been made for a few pure metals and alloys by assuming their mating surfaces to have specific idealized topographies such as triangular ridges. Results of a set of such calculations for Fe are presented in Figure 3.20. These data

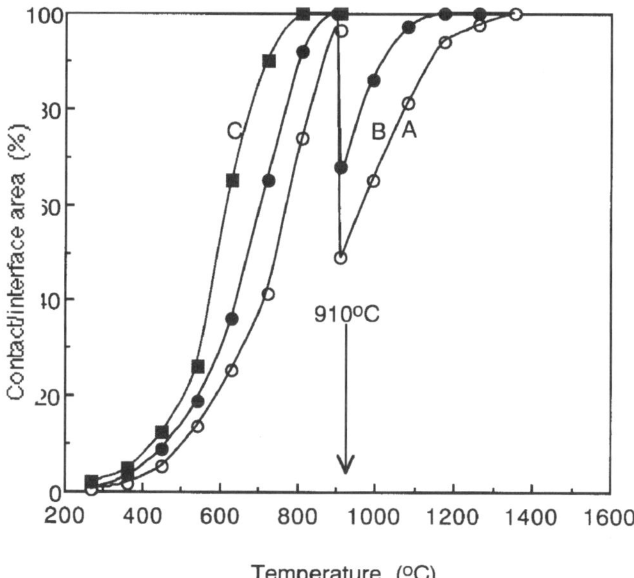

Figure 3.20 Predicted contact/interface areas for Fe diffusion bonded for:
A, 10 minutes using a pressure of 7 MPa,
B, 60 minutes using a pressure of 7 MPa and,
C, 60 minutes using a pressure of 15 MPa. After Derby and Wallach (1984)

suggest that both time and pressure can have a significant effect on the extent of bonding but that the most dramatic effects are produced by varying the temperature. Increasing the temperature up to 910 °C should cause a rapid increase in the bonded area up to 100%. Similarly, increases in temperature above 910 °C should also enhance bonding. However, the most notable feature of the figure is the prediction that the bonding achieved in a particular time at temperatures just above 910 °C will be less than that achieved in a similar time at a slightly lower temperature. This startling prediction reflects the fact that Fe transforms from a body centred cubic to a face centred cubic lattice structure when the temperature is increased above 910 °C and diffusion rates in the high temperature structure are relatively slow. Thus the D_V value at 1382 °C, $0.9\,T_M$, obtained by extrapolation of low-temperature body centred cubic structure data is $9.7 \times 10^{-13}\,m^2s^{-1}$ while that actually measured for the face centred cubic material is $0.4 \times 10^{-13}\,m^2s^{-1}$.

These and data for other systems predict that the achievement of nearly complete diffusion bonding requires a high fabrication temperature, typically greater than $0.6\,T_M$, which is to be expected since the model mainly involves operation of high-temperature mechanisms. However, the other main predictions that fabrication times of usually several hours will be required but that pressures need be only modest are not automatic results of the choice of mechanisms. It should be noted, however, that the modesty of the pressure requirement depends on the yield stress of the material which is in turn a function of the temperature.

Calculations of the growth of the bonded areas also predict which is the dominant mechanism and how this dominance varies with the assumed fabrication conditions and material characteristics. Such predictions often suggest that surface diffusion is dominant at low and moderate temperatures, and this is the case for the data for Fe plotted in Figure 3.20, and that volume diffusion becomes more important at high temperatures but that vapour phase transport is seldom important.

The models described that require such data refer to very idealized conditions in which mating surfaces are perfectly flat and uncontaminated and the materials being joined are identical and pure, but some additional modelling has been done to take account of these complications. Flat mating surfaces are rarely achievable with real surfaces being populated by long wavelength undulations whose presence causes the mating surfaces to contact each other at a few places and to arch away from these to create large voids, as illustrated in Figure 2.4. The collapse of such arches must be realized before bonded area growth can commence over the entire mating surface and this collapse has been attributed to plastic deformation and creep. Regardless of which mechanism is dominant, the necessity for collapse of the arches can significantly lengthen the time needed to achieve complete bonding of the mating surfaces.

The quantitative effects of contamination such as the presence of oxide films on the surfaces of metals on the operation of the diffusion bonding mechanisms has yet to be modelled rigorously but attention has been paid to the effects of using impure or dissimilar materials. If both of the blocks being bonded are identically impure, expressions describing the growth of bonded areas between pure materials can be used with input data that reflect any tendency of the impurities to evaporate more easily or to segregate to and diffuse rapidly along surfaces and interfaces. The use of dissimilar materials complicates modelling because it introduces additional chemical driving forces and destroys the symmetry of void closure processes. Thus while a compositional difference may promote diffusional flow and hence speed up the bonding process, the imbalance of the flow can also cause detrimental formation of Kirkendal porosity in already bonded interface areas and migration of the interface to leave voids that cannot then be closed by the operation of surface and interface diffusion mechanisms.

The differing mechanical properties of dissimilar blocks is also of importance and has been modelled by Derby (1986) for the case of a completely rigid and refractory block being bonded to one that is ductile and has relatively large diffusion coefficients, such as when a ceramic is bonded to a metal. The modelling assumed that the mating surface of the rigid block was populated by a cubic array of cones and that there was no material transfer from the rigid block by creep or diffusional flow. The rigid block merely indented its partner and closure of voids situated at the bases of the conic array was exclusively by material transfer from the ductile block. Closure occurred in two stages, the first being growth of isolated contact/interface areas at a rate described by

$$dA_B/dt = [2\delta.D_{int}.V_m/kT].P_{eff}.[S^3/(S^2 - A_B^2)A_B^2h_V]^{-1} \qquad \{3.51\}$$

when the conical geometry of the indenting mating surface is taken into account and where δ is the width of the interface, V_m is the atomic volume of the ductile material, h_V is the void height and P_{eff} is P_{ext} plus a contribution due to curvature terms and k is Boltzman's constant. Similarly the second stage during which shrinkage and closure of isolated voids occurs can be described by

$$dr_v/dt = -[\delta.D_{int}.V_m/kT]S^3P_{eff}.[(S^2 + r_v^2)In_e(S/r_v) - (S^2 - r_v^2)r_v^2h_V]^{-1} \qquad \{3.52\}$$

where r_v is the interface plane radius of the voids, equal to $(S\text{-}A_B)$.

These expressions are daunting but once again they lead to the conclusion that the growth of bonded areas is very dependent on the details of the assumed surface topography and diffusion characteristics. No one has yet developed a mathematical model for the even more complex case for the bonding of dissimilar materials when each can contribute to the growth of bonded areas.

3.2.2 Behaviour of real metal–metal systems

Assessments of the extent of interface creation when bonding metals to metals are usually based on qualitative interpretation of indirect characteristics. Thus interfaces that are gas tight have at least reached the stage of being populated only by isolated voids, while those that have near parent metal strengths are considered to be substantially bonded. Such qualitative assessments are common and support the conclusions that temperature is the processing parameter of prime importance but that little benefit is gained by prolonging fabrication times from a few hours to tens of hours.

Quantitative studies of joint formation when diffusion bonding metals are sparse, not least because defining the shape and size of populations of the original topographic features of the mating surfaces and of the small voids strung out along the bonded interfaces can be difficult, uncertain and tedious. The mating surfaces are seldom populated by the regular arrays of simple geometrical shapes assumed by the models but rather by a random distribution of small asperities of no fixed geometry, but with typically an average amplitude of 0.1 to 1.0 μm and a width of 10 to 100 μm. Characterizing the void populations is also difficult. Non-destructive evaluation techniques generally are unable to characterize the small voids present at interfaces between the relatively thick metal blocks used when diffusion bonding, so shape and size information has to be gathered by examination of the separated faces of samples whose fracture paths ran along bonded interfaces or by optical and scanning electron microscopy of cross-sections.

Despite these difficulties, some quantitative data are available and these show that the models provide guidance to the behaviour of real materials. Thus interface creation was characterized for the diffusion bonding of Fe and a low alloy steel using microscopy of cross-sectioned samples. The data obtained for both materials were in fair agreement with the predictions of a model that assumed the mating surfaces to be populated by triangular ridge. As shown in Figure 3.21, vacuum bonding of Fe at 700 °C using mating surfaces with an average roughness amplitude of 0.4 μm and a width of 50 μm achieved 65–75% of the predicted bond area growth with time.

The data plotted in Figure 3.21 for bond areas of up to 98% can be described by a simple

$$A_M = A^*.t^n \qquad \{3.53\}$$

relationship where A_M is the measured bond area and A^* and n are constants. Fitting the theoretical data yields values of 36.4% for A^* and 0.25 for n, while for the maximum Fe-Fe data the values are 27.3 and 0.28 respectively. Similar behaviour has been reported for the bonding of

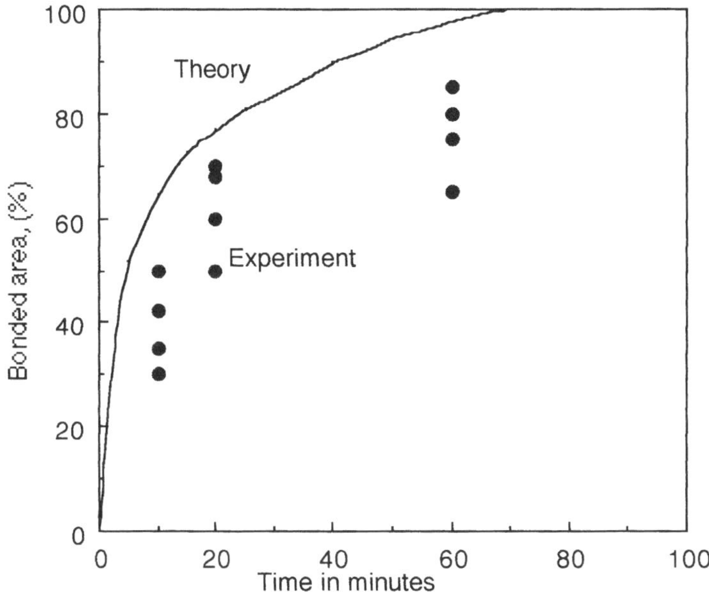

Figure 3.21 The bonded areas of blocks of Fe diffusion bonded at 700 °C using a pressure of 4 MPa. After Derby and Wallach (1984)

Ti-6Al-4V samples in vacuum at 850 °C using a pressure of 4 MPa. Examination of fractured interfaces showed the bonded area increased from 63% after 20 minutes to 89% after an hour and a plot of these data yields an A^* of 28.1% and an n of 0.27.

Experimental examination of bonded Fe–Fe and steel–steel samples showed that the voids along the bonded interface became more rounded as bonding proceeded, confirming the model prediction that surface diffusion is an important material transport process. However, there are obvious deviations between prediction and observation because the bonded areas measured for Fe, steel and Ti-6Al-4V samples were all somewhat smaller than the predicted values. This shortfall has been attributed to two factors, the presence of undulations on the mating surfaces that delayed the onset of void closure processes and contamination by tenacious oxide films.

The effects of undulations on bonding have been studied by pressing samples of Ti-6Al-4V with mating surfaces covered by triangular ridges 3.2 mm high and 6.4 mm wide against a flat tool steel plate, Garmong *et al.* (1975). The model described earlier predicts that the flattening rate of asperities, defined as $d\ln_e w^*/dt$ where w^* is the asperity width at a particular location, will be independent of their size, and hence the behaviour of these experimentally convenient artificial asperities, a hundred to a thousand times greater than those normally encountered in

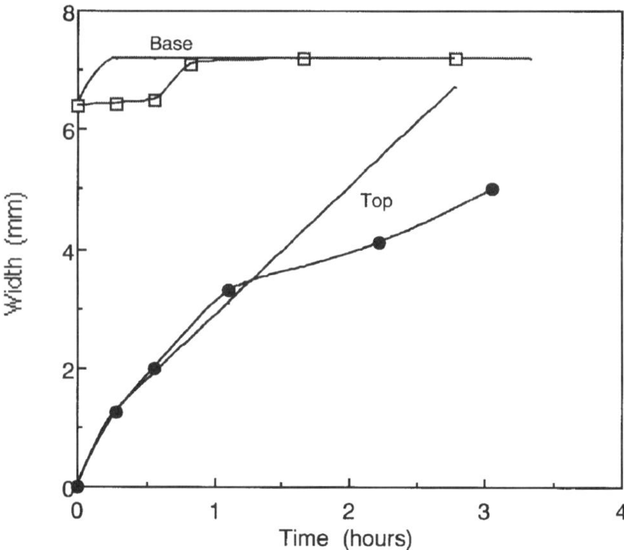

Figure 3.22 Comparison of predicted and observed widths at the top and bottom of 3.2 mm high 6.4 mm wide triangular asperities on the surfaces of compressed Ti-6Al-4V samples. After Garmong *et al.* (1975)

practice, provided a test of the model validity. Comparison of observation and prediction showed a reasonable accord for samples pressed at 1027 °C, Figure 3.22, the deviation of predicted and observed top widths at the longer pressing times being attributed to friction between the steel plate and the Ti-6Al-4V alloy.

This Ti-6Al-4V alloy did not bond with the plates against which it was compressed, probably because of the presence of an Al_2O_3 film on its surface and such films are detrimental for a number of reasons. They prevent the desired metal–metal contact and the surface energy driving force for growth of bonding between the contaminated surfaces is less because γ_S values for oxides typically lie within the range 0.3–1.0 J.m^{-2} as compared to the 0.5 to 3.5 J.m^{-2} for most metals. While the γ_S values for oxides may be low, the interfacial energies for metal oxide systems are usually large so the driving force will be diminished even if only one mating surface is contaminated. Further, the presence of oxides degrades the kinetics of interface creation because oxides are almost always more refractory than the metals from which they form and hence the creep and diffusion processes that could promote bonding will progress slowly at fabrication temperatures chosen for metal–metal systems.

If oxide films are not present, diffusion bonding can be achieved relatively readily. Diffusion bonding of Au and Ag is an ancient practice and strong bonding of even Al–Al couples has been achieved at room temperature using an ultra-high vacuum facility and mating surfaces cleaned by ion bombardment. In practice, less sophisticated techniques are usually employed. Fluxing has not found applications in diffusion bonding practice but most bonding is conducted in vacuum or high-quality inert atmospheres to impede the formation and growth of oxide films and promote dissociation and dissolution in the substrate metal of any already present. Thus bonding of Ni and some Ni alloys can be achieved in a good-quality vacuum at temperatures above 1000 °C because of dissociation of NiO, and the development of diffusion bonding practices for Ti and its alloys owe much to the 33 atom% solubility of O at fabrication temperatures of 800–1000 °C. Another practice is to coat the mating surfaces with a metal, usually Ag, that has a low affinity for O_2. These coatings can vary in thickness, from a fraction of a micrometre to tens of micrometres and this influences the bonding conditions. If the film is thin, i.e. has a thickness less than the amplitudes of the surface roughness, the fabrication temperature, pressure and time will be that expected to be suitable for the substrate metal, but conditions suitable for diffusion bonding the coating material will be needed if the coating has a thickness of several times the amplitude of the surface roughness. An extreme example of this latter practice is the use of ductile foil interlayers to promote bonding which has found application in the bonding of ceramic–metal systems.

3.2.3 Behaviour of real ceramic–metal systems

The systems to be discussed under this heading are those in which a metal interlayer is used to promote bonding between ceramic and metal components or between pairs of ceramic components and attention will be paid only to the creation of ceramic–metal interlayer interfaces. The interlayers used are relatively soft and ductile metal foils or sheets that are much wider than they are thick and this constrains their ability to deform when compressed between the effectively rigid components. The pressure that must be exerted to achieve flow increases linearly as the width/thickness ratio increases, in accord with the von Mises criteria, from $(2/\sqrt{3})\sigma_S$ for a ratio of 1/1 to, for example, $5.75 \times (2/3)\sigma_S$ for a ratio of 20/1. Even larger pressures are used sometimes because the massive extension in surface area caused by a gross compression of the interlayers will disrupt and disperse oxide films that prevent ceramic–metal contact.

The diffusion bonding of ceramics to metals exhibits a number of scientifically interesting features, the most notable of which is that bonding has been achieved for a wide range of material combinations and none has been classified as unbondable. This is in sharp contrast with the inability of solid ceramic–liquid metal systems to create interfaces due to their lack of wetting because energies of the solid ceramic–liquid metal interfaces are

Table 3.11 The solid surface and interfacial energies of some Al_2O_3-metal systems, $J.m^{-2}$. The solid surface energy of Al_2O_3 is 0.9 $J.m^{-2}$

Metal	Ag	Al	Cu	Fe	Ni
Interfacial energy	1.55	1.8	2.2	2.2	2.75
Surface energy	1.1	1.4	1.78	2.1	2.28

larger than those of the surfaces of the solid ceramics. What few values are available for solid ceramic–solid metal interfaces also show their energies to be large, and as can be seen from the data for Al_2O_3-solid metal interfaces assembled in Table 3.11 the values for some systems approach (γ_S ceramic + γ_S metal). Thus the surface energy driving force for diffusion bonding is far less than that assumed by the modelling of idealized systems and indeed for some real ceramic–metal systems may be negligible. Nevertheless there seems no significant difficulty in diffusion bonding ceramic–metal systems, so the modelling of idealized systems may not have taken into account all the factors that promote the creation of interfaces.

The diffusion bonded areas of ceramic–metal interfaces have been measured quantitatively for only a few systems but the behaviour observed is similar to that found with metal–metal systems. Thus Figure 3.23 shows

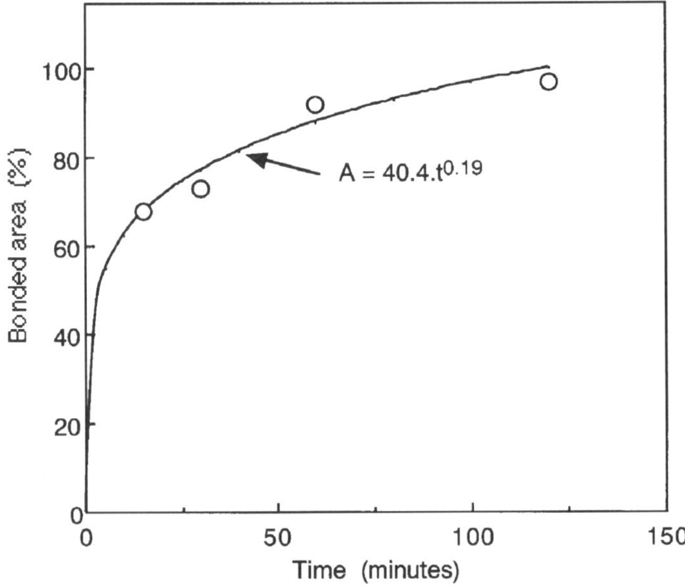

Figure 3.23 The effect of time and temperature on the area of ZrO_2-Pd interfaces bonded at 1100 °C using a pressure of 10 MPa. After Qin *et al.* (1992)

the bonded areas of Pd–ZrO_2 interfaces derived by examination of fracture surfaces to approach 100% after about two hours. These data can be fitted to the simple $A^*.t^n$ relationship found to be adequate for metals, but the n values are lower, 0.19 in this case, and this difference has yet to be fully understood. Because measurements of bond areas are rare, insight into bond area growth has been sought also from the effects of fabrication parameters on bond strengths. Thus attention has been drawn by Derby (1986) to the fact that the effect of fabrication temperature on the tensile strengths of diffusion bonded Al_2O_3-Al-stainless steel samples that failed at their Al_2O_3-Al interfaces can be described by a relationship

$$\text{Bond strength} = 3.14 \times 10^6 \exp^{-Q/RT} \tag{3.54}$$

where Q is 83.3 kJ.mol^{-1}, close to the 84 kJ.mol^{-1} activation energy for grain boundary diffusion of Al.

The systems mentioned so far are nominally inert, but there is evidence suggesting a similarly important role for diffusion processes within the metal interlayer for chemically reactive systems. Thus the strengthening of diffusion bonded samples of Al_2O_3-(Al-4.5Mg)-Al_2O_3 by post-fabrication heat treatment has been attributed to the formation of the $MgAl_2O_4$ spinel as reaction product layers that bridge the ceramic–metal interfaces. It is notable, therefore, that correlation of the times and temperatures needed to achieve maximum strengthening of such samples yields an apparent activation energy of 120 kJ.mol^{-1} that is close to the 116–119 kJ.mol^{-1} reported for volume diffusion of Mg in the alloy. Similarly, the formation of a CrN reaction product layer has been associated with the creation of strong joints in diffusion bonded Si_3N_4-(Ni-20Cr)-Si_3N_4 samples and the activation energy of 265.6 kJ.mol^{-1} for the growth kinetics of this layer has been related to the 264 kJ.mol^{-1} for volume diffusion of Cr in the alloy.

There is, therefore, circumstantial evidence that the creation of ceramic–metal interfaces not only involves some of the processes that dominate the behaviour of metal–metal systems but that the formation of reaction products can play an important and beneficial role. However, the mechanism whereby chemical reactions cause this enhancement is not yet clear but a number of possibilities exist. Thus lateral growth of reaction product layers could enhance interface formation as it does when active metal brazing or the free energy released by the reaction could help make diffusion bonding energetically favourable in ceramic–metal systems that have inherently high interfacial energies.

3.3 HYBRID PROCESSES

Transient liquid phase bonding, TLPB, and partial transient liquid phase bonding, PTLPB, are hybrid brazing/diffusion bonding processes in which interfaces are created by the flow of a liquid metal but which can result in joint structures similar to those produced by diffusion bonding as was illustrated schematically in Figure 2.9. The usual practice is to place a thin metal interlayer between the components that forms a liquid when heated but then resolidifies isothermally to produce a permanent joint. Continuing interdiffusion between the joint seam and the components then produces a chemical homogenization of the joint and the components. The difference between the two hybrid processes is that all the interlayer becomes liquid during TLPB but only the surface region does during PTLPB. Like the other joining processes, the prime driving force for interface creation is a reduction in the energy of the whole system produced by replacing pairs of mating surfaces with interfaces, while the predominant mechanism that effects this change in configuration is interdiffusion of the component and joining materials.

Scientific studies of these processes are relatively recent and so far have concentrated on mainly phenomological assessments of the TLPB of metal–metal systems. These have identified five sequential stages in the bonding of an idealized system in which an interlayer of pure A is used to bond components of pure B, the understanding of which can be assisted by reference to the phase diagram for the A–B system shown as Figure 3.24.

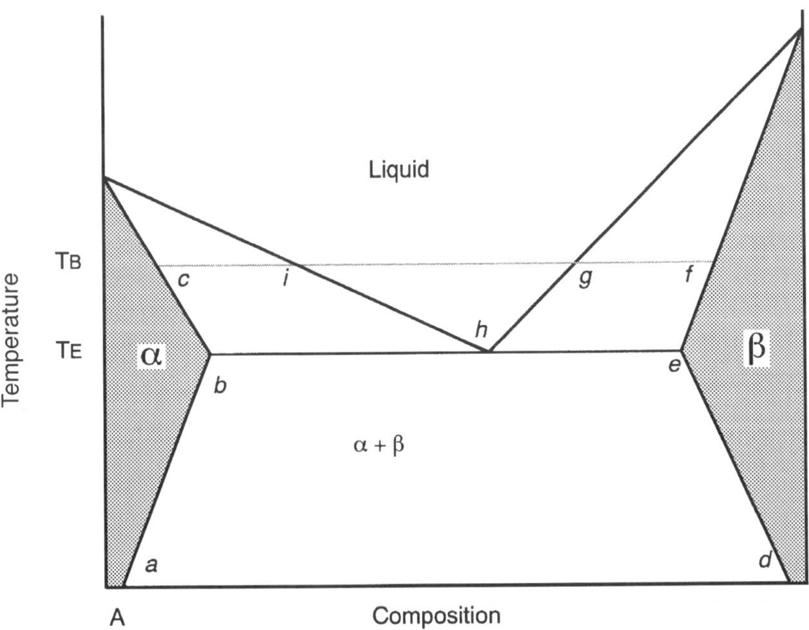

Figure 3.24 Phase diagram of the A–B system

1. The initial assembly of the components and interlayers prior to the bonding cycle causes plastic deformation of asperities to produce small isolated areas of solid–solid contact and interface formation.
2. Raising the assembly temperature to T_E causes additional plastic deformation of the asperities and hence growth of the isolated interface islands because the yield stress decreases with temperature. This heating also allows interdiffusion to occur at an increasingly rapid rate to change the compositions of the interlayer and the components as the assembly is heated to T_E, the surface region of the interlayer of A becoming alloyed to produce an α phase which contains an increasing amount of B, shown as $a \rightarrow b$ in Figure 3.24, while the surfaces of the components of pure B are alloyed to produce a β phase that contains an increasing amount of A, shown as $d \rightarrow e$.
3. Increasing the temperature from T_E to the T_B chosen for bonding causes the B content of the α phase and the A content of the β phase to decrease, from $b \rightarrow c$ and from $e \rightarrow f$, and a eutectic liquid to form with a composition lying within the *ghi* region of the figure. This liquid wets and spreads over the solid surfaces of the components and the interlayer to increase their contact area and hence the rate of alloying and liquid formation caused by interdiffusion.
4. Holding the assembly at the bonding temperature T_B causes a liquid layer to spread over the whole of the mating surfaces and initially increase in thickness as interdiffusion continues until the whole of the original interlayer is consumed. Continued interdiffusion then causes a progressive change in the composition of the liquid from the *h* to the *g* shown in the figure. Thereafter, isothermal resolidification of the liquid occurs to cause thinning and ultimately the complete disappearance of the liquid layer.
5. Continued interdiffusion between the now solid phases present in the bonded assembly causes the A content of the joint region to decrease and that of the components to increase until, ultimately spatial variations of composition become trivial and the assembly structure and chemistry resembles that of diffusion bonded components of an alloy of B containing a small amount of A.

The creation of an interlayer/component interface completes the joint formation process when brazing or diffusion bonding, but TLPB is complete only when the bonded joint/component assembly is homogenized to produce a structure akin to that achievable by diffusion bonding components without the use of an interlayer. Thus the TLPB process involves all five stages and not merely the first three in which a liquid interlayer/component interface is created. It can be argued that the production of a brazed joint should also be regarded as completed only when the bonded assembly has been equilibrated by post-fabrication heat treatment to produce homogenization of the joint and component compositions, but such practice is exceptional and therefore was not considered in section 3.1. There is no generally accepted complete model of TLPB that describes all five stages but analytical descriptions have been advanced for certain aspects of both ideal and real systems.

3.3.1 Modelling TLPB of ideal systems

Modelling of idealized and real systems, Zhou *et al.* (1995), has been concerned mainly with configurations in which a thin interlayer of pure A is used to bond components of pure B, the thinness of the interlayer being such that the A and B contents of the assembly lie within the β phase region shown in Figure 3.24. There are no barriers in an ideal system to direct contact between the materials being bonded, contact and interface formation are equivalent. Thus interface formation at contacting asperity tips commences immediately when the components and interlayers are assembled, and the extent of this initial formation depends on the stress applied to the assembly and the yield characteristics of the deforming asperities, usually assumed to be only those on the surface of the interlayer. As in the case of diffusion bonding, isolated islands of interface are formed almost instantaneously by plastic yielding to accommodate the external pressure in accord with the expression

$$A_F = P_{ext}/\text{Constant}.\sigma_Y \qquad \{3.55\}$$

where A_F is the fractional interfacial area, (actual/nominal), and the constant approaches a value of 3 for asperities with triangular cross-sections. In practice the P_{ext} applied is usually modest and intended only to hold the assembly in place while being heated to the bonding temperature. Thus the interfacial area created by this plastic deformation also will be modest until the yield stress becomes small at temperatures above about $0.6\,T_M$. At these high temperatures, slow continuing creep deformation also can cause growth of the contact/interface area in accord with the expression

$$d\varepsilon_{SS}/dt = A(\sigma/E)^n \exp(-Q_V/RT) \qquad \{3.47\}$$

but this contribution to the growth of the interfacial area will be relatively small because of the inherent slowness of creep deformation and the short time spent at high temperatures. Once a liquid is formed, its flow becomes the dominant mechanism of interface creation.

The formation of a liquid eutectic at temperatures of T_E and above caused by interdiffusion of interlayer and component materials, A and B, results in a net transport of solid material from the interlayer. The diffusion behaviour of an assembly of a thin interlayer sandwiched between two large components has been analysed using the 'thin film solution' which is valid when the components are at least $4(D_{\underline{A}}t)^{1/2}$ thicker than the interlayer, where $D_{\underline{A}}$ is the coefficient of diffusion of A in B. For such a configuration, the concentration of the interlayer material A at different locations can be described by the expression

$$N_A = w.\exp(-x^2/4D_{\underline{A}}t)(4\pi D_{\underline{A}}t)^{-1/2} \tag{3.56}$$

where w is the width of the layer and x is the distance from the centre of the layer. The resulting concentration gradient embraces the entire range from pure A to pure B and hence results in the phase sequence $\alpha \rightarrow$ liquid $\rightarrow \beta$. As time increases, so will the depth of interdiffusion, while the concentration of A at the centre of the interlayer, $x = 0$, will decrease as a function of $t^{1/2}$. If the intermediate phase was a solid, it would at first thicken with $t^{1/2}$ until all the interlayer is consumed and then waste away, again as $t^{1/2}$, as continuing interdiffusion replaced more and more A with B.

The behaviour of even an ideal system will be complicated by the facts that the coefficient of diffusion and the concentrations of A at which phase changes occur change as the temperature is raised to T_B, requiring the use of 'effective' coefficients and saturation solubilities. Further, while a solid intermediate phase will be nucleated at discrete sites and grow laterally as well as thicken to form a continuous layer, formation of a liquid intermediate phase will change the growth process by spreading over the surfaces of the solid α and β phases, causing an early thinning of the intermediate layer and increasing the area across which interdiffusion can occur and hence the dissolution rate of the interlayer.

Modelling the flow of wetting liquids has been described earlier and it is sufficient to remember at present that this should be both very extensive and very rapid for ideal metal–metal systems so that interface creation should be complete shortly after the temperature exceeds T_E. However, the subsequent interdiffusion induced resolidification of a liquid that has spread all over the mating surfaces should again be a function of $t^{1/2}$ as should the final decrease in the concentration of A to a level close to that for the whole bonded assembly. The times required for completion of these two changes will be longer than for the formation of a liquid layer because diffusion has to take place over longer distances, and hence it is this effective conversion of the joint from a brazed to a diffusion bonded structure that largely determines the process time for joining by TLPB.

3.3.2 TLPB of real metal–metal systems

TLPB has been used successfully to join many metallic materials, including superalloys based on Fe and Ni, stainless steels, Ti and Al alloys and also Ti and Al matrix composites, and the experience gained during this work contributed to the development of models for the idealized systems. It is not surprising, therefore, that the behaviour observed during the TLPB of metal–metal combinations is generally in accord with that predicted by models of ideal systems. Nevertheless there are some differences between real and ideal systems that can be important. In particular, the behaviour of real materials is influenced by more factors than are taken into account by the idealized models and real materials do not progress through a sequence of distinctively separate stages when being joined by TLPB.

Reference has been made already to the fact that the liquid produced by interdiffusion can spread over component surfaces, decreasing its thickness but increasing the area across which further interdiffusion can occur. These effects will distort stages (2) and (3) of the TLPB process and it is perhaps not surprising therefore that the dissolution of real interlayer materials cannot be described by a simple $t^{1/2}$ relationship and is often somewhat faster than predicted. However, the liquid phase bonding stages of TLPB are of relatively short duration and hence these kinetic differences do not have a great effect on the total processing times. Further, examination of Ni/Ni-11P/Ni assemblies has shown that the maximum extent of component dissolution, which is related to the volume of liquid phase formed, is affected by process parameters such as the initial thickness of the interlayer, the heating rate and the bonding temperature, all of which are in qualitative accord with theoretical expectations.

Agreement between theory and practice can be in reasonable quantitative accord once a stable configuration has been achieved. Thus thinning of the liquid layer by isothermal resolidification is found to be a temperature dependent function of $t^{1/2}$ when bonding Ni components with a Ni-B interlayer in accord with model predictions. Again, as shown in Figure 3.25, while the observed increase of a component solute at a joint centre-line during the fifth and final stage of TLPB is faster than model expectations, the form of time dependence is similar.

The presence of oxide films on the surfaces of most metals and of grain boundaries in real component substrates can have profound effects on their TLPB behaviour. Surface oxide films prevent direct metal–metal contact and this can impede joint formation when joining by TLPB. The oxide films on interlayer materials may be disrupted when melting occurs but those on the surfaces of the solid components can be significant barriers. Thus the dissolution of MM007, Ni-10Co-8Cr-6Mo-6Al-4Ta-1.3Hf-0.1C-0.02B, super alloy components at 1123 °C when using a MBF-80, Ni-15.5Cr-3.7B, interlayer is initially impeded by oxide film: a 0.13 μm thick film halves the normally observed dissolution rate for the first 250 seconds of contact and a 0.23 μm layer decreases the rate tenfold for more than 600 seconds, Nakao *et al.* (1990). This slow start to component dissolution was attributed to the need for the liquid constituents to diffuse through the oxide films, and presumably then to disrupt

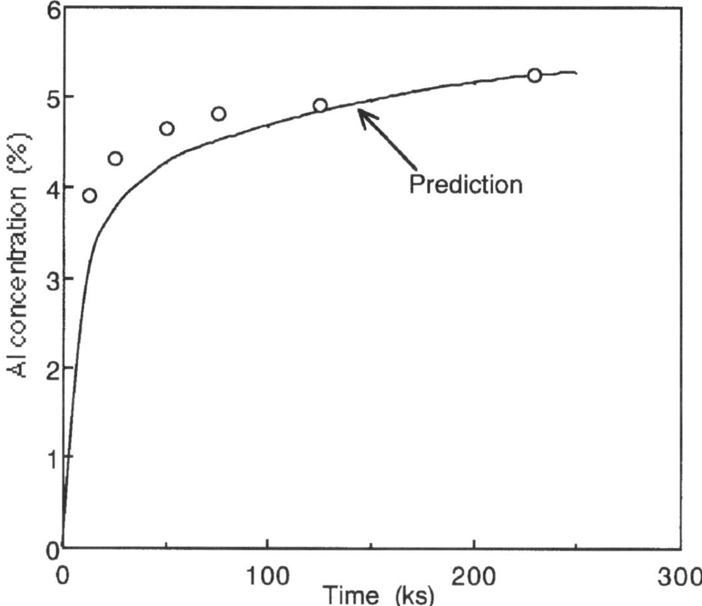

Figure 3.25 Build up of Al at the centre-line of a TLPB assembly of MM007 components joined at 1150 °C using a MBF-80 interlayer. After Nakao *et al.* (1990)

and disperse it by causing the substrate to melt. However, it is also possible that the mechanisms that permit brazes to make contact with oxidized components may be relevant: dissociation in low O_2 environments, dissolution in the substrate, reduction by braze or substrate constituents and undermining by substrate dissolution initiated by liquid flow through tears or cracks in the film.

In contrast, the presence of grain boundaries in the component substrates is generally beneficial because they offer opportunities for accelerated interdiffusion. This arises because grain boundaries that intersect liquid–solid interfaces form grooves, subtending an angle ϕ where $\cos\phi/2$ equals $\gamma_{GB}/2\gamma_{SL}$ and hence increase the area across which interdiffusion can occur. Additionally, some solutes segregate to grain boundaries, and most importantly grain boundaries are high diffusivity paths. D_{GB} for self-diffusion is typically 10^4 larger than D_V at $0.9\,T_M$, Table 3.10, and the disparity will be even larger at lower temperatures because Q_V is larger than Q_{GB}. The amount of material transported along grain boundaries is limited by their small width, typically about 1 nm, but nevertheless the contribution is significant and it has been calculated that such a change should decrease the resolidification time by 25% for a material with a 33 μm grain size, and by 50% for one with a 10 μm grain size. Assuming such a grain boundary enhancement is also applicable to solute diffusion would account for the faster resolidification during the TLPB of fine-grained Ag or Ni components. Thus Figure 3.26 shows resolidification of the liquid layer formed decreased linearly as a function of $(\text{time})^{1/2}$ when a Ni base super alloy was joined with a Ni-B interlayer, indicating that diffusion was the controlling process. Further, the time for complete resolidification decreased by 27% when the grain size was reduced from several millimetres to 33 μm. Similarly, such grain boundary enhancement of material transport by diffusion could account for the faster than predicted build-up of Al at a joint centre-line shown in Figure 3.25.

Finally, the behaviour of real systems can sometimes display unexpected microstructural effects when some but not all of the solutes can diffuse very rapidly. Thus particles of Ni_3B but not of silicides were precipitated in Ni near the liquid–solid interface during TLPB with a Ni-4.5Si-3.2B interlayer at 1065 °C, and this was due to two effects, the high coefficient of diffusion, D_V, for B in Ni, $10^{-10}\,m^2s^{-1}$ as compared to $10^{-14}\,m^2s^{-1}$ for Si, and the decrease in the solubility of B in γNi in the presence of Si. Thus

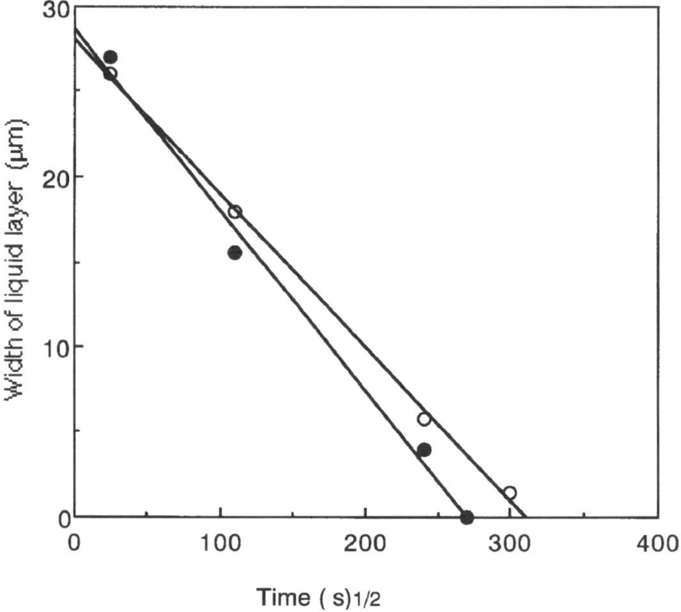

Figure 3.26 The effect of time on the thickness of the liquid layer when joining a Ni base super alloy with a Ni-B interlayer. The hollow and full symbols relate respectively to the joining of coarse- and fine-grained components

B diffused readily into the Ni and continued to do so until the concentration at the joint centre line fell to a very low level, however the slower diffusion of Si into the γNi caused the local solubility limit for B to decrease and Ni_3B to be precipitated.

3.3.3 Real ceramic–ceramics systems

The joining of ceramics by TLPB is impeded by the non-wetting behaviour of most ceramic–metal systems. Thus an active metal braze alloy could be used as the interlayer, but that process would not be TLPB because the isothermal resolidification and homogenization stages would not occur. However, the partial transient liquid phase bonding variant of TLPB has been developed specifically to produce ductile joints for ceramic components that must withstand high service temperatures.

The essential stages in PTLPB are illustrated schematically in Figure 3.27. The interlayer used is more complex than those employed when joining by TLPB. It consists of a high melting temperature metal or alloy core that has a thickness of typically 25–125 μm whose surface is coated with a 1 μm or so layer of a low melting temperature metal or alloy. When heated to the

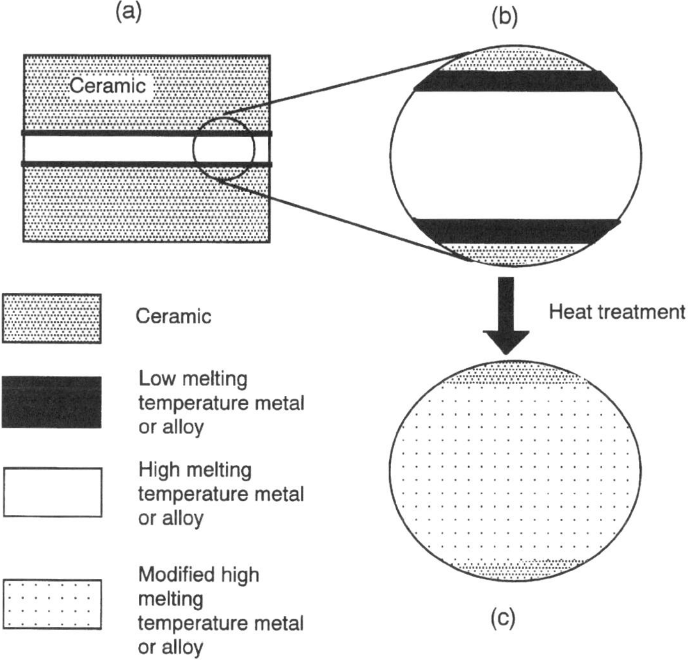

Figure 3.27 The principles of PTLPB joining of ceramics

bonding temperature, the coating melts either because its liquidus has been exceeded or because interdiffusion between it and the interlayer core produces a liquid eutectic composition that is formed initially within the interlayer at the core/coating interface. Ultimately the surface of the interlayer is covered by a liquid film which fills the capillary gap between the ceramic component and the interlayer. When isothermal resolidification of the liquid starts, it is due to interdiffusion with the core of the interlayer as well as interaction with the component. Finally, continued heat treatment after the liquid layer has resolidified homogenizes the interlayer to produce a composition little different from that of the original core.

PTLPB is still a laboratory process and most of the scientific aspects that must be considered if it is to be used as a production process have been mentioned already when discussing TLPB. However, the successful laboratory use of PTLPB to join Al_2O_3 using interlayers of Pt, Ni and Nb coated with Cu and Si_3N_4 and SiC using Ni-20Cr interlayers coated with Au or Ti and Au-Cu has revealed a number of additional scientific aspects that must be considered.

Successful application of TLPB depends on the ability of the liquid to wet the metal components and this requirement usually can be satisfied quite easily. The equivalent requirement when using PTLPB to join ceramic components is more complex because they are wetted by few metals. However, filling the capillary gap between the ceramic and the interlayer does not require the ceramic to be wetted but merely that the sum of the contact angles between the liquid and the ceramic and interlayer should be less than 180°. This easier requirement arises because the gap will fill if the energy change δE caused by a small movement of the liquid front is negative. Thus at equilibrium

$$\delta E = (\gamma_S - \gamma_{SL})_{metal} + (\gamma_S - \gamma_{SL})_{ceramic} = 0 \quad \{3.57\}$$

$$\gamma_L \cos\theta_{metal} + \gamma_L \cos\theta_{ceramic} = 0 \quad \{3.58\}$$

$$\cos\theta_{metal} = -\cos\theta_{ceramic} \quad \{3.59\}$$

$$\theta_{metal} = 180^\circ - \theta_{ceramic} \quad \{3.60\}$$

$$\theta_{metal} + \theta_{ceramic} = 180^\circ \quad \{3.61\}$$

Since uncontaminated metal components are well wetted, $\cos\theta \rightarrow 1.0$, by a wide range of liquid metals and alloys, many liquids should be able to fill capillary gaps between ceramics and metal components. However, it should be beneficial if the liquid does wet the ceramic because this would enable small crevices to be filled whose presence as voids could degrade the mechanical properties of the joints. It is interesting to note, therefore, that the strongest PTLPB joints reported so far have been for the joining of Si_3N_4 using interlayers of (Ti-Au-Cu)-Ni-(Cu-Au-Ti) and Au-(Ni-20Cr)-Au in which a reactive metal has been available to promote reactive wetting.

First estimates of the times needed for joint formation by PTLPB can be derived from knowledge of saturation concentrations and diffusion characteristics. Thus PTLPB of Si_3N_4 using an interlayer of (2.5 μm Au/ 125 μm Ni-20Cr/ 2.5 μm Au) at 1000 °C resulted in isothermal resolidification of the coating within 4 hours. The melting temperature of Au is higher than 1000 °C but an Au-5Ni alloy will melt at 1000 °C and resolidify when the Au concentration falls to 55%. The diffusion coefficient of Au in Ni-20Cr is not known but if the coating is regarded as half of a thin film in contact with a semi-infinite block, the observed resolidification time suggests a coefficient of at least $1.9 \times 10^{-15}\,m^2s^{-1}$ and this is reasonably close to the $4.5 \times 10^{-15}\,m^2s^{-1}$ reported in the literature for the volume diffusion coefficient of Au in Ni. More sophisticated analysis would take into account the effect of ingress of Si from the ceramic which will tend to keep the coating liquid and the contribution of grain boundary diffusion within the interlayer core which could result in earlier isothermal resolidification, but the similarity of observation and prediction suggests that even approximate models and input data can provide some useful guidance to process parameters and insight into the controlling mechanisms.

REFERENCES

Ambrose, J. C., Nicholas, M. G. and Stoneham, A. M. (1993) Dynamics of liquid drop spreading in metal-metal systems, *Acta Metallurgica et Materialia*, **41**, 2395–402.

Derby, B. and Wallach, E. R. (1982) Diffusion bonding: development of a theoretical model, *Metal Science*, **18**, 427–32.

Derby, B. and Wallach, E. R. (1984) Diffusion bonds in iron and low alloy steel, *Journal of Materials Science*, **19**, 3149–58.

Derby, B. (1986) The influence of surface roughness on interface formation in metal/ceramic bonding. In J. A. Pask and A. G. Evans (eds), *Ceramic Microstructures 86*, Plenum Press, New York, 319–28.

Garmong, G., Paton, N. E. and Argon, A. S. (1975) Attainment of full interfacial contact during diffusion bonding, *Metallurgical Transactions A*, **6A**, 1269–79.

de Gennes, P. D. (1985) Wetting: statics and dynamics, *Reviews of Modern Physics*, **57**, 828–64.

Hitchcock, S. J., Carroll, N. T. and Nicholas, M. G. (1981) Some effects of substrate roughness on wettability, *Journal of Materials Science*, **16**, 714–32.

Johnson, R. E. and Dettre, R. H. (1963) Contact angle hysteresis: 1 A study of an idealised rough surface, *Advances in Chemistry*, **43**, American Chemical Society, 112–36.

Latin, A, (1946) Capillary flow in the soldering process and some measurements of the penetration coefficients of soft solders, *Journal of the Institute of Metals*, 265–82.

Miedema, A. R. and den Broeder, F. J. A. (1979) Interfacial energy in solid-liquid and solid-solid combinations, *Zeitschrift für Metallkunde*, **70**, 14–25.

Naidich, Y. V. (1981) The wettability of solids by liquid metals, *Progress in Surface and Membrane Science*, **14**, 354–485.

Nakao, Y., Nishimoto, K., Shinozaki, K. and Kang, C. Y. (1990) Transient liquid insert metal diffusion bonding of nickel-base superalloys. In T. H. North (ed.), *Advanced Joining Technologies*, Chapman & Hall, London.

Nicholas, M. G. and Crispin, R. M. (1986) Some effects of anisotropic roughening on the wetting of metal surfaces, *Journal of Materials Science*, **21**, 522–8.

Pauling, L. (1960) *The Nature of the Chemical Bond*, Cornell University Press, Ithaca, NY.

Qin, C-D., James, N. A. and Derby, B. (1992) Pd-ZrO_2 diffusion bonding: mechanical properties and interface reactions, *Acta Metallurgica and Materialia*, **40**, 925–32.

Shuttleworth, R. and Bailey, G. L. J. (1948) The spreading of a liquid over a rough surface, *Discussions of the Faraday Society*, **3**, 16–35.

Zhou, Y., Gale, W. F. and North, T. H. (1995) Modelling of transient liquid phase bonding, *International Materials Reviews*, **40**, 181–97.

Selection of materials | 4

There is a continuing demand for more economic joining processes that produce components capable of satisfying performance specifications that are becoming ever more stringent. These requirements make it essential that implementation of the joining processes is optimized, and many and varied decisions must be made before joining components by brazing, diffusion bonding or hybrid-bonding processes. While numerous and varying in individual importance from application to application, the decisions that have to be made can be grouped into sets. Those that relate to materials selection are described in this chapter, while those concerned with process parameters and the design of joints are described later.

It is relatively rare for the fabricator to be involved in the specification of component materials even though this may include characteristics such as resistance to chemical attack or stiffness that can restrict the choice of joining process and materials. The purpose of this chapter is to describe how the fabricator can accommodate the differing component material characteristics summarized in the next section by careful selection of joining materials.

4.1 CHARACTERISTICS OF COMPONENT MATERIALS

4.1.1 Metals

Metallic components that need to be joined are usually of alloys based on one or more of the twelve metals for which some characteristic properties are listed in Table 4.1. The table divides the metals into four groups, three of which emphasize a shared and relatively extreme characteristic – the low density of Al, Be and Ti, the high thermal and electrical conductivities of Ag, Au and Cu or the refractoriness of Mo and W – while the fourth grouping of Co, Fe and Ni reflects the importance of steels and Ni or Co super alloys in a vast range of technical applications rather than a particular shared property.

Table 4.1 Some properties of pure metals

Metal	*Melting temp.* °C	*Density* $Mg.m^{-3}$	*Expans. coeff.* $10^{-6} K^{-1}$	*Electrical cond.* $S.m^{-1}$	*Thermal cond.* $W.m^{-1} K^{-1}$	*Young's modulus** GPa	*Yield strength*† MPa	*O affinity* kJ/mole of O
Al	660	2.70	23.5	4×10^7	238	70.6	20	−528
Be	1287	1.85	12.0	3×10^7	194	318	240	−580
Mg	649	1.74	26.0	2×10^7	156	44.7	69	−571
Ti	1667	4.50	8.90	2×10^6	21.6	120	103	−509
Ag	962	10.5	19.1	6×10^7	425	82.7	–	−11
Au	1064	19.3	14.1	5×10^7	316	78.5	–	+
Cu	1085	8.96	17.0	6×10^7	397	130	48	−127
Co	1494	8.9	12.5	2×10^7	96.0	211	345	−212
Fe	1535	7.87	12.1	1×10^7	78.2	211	346	−241
Ni	1455	8.90	13.3	2×10^7	88.5	199	60	−213
Mo	2615	10.2	5.1	2×10^7	137	325	345	−266
W	3387	19.3	4.5	2×10^7	174	411	550	−255

* Young's modulus of elasticity.

† Defined as the (load/original cross-sectional area) needed to cause a permanent deformation of 0.2% of annealed, fully softened, material.

Many but not all properties are linked by their dependence on electronic and lattice structures. Thus the group of low-density metals has high chemical affinities for O and quite high specific stiffnesses (elasticity/density) ratios that can cause problems for the joint fabricator, and three of the four also have high electrical and high thermal conductivities. The group of high conductivity metals has colours that make them attractive for use in jewellery and other high value decorative items and moderate or negligible affinities for O. The refractory metals have low coefficients of expansion and high modulus of elasticity values, but their high densities means that their specific stiffnesses are not particularly high. Finally, it can be seen from the table that Co, Fe and Ni are quite refractory and have fairly high specific stiffnesses.

When the component material is an alloy rather than a pure metal, the joint fabricator must pay attention to the effects of the additions on physical, chemical and mechanical properties. An immense number of alloy compositions have been developed, examined or suggested that could be used as component materials, but guidance as to what sort of compositions will be most commonly encountered can be gained by resorting to standards specifications and reference tabulations such as those in the *ASM Metals Handbook* or *Smithells Metals Reference Book*. These reveal that some alloying additions made to enhance service performance will also have detrimental effects on joining behaviour. Thus they can decrease the melting temperature or change the chemistry of the oxide films on the component surfaces. Similarly, the mechanical properties of alloys can be changed by cold working and also by heat treatments that take alloying elements into solution in the solvent metal matrix or cause them to be rejected and change the size, shape or distribution of the alloy phase precipitates.

Al and its alloys

The low melting temperature and electrode potential of Al and Al alloys compared to braze metals such as Ag, Cu or Ni restricts the range of materials with which it can be joined effectively to Al alloys such as the Al-Si eutectic brazes. Commercially developed Al alloys have been standardized and divided into eight groups relating to their principal alloying element, examples of which are listed in Table 4.2. Using brazes developed for pure Al, the easiest alloys to join are commercially pure 1XXX series, the Mn rich 3XXX series of alloys and Mg lean members of the 5XXX series. Problems can be encountered with the 2XXX series and the 7XXX because of their low melting temperatures, an addition of 5Cu decreasing the melting temperature of Al to 548 °C. Alloys in the 5XXX and 6XXX series generally contain a percent or more of Mg and this can both thicken and change the chemistry of the surface film of oxide. Thus care should be taken when annealing such materials to do so in a low O

Table 4.2 Examples of Al alloy component materials

Alloy	*Composition, wt% of additives*	*Comments*
1060	< 0.40	Solidus 660 °C
1100	< 1.0	Solidus 660 °C
2004	6.0Cu, 0.40Zr	Superplastic
2024	4.5Cu, 1.5Mg, 0.60Mn	Solidus 502 °C
3003	1.2Mn	Core of brazing sheet
3004	1.2Mn, 1.0Mg	Solidus 629 °C
5050	1.2Mg	Solidus 627 °C
5056	5.2Mg, 0.10Mn, 0.10Cr	Solidus 568 °C
6061	1.0Mg, 0.60Si, 0.25Cr, 0.20Cu	Solidus 552 °C
6151	1.0Si, 0.60Mg, 0.20Cr	Solidus 582 °C
7016	4.5Zn, 1.1Mg, 0.75Cu	
7075	5.5Zn, 2.5Mg, 1.6Cu, 0.25Cr	Solidus 476 °C
8090	2.5Li, 1.3Cu, 0.80Mg, 0.12Zr	Superplastic

activity environment. On the other hand, a moderate concentration of Mg can assist joining in evacuated chambers by evaporating to disrupt the surface oxide. The additions of Zn present in 7XXX alloys can also render the surface oxide unstable, as can the Li present in the 8090 whose superplastic characteristics can be exploited in the diffusion bonding of complex thin wall structures such as heat exchangers.

Mg and its alloys

Similar restrictions to the choice of joining materials for Mg apply as for Al, and just as Al is alloyed with Mg so Mg is alloyed with Al (Table 4.3). Other important solutes are Th, Zn and Zr. These additives all have high affinities for O, but not significantly more so than that of Mg itself and, therefore, do not have marked effects on surface chemistries. However, the additions of Al, Th or Zn present in some Mg alloys are sufficient to decrease the already low melting temperature of Mg by between 25 and 55 °C, so restricting even more the choice of suitable brazing alloys.

Ti and its alloys

The lattice structure of Ti changes from that of a close packed hexagonal α phase to that of a body-centred cubic β phase when its temperature rises

Table 4.3 Some Mg alloy component materials

ASTM alloy	*Composition, wt% additives*	*Comment*
M1A	1.5Mn, 0.09Ca	
AZ31B	3.0Al, 1.0Zn, 0.20Mn	Solidus 605 °C
AZ61A	5.5Al, 1.0Zn, 0.20Mn	Solidus 525 °C
AZ80A	8.5Al, 0.50Zn	Solidus 490 °C
ZK20A	2.3Zn, 0.55Zr	
ZK60A	5.5Zn, 0.50Zr	Solidus 520 °C
HM21A	2.0Th, 0.50Mn	Creep resistant
HM31A	3.0Th, 1.2Mn	Creep resistant

through 882 °C, and alloys have been developed that stabilize both phases as well as $\alpha + \beta$ microstructures. Alloying to stabilize the β or $\alpha + \beta$ requires complex additions that also produce stronger alloys as indicated in Table 4.4, but the existence of these different structures also affects the selection of possible joining materials. Thus brazing of the high temperature $\alpha + \beta$ and β alloys should be conducted at temperatures below the $\alpha \rightarrow \beta$ transition temperature of the alloy if ductility is not to be impaired by the abrupt lattice change or excessive chemical interaction with the braze alloy. On the other hand, the superplasticity that can be displayed by the $\alpha + \beta$ IMI 318 alloy at 750–1000 °C can be exploited in diffusion bonding processes.

Despite the chemical affinity of Ti for O, the pure metal is relatively easy to join in evacuated or inert environments because the high solubility of O, 33 atom% in αTi at 800 °C, assists the formation of clean bonding surfaces. However, alloying additions of Al can cause formation of substantial surface films of relatively insoluble Al_2O_3 in the environments to which the components are exposed prior to joining. Awareness of the chemical affinity of Ti for other elements is also of importance to the joint

Table 4.4 Examples of Ti alloy component materials

Alloy	*Composition, wt% of additives*	*Structure and room temperature UTS* (MPa)
IMI 125	< 0.8	αTi, 480
IMI 317	5Al, 2.5Sn	αTi, 920
IMI 318	6Al, 4V	$\alpha + \beta$Ti, 1040
IMI 550	4Al, 2Sn, 4Mo, 0.5Si	$\alpha + \beta$Ti, 1220
IMI 685	6Al, 5Zr, 0.5Mo, 0.25Si	βTi, 1560
IMI 829	5.5Al, 3.5Sn, 3Zr, 1Nb, 0.3Mo, 0.3Si	βTi, 1040

Table 4.5 Typical Cu alloy component materials

Alloy	*Composition, wt% of additives*	*Yield strength,* (MPa)
OFHC Cu	< 0.05	48
Deoxidized Cu	0.04P	54
Chromium copper	0.6Cr	324
Beryllium copper	1.85Be, 0.25Co	224
Gilding metal	10Zn	100
Brass	30Zn	115
Brass	40Zn	309
Aluminium brass	22Zn, 2Al	139
Naval brass	37Zn, 1Sn	170
Nickel silver	28Zn, 10Ni	100
Nickel silver	18Zn, 25Ni	124
Phosphor bronze	3.5Sn, 0.1P	124
Phosphor bronze	7Sn, 0.1P	139
Silicon bronze	3Si, 1Mn	77
Aluminium bronze	5Al	147
Aluminium bronze	8Al	170
Cupronickel	5.5Ni, 1.2Fe, 0.5Mn	116
Cupronickel	31Ni, 1Fe, 1Mn	170

fabricator if detrimental effects such as embrittlement due to pick up of H_2 from reducing atmospheres and the formation of brittle intermetallic layers by reaction with joining materials are to be avoided.

Cu and its alloys

Cu components are relatively easy to join because of their good ductility and the low chemical stability of their surface oxides, but some of the wide range of alloying additions for Cu identified in Table 4.5 can cause problems. Thus alloying Cu to increase its strength can make successful diffusion bonding more difficult to achieve. Further, many Cu alloys have to be stress relieved at temperatures of about 300 °C before being brazed to avoid cracking, liquid metal embrittlement, induced by contact with the molten filler metal. Alloying can also cause chemical effects of concern when making a joint, for example, additions of Al, Be, Cr or Zr can cause component surfaces to be coated with chemically stable and physically tenacious oxide films whose presence impedes the achievement of high integrity interfaces. This detrimental effect of surface oxides occurs with

Al bronzes containing more than about 8Al, but normally not with bronzes incorporating Sn or Si. Other detrimental chemical effects that may be of concern are the loss of Zn and the consequent mechanical degradation of brasses that can occur when using high joining temperatures, and the formation of surface dross by alloys to which Pb has been added to enhance machinability.

Ni and its alloys

The joining of Ni components by brazing or diffusion bonding does not present great problems for the fabricator, and indeed Ni coatings can be used to promote the joining of ceramics to metals and of metal alloys covered by surface films of stable oxides. Ni alloys are also joined successfully by both brazing and diffusion bonding, but their characteristics do present some problems.

Technically important families of Ni alloys have been developed by Inco and other companies that are based on the incorporation of substantial concentrations of Cu (Monels), of Cr and Fe (Inconels, Incoloys, and some Hastelloys and Nimonics), Cr or Cr and Co or Mo (some Hastelloys and Nimonics), but many of the compositions quoted in Table 4.6 also

Table 4.6 Examples of Ni alloy component materials

Alloy	*Composition, wt% of additives*	*Yield strength,* (MPa)*
Monel 400	30Cu, 1.5Fe, 1Mn	230 (570)
Monel K500	29Cu, 2.8Al, 0.5Ti	600 (850)
Inconel 600	16Cr, 7Fe, 1Mn, 0.3Si, 0.2Ti, 0.16Al	260 (700)
Inconel 625	22Cr, 4Nb, 9Mo	352
Incoloy 800	45Fe, 21Cr, 0.4Ti, 0.4Al	310 (700)
Incoloy 825	32Fe, 21Cr, 3Mo, 2Cu, 1Ti	340
Hastelloy C4	16Mo, 16Cr	416
Hastelloy X	9Mo, 21Cr, 18Fe	350
Nimonic 75	20Cr, 0.4Ti	350
Nimonic 80A	20Cr, 2Ti, 1.5Al	(800)
Nimonic 90	20Cr, 17Co, 2.4Ti, 1.4Al	(800)
Nimonic 105	15Cr, 20Co, 5Mo, 5Al, 1.2Ti	(800)
Nimonic PE16	32Fe, 16Cr, 3Mo, 1Ti, 1Al	(460)
Udimet 700	18Co, 15Cr, 5Mo, 4.2Al, 3.5Ti, 1Fe	985

* The bracketed figures refer to alloys that have been cold-worked or heat-treated to achieve an optimum precipitate distribution. The other values are for alloys that have been annealed or heat treated to take the alloying elements into solution in the metal matrix.

contain significant concentrations of Al or Ti or both. Careful processing control of some of the most chemically complex alloys produces multiple phase microstructures that confer the excellent high temperature mechanical properties needed for aircraft engine manufacture. However, the disadvantage with such compositions is that many of the high strength Ni alloys are susceptible to liquid metal embrittlement caused by contact with low melting temperature constituents of brazes, such as Bi, Sb or Zn and the use of high temperature brazes can result in coarsening of the microstructures and consequent mechanical degradation unless great care is taken in the selection of the brazing alloy. The stiffness and high yield strengths of some Ni alloys also cause problems when joining is to be achieved by diffusion bonding, and it is fairly common practice to use ductile interlayers to produce joints between high-strength Ni alloys.

Alloying Ni also affects its chemical behaviour. Thus the high O affinity of additions of Cr, Al or Ti can produce surface films that make it difficult to achieve the metal–metal contact necessary to form high-integrity interfaces. Thus growth of Al_2O_3, Cr_2O_3, TiO_2 and other stable oxides must be prevented whenever possible, for example by plating the surfaces with a non-reactive metal. If this is not feasible on technical or economic grounds, the surface films have to be penetrated during the bonding process by use of very reactive brazing materials or a high-quality vacuum and a high temperature to exploit possibilities such as evaporative disruption or dissolution due to increased solubility of O in the metal substrate.

Fe and its alloys

Fe undergoes two phase changes when heated to its melting temperature, from a body-centred cubic αFe, ferrite, phase to a face-centred cubic γFe, austenite, phase at 912 °C and back to a body-centred cubic δFe phase at 1394 °C. The change from γFe to δFe occurs at too high a temperature to be of concern to the joint fabricator using brazing or diffusion bonding, and even the volume change that accompanies reversion to αFe during cooling of joints made at temperatures above 912 °C is seldom the cause of serious stressing of the joint because of the ductility of the joining materials.

A vast range of individual Fe alloys may require joining, but they can be systematized in terms of compositions and microstructures. The most important alloying element is C, whose addition leads to the formation of cast iron or steel. Cast iron typically contains 2.5–3.5 wt% C, and Figure 4.1 shows this near eutectic composition to depress the liquidus to about 1200 °C. The equilibrium microstructure of such Fe-C alloys, grey cast iron, has a matrix of αFe in which excess C is embedded as flakes of graphite. However, fast cooling can produce white cast iron with a metastable microstructure in which excess C is present as Fe_3C, cementite, as primary phases or a constituent of the αFe-Fe_3C pearlite

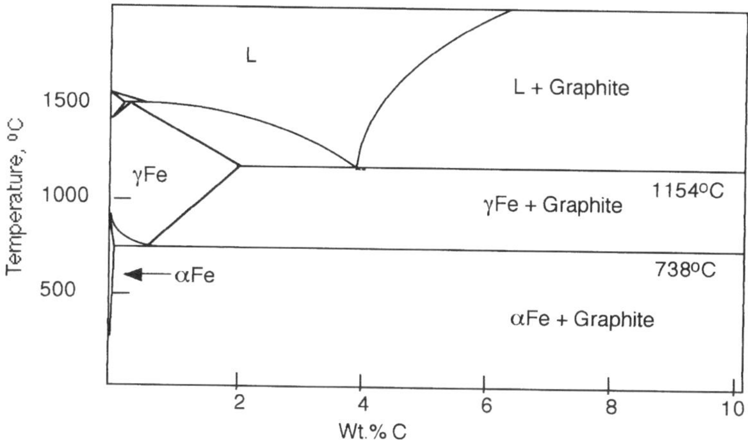

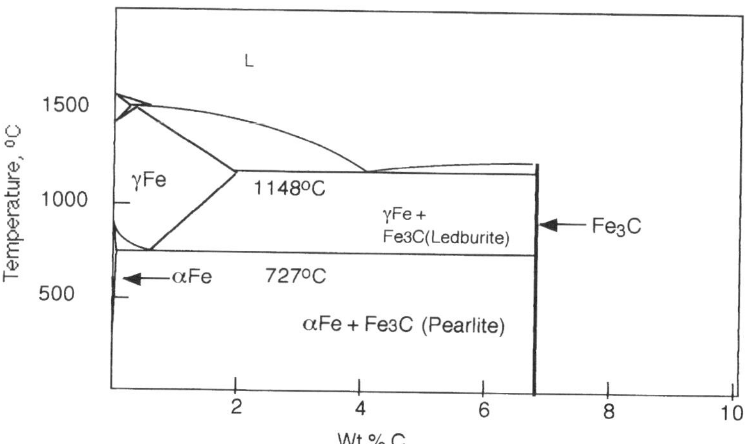

Figure 4.1 Phase diagrams of the Fe-graphite and Fe-cementite systems

eutectoid mixture. Steels differ from cast irons in containing less C, typically less than 0.6 wt%, and hence there is no opportunity for the rejection of C until the alloy has cooled to the γFe $\rightarrow$ αFe transformation temperature, which is too low to permit nucleation and growth of graphite. Thus Fe-C steels have metastable microstructures, mixtures of ferrite and pearlite for low C steels and cementite and pearlite for high C steels as indicated in Figure 4.1.

In practice, neither cast irons or steels are binary alloys but contain additional elements that are introduced to enhance one or more desired characteristics, examples of which are presented in Table 4.7. Thus grey

Table 4.7 Some cast-iron and steel component materials

Type		*Alloying addition, wt%*					
		C	Cr	Mn	Ni	Si	*Other*
					Cast irons		
White		3.5	1	0.60	–	0.7	< 0.3P, < 0.15S
Grey		3.1	–	0.60	–	2	0.20P, 0.10S
Nodular		3.6	–	0.40	1.1	2.4	0.02P, 0.10Cu, 0.05Mg
					Steels		
C steels							
Low,	AISI 1006	0.06	–	0.35	–	–	< 0.04P, < 0.05S
	AISI 1021	0.21	–	0.45	–	–	< 0.04P, < 0.05S
Medium,	AISI 1035	0.35	–	0.75	–	–	< 0.04P, < 0.05S
	AISI 1043	0.44	–	0.85	–	–	< 0.04P, < 0.05S
High,	AISI 1078	0.80	–	0.45	–	–	< 0.04P, < 0.05S
	AISI 1085	0.87	–	0.85	–	–	< 0.04P, < 0.05S
Alloy steels							
Low,	BS 4360-40B	0.10	–	1.0	–	–	
	ASTM A242	0.10	0.50	0.70	–	0.70	0.10Mo, 0.10Zr

Low alloy, high strength	ASTM A533	0.20	–	1.2	0.50	–	0.50Mo
	ASTM A517F	0.15	0.50	1.3	0.85	–	0.50Mo, 0.30Cu, 0.05V, B
Tool steels	BS BM15	1.50	4.8	0.30	–	0.20	6.5W, 5V, 5Co, 3Mo
	BS BT1	0.80	3.9	0.20	–	0.30	18W, 1.1V, 0.50Mo, 0.40Co
Austenitic, stainless	BS 304	0.04	18.0	1.0	10.0	–	
	BS 316	0.02	17.5	1.2	12.5	–	2.7Mo
Ferritic, high Cr	BS 405	0.04	13.0	0.70	0.40	–	0.20Al
	DINX10CrSi29	0.10	25.0	0.80	–	1.50	
Martensitic	BS 416	0.10	12.0	1.0	0.50	0.30	< 0.60Mo, 0.20S
	BS 431	0.15	16.0	0.50	2.0	–	
Low expansion	Nilo K†	–	–	–	29.0	–	17Co
	Invar†	–	–	–	36.0	–	
High-temperature oxidation resistant	Fecralloy†	–	6.0	–	–	–	4Al, 0.25Y
	Kanthal†	–	25.0	–	–	–	5Al, 3Co

AISI American Iron and Steel Institute
ASTM American Society for Testing Materials
BS British Standards Institution
DIN Deutsche Industrie Norm
† Trade names

cast irons invariably also contain Si to promote formation of graphite while additions of trace amounts of Mg change the configuration of graphite from flakes to nodules. Similarly, additions of Cr to steels promote stabilization of the ferrite phase and carbide formation, can harden the steel and can promote formation of chemically stable and physically tenacious oxide films that improve corrosion resistance. Additions of Ni stabilize the ductile austenite phase in steels and can promote its transformation when rapidly cooled to tough or hard forms of distorted ferrite that contain excess C, such as bainite or martensite. Very large additions of Ni or of Ni and Co drastically decrease the coefficient of thermal expansion of Fe, while simultaneous additions of Ni and Cr can produce stainless steels that have ductile austenitic lattice structures and are corrosion resistant.

The joining of steels or cast irons can present some problems. The characteristics of each individual material will influence the choice of the joining process, but some problems that have to be solved are common to many of the materials. Thus the joining characteristics of cast irons are usually dominated by the presence of graphite flakes or nodules that are not wetted by many braze constituents, are physically weak and provide a reservoir of poorly bonding material that can be spread about during preparation of component surfaces. With steels, the most important surface contaminant is usually the presence of chemically stable and physically tenacious oxide films such as those formed by the Cr and Al additions in high Cr steels, stainless steels and Fecralloy, which act as diffusion barriers and are often unwetted by common braze alloys. Such oxide films have to be disrupted and dispersed, for example by using very reactive brazing materials that can penetrate beneath to the metal substrate through adventitious gaps such as tears at grain boundary triple points and thereafter cause undermining by chemical or electrolytic interactions at the oxide-substrate interface. In a few cases, Cr_2O_3 films are disrupted during the joining process by substrate effects, such as the evaporation of MoO_3 from AISI 315 stainless steel or by reactions with C in the substrate to cause dissociation and the formation of CO.

Fe_3C, Cr_3C_2 and other alloy carbides present in steels are wetted by some braze alloys, but the very close control of the heat treatment cycles required to produce their optimum distributions can be incompatible with simultaneous joining by either brazing or diffusion bonding. The temperatures used during the heat treatments to achieve the desired mechanical properties depend on the chemistry of the alloy but in many cases lie within the broad range of 540–820 °C, which is also the operational range of many braze alloys. Similar restrictions can be caused by the need to avoid carbide precipitation in some stainless steels held at 450–800 °C that depletes their grain boundaries of Cr and renders them susceptible to corrosion.

Problems can arise also due to reactions with constituents of joining materials. Thus care has to be taken when joining at high temperatures with Ni brazes to avoid excessive ingress of Si which can embrittle, and of B into steel components for nuclear applications because it could be transmuted into He bubbles. Similar embrittlement problems can arise when joining at temperatures low enough for them to have no effect on the microstructure of the steel since ingress of Bi, Cd, In, Sb, Zn and a number of other braze melting temperature depressants can result in liquid metal embrittlement of high strength steels, and Al and Al alloy brazes and diffusion bonding interlayers readily form brittle intermetallics with Fe and steel components such as Cr and Ni.

In principle, these interaction problems can be avoided by diffusion bonding steel components to each other at low temperatures without the need for interlayers, but this requires the preparation of exceptionally flat and smooth bonding surfaces if intimate contact is to be achieved at loads insufficient to cause gross deformation of the components. However, the superplasticity exhibited by some steels and a few cast irons at temperatures as low as 550 °C can be exploited to achieve diffusion bonding without the use of interlayers, but major applications have yet to emerge.

Other metals and their alloys

A very large number of other metals and alloys can be required to be joined, and each individual material and application must be considered separately when considering the suitability of joining processes, materials and conditions. Component material characteristics of importance to joint fabrication usually include the melting temperature, phase stability, chemical reactivity, stiffness and ductility, but others can be of importance and even of predominant importance. Thus the high affinity of Be for O and the physical tenacity of BeO pose considerable problems, but of even greater importance is the need to be aware of the toxicity of Be ingested as fine oxide particulates, which can cause irreversible and sometimes fatal respiratory damage even though the effects may not manifest themselves until many years after exposure.

In contrast, the precious metals Ag, Au and Pt pose no major problems for the joint fabricator. All three metals have negligible affinities for O and are ductile, so they can be brazed or diffusion bonded with relative ease, and have been treated thus for millennia in the case of Ag and Au. However, this does not mean that no thought need be given to the selection of joining materials, processes or conditions. Thus, attention may have to be given to a somewhat unusual consideration when brazing to fabricate jewellery items: ensuring that the colour of the joint is close to that of the component metal so that it does not flaw the beauty of the product.

The moderate O affinity of the refractory metals Mo, Nb, Ta and W means that bonding inhibition due to oxide barriers is not a major problem. However, other characteristics of these metals impose restrictions on the joining processes that can be used. Thus Mo and W are brittle at room temperature and hence components to be joined must be in a stress-free condition. Similarly, the use of high-temperature joining processes can be restricted by the onset of embrittlement due to recrystallization of the refractory metal at moderate temperatures within the range $0.4 \rightarrow 0.54\,T_M$: 1000–1150 °C for Nb, 1100–1400 °C for Ta, 1150–1200 °C for Mo and 1200–1650 °C for W. However, some increases in these restrictive temperatures are achievable by close control of grain boundary chemistries and it is noteworthy that recrystallization of a Mo–0.5Ti–0.7Zr alloy, TZM, does not occur until about 1480 °C. Embrittlement can also be caused by interdiffusion with the joining material to form, for example, layers of silicide or aluminide intermetallics, by contact with graphite holding jigs or by the absorption of furnace gases: Nb and Ta being embrittled by H_2 absorbed at low temperatures and Mo, Nb and Ta by N_2 absorbed at high temperatures.

4.1.2 Ceramics

The need to join ceramics relates mainly to their developing importance in engineering as designers seek to exploit common characteristics that differentiate them from metals, such as a lower density or greater refractoriness and strength than metals, and Table 4.8 expands on the presentation made in Chapter 1 of typical property values for a number of ceramics of current importance.

Some of these and other characteristics are of particular note. Thus the refractoriness of ceramics means that joining temperatures are seldom limited by coarsening and mechanical degradation of the component microstructures, but other characteristics that can cause problems include:

- generally poor wettability, which restricts the choice of materials that can be used as brazes or PTLPB interlayers;
- frequently exceptionally high stiffness and hardness, which can make the preparation of bonding surfaces arduous;
- low coefficients of thermal expansion, which can result in excessive strains and stresses during the cooling of joints made to metal;
- phase changes, the avoidance of which during cooling restricts the range of possible joining materials.

The most commonly joined ceramics are Al_2O_3 and C and have been so for many years but the range has widened in the last few decades with the introduction of new engineering ceramics, such as Si_3N_4 and SiC. At the same time, the requirements for the service performance of joints have

Table 4.8 Properties of some ceramics

Material	*Melting temp.*, °C	*Density*, Mg.m^{-3}	*Thermal expans.*, $\times 10^{-6}$ K^{-1}	*Strength**, MPa	*Elasticity*, GPa	*Thermal cond.* Wm^{-1} K^{-1}	*Elect. cond.* Sm^{-1}
C, graphite	3650	2.25	***	–	–	25	2.7×10^6
SiC	2700	3.2	4.3	141	211	50	10^3
TiC	3140	4.9	7.2	703	422	36	2×10^7
WC	2777	15.8	5.2	598	704	84	–
AlN	2197	3.3	4.6	400	310	165	10^{-12}
BN, hexagonal	3000	2.3	****	45	23	50	2×10^{-12}
Si_3N_4	1900**	3.2	3.2	750	280	35	2×10^{-9}
TiN	2900	5.3	7.5	–	–	29	–
Al_2O_3	2050	4.0	7.6	490	363	32	10^{-14}
BeO	2530	3.1	7.4	246	401	210	10^{-10}
MgO	2800**	3.6	11.6	281	394	62	10^{-12}
ZrO_2	2690	5.8	6.5	176	140	2	3×10^{-5}

* Four point bend test data.

** Does not melt but sublimes, decomposes or vaporizes.

*** 1×10^{-6} parallel to and 5×10^{-6} perpendicular to the grain for polycrystalline material.

**** 0.8×10^{-6} parallel to and 7.5×10^{-6} normal to the basal plane.

become more stringent and this has intensified the need for the fabricator to optimize choices of joining materials and processes. Each choice will be influenced by the specific characteristics of the individual component materials, but some characteristics are familial.

Oxide ceramics and glasses

Al_2O_3 is the most widely used of the oxide ceramics, finding many applications that include electronic substrates, electrical insulators, leak-tight containers for lighting and reactor technology, cutting and forming components of machine tools and high-temperature containment of liquid metals. Its ability to fulfil so many functions derives from the ionic nature of oxide lattices which confer characteristics such as low electrical and thermal conductivity and thermal expansivity, refractoriness, hardness, strength and stiffness, as exemplified by the data quoted in Table 4.8 for Al_2O_3 and a number of other commercially pure oxides. Exceptions to such typically ionic characteristics also lead to applications, such as the exploitation of the high thermal conductivity of BeO in heat-dissipating electronic substrates and of ZrO_2 in gas sensors.

The data quoted in Table 4.8 are typical of technical grades of oxide ceramics that contain small additions of sintering aids, such as SiO_2 or MgO. However, some technically important oxide ceramics contain deliberate and substantial additions of foreign oxides made to enhance desirable characteristics. As shown by the data assembled in Table 4.9, the thermal shock resistance of mullite, $3Al_2O_3.2SiO_2$, is much better than that of Al_2O_3, as that of zircon, $ZrO_2.SiO_2$, is better than that of ZrO_2. Similarly, a substantial toughening is achieved by the addition of CaO, MgO or Y_2O_3 to ZrO_2 to produce Partially Stabilized Zirconia, PSZ, that does not suffer fully the mechanically degrading tetragonal → monoclinic phase change on cooling through a temperature of about 1000 °C. Further, the phase change from liquid SiO_2 to solid crystalline quartz can be suppressed by rapid cooling to produce a glass, which also decreases the thermal expansivity from 22×10^{-6} to $0.5 \times 10^{-6}\,K^{-1}$. This suppression can be promoted by the addition of other oxides such as Na_2O, K_2O and B_2O_3 (Table 2.1), and recent years have seen the development of glass–ceramics in which partial recrystallization of a glass is allowed to enhance toughness and permit tailoring of expansion coefficients.

The joining of oxides present problems typical of most ceramics. Thus surface preparation prior to joining can be difficult. Chemical cleaning can involve heating at high temperatures to remove adsorbed hydroxyl groups. Again, their hardness makes arduous the achievement of flat bonding surfaces by mechanical finishing and their brittleness can cause

Table 4.9 Properties of four complex ceramics

Material	*Density*, $Mg.m^{-3}$	*Thermal expans.*, $\times 10^{-6}\,K^{-1}$	*Flexural strength*,* MPa	*Elasticity*, GPa	*Thermal cond.*, $Wm^{-1}\,K^{-1}$	*Fracture toughness*,* K_C, $MPa.m^{1/2}$
Mullite, $3Al_2O_3.2SiO_2$	3.15	5.5	355	–	7.1	–
Partially stabilized Zirconia, $ZrO_2 + MgO$	5.7	10.2	520†	210	2.5	8.1#†
Zircon, $ZrO_2.SiO_2$	4.6	3.8	–	–	6.1	–
Sialon, $Si_3N_4 + Al_2O_3$	3.25	3.0	945	286	21.3	4.55#

* The derivation of these characteristics is described in Chapter 7.

\# For comparison, the fracture toughness of Al_2O_3 is 2.3 $MPa.m^{1/2}$ and that of Si_3N_4 is 5.6 $MPa.m^{1/2}$, while the values for glasses are generally less than 1 $MPa.m^{1/2}$ and those for ductile metal alloys are several tens of $MPa.m^{1/2}$.

† ZrO_2 stabilized with Y_2O_3 has a strength of 950 MPa and a fracture toughness of 10.5 $MPa.m^{1/2}$.

chipping during such preparation. Even when clean and smooth, oxide surfaces are not wetted by many conventional brazes so the choice of joining materials is limited. Again the low-expansion coefficients and high-elastic moduli of oxides can complicate the joining of ceramics to metal components

Carbon ceramics

Carbon exists in three allotropic modifications that have markedly different characteristics: graphite, diamond and fullerene. Each allotrope has a distinctive lattice structure: thus the atoms in graphite are covalently bonded to their nearest neighbours in hexagonal arrays to produce layers that are held together by Van der Waals forces, while the C atoms in diamond are each at the centre of a covalently bonded tetrahedron. The Van der Waals forces are much weaker than those exerted by covalent bonds, and this is reflected in the 0.34 nm layer–layer distance of graphite as compared to the 0.14 or 0.15 nm C–C distances in graphite and diamond. In the case of the recently discovered fullerene structure, covalently bound pentagonal and hexagonal rings of C atoms form the facets of hollow spheres or closed tubes of 60, 70 or 84 atoms in an arrangement that is similar to the geodesic architectural forms developed by R Buckminster Fuller.

Carbon has many properties that make it an important technical material. Both graphite and diamond are very refractory, sublimation rather than melting occurring at 3650 °C and 3550 °C respectively. The hardness of diamond is exploited in cutting tools while its high thermal conductivity, typically more than $1000\,W.m^{-1}K^{-1}$, makes it a technically attractive heat dissipating electronic substrate. Similarly, graphite is a good solid lubricant and electrical conductor that is used as a contact material in electric motors, while very fine-grained, amorphous, graphite can be formed as carbon fibres, with uniaxial tensile strengths of up to 500 MPa, or a pore free and relatively hard vitreous carbon. The introduction of K atoms can make fullerene a superconductor.

As with oxides, C ceramics are wetted by few braze alloys and the joint fabricator must give careful thought to their cleaning before brazing or diffusion bonding. The presence of porosity and residual ash can result in enhanced and detrimental high temperature degassing of graphite components, but the mechanical preparation of smooth flat surfaces on graphite components is relatively simple. The mechanical weakness of graphite and its low coefficient of thermal expansion can also cause problems when components have to be joined to otherwise metal structures.

Other non-oxide ceramics

A number of non-oxide ceramics, but principally carbides and nitrides, are of considerable technical importance because of their differing characteristics. Thus, carbides are predominantly covalent rather than ionic but also can display some metallic characteristics, as demonstrated by the data assembled in Table 4.8. Thus the refractoriness and hardness of SiC, TiC and WC that are exploited in their application as cutting, forming and wear resistant materials for machine tool components are typical of covalent ceramics. However, their metal-like behaviour is shown by their high electrical conductivities and the fact that many, such as Cr_3C_2, HfC, Mo_2C, NbC, TaC, TiC, WC or ZrC are wetted by Ag, Cu or Ni and alloys developed for metal–metal joining.

In contrast, nitride ceramics are generally more ionic, albeit not as much as oxides. However, like oxides they are usually refractory, strong, have low thermal and electrical conductivities and are unwetted by braze metals and alloys developed for metal–metal joining. Thus TiN is exceptionally good as a wear resistant coating and diffusion barrier while Si_3N_4 is one of the new engineering ceramics that offers a high strength of 750 MPa that can be further enhanced to about 950 MPa by the introduction of Al_2O_3 to form a complex Sialon, Si-Al-O-N, ceramic. Nevertheless, as might be expected with an intermediate group of materials, there are exceptions to these generalizations such as AlN whose high thermal conductivity has led to applications as a heat dissipating electronic substrate and BN which has a low ionicity and behaves somewhat like C, having graphitic and diamond-like allotropes.

For the joint fabricator, carbides are a particularly notable group of ceramics because of their readier wettability, but otherwise the non-oxide ceramics present the familiar joining problems of ceramic components that result from their hardness, stiffness and low thermal expansivity.

4.1.3 Composites

The development and application of ceramic and metal matrix composites has become of growing importance in recent decades. Thus Al_2O_3, SiC, Si_3N_4 and other ceramic matrices have been strengthened or toughened by the introduction of Al_2O_3, C, SiC, Si_3N_4 and other fibres to produce both dissimilar and chemically homogeneous composites, such as SiC fibres in a Si_3N_4 matrix or SiC fibres in a SiC matrix. Similarly, Al, Ni and Ti and other metal matrices have been reinforced successfully with fibres such as Al_2O_3, B, C and SiC to achieve room temperature mechanical properties values approximating to rule of mixtures expectations when optimized processing conditions have been used. However, the high strengths of composites can be degraded by the onset of fibre–matrix reactions, which limit both service and joining temperatures so that, for example, brazing or diffusion bonding of Al alloy matrix composites reinforced with B or SiC fibres must be achieved at temperatures of less than about 450 °C.

Fibre-reinforced composites, however, are not the only ones of importance. Thus, metal matrix materials containing dispersions of oxides, such as thoriated W or sintered aluminium powder (SAP), are well-established component materials, and indeed it can be argued that graphitic cast irons should also be regarded as composites. A number of non-fibrous ceramic matrix composite materials are also of technical importance, particularly WC-Co mixtures in which typically 5 to 20% by volume of Co provides the matrix for well-bonded angular grains of WC to produce a material that finds numerous machine tool applications that require toughness as well as hardness, strength and wear resistance. Another example is provided by the development of reaction bonded SiC ceramics in which the presence of minor volume fractions of elemental Si promotes bonding of the ceramic grains and hence toughens the ceramic.

While advantageous for the end user, the joining of composite materials can pose considerable problems for the joint fabricator, particularly in the case of metal matrix composites. Temperature limitations on joining processes have been mentioned already, but additionally the stiffening of metals by the introduction of ceramic fibres can increase the stresses generated during the cooling of joints between materials with mismatched thermal expansivities. Further, the presence of substantial additions of ceramic fibres or particulates can make it more difficult to achieve the flat and smooth surfaces required for diffusion bonding. Of most importance to the joint fabricator, however, can be the chemical complexities of composite materials. Preparation of the bonding surfaces will often expose some of the introduced material and therefore a high integrity interface will have to be formed simultaneously to materials with markedly different bonding behaviour. Exposure of the second, minor, phase will be generally beneficial when it is a metal, as in the case of WC-Co, but difficulties can arise when the second phase is a ceramic. Thus fibres partially exposed at the bonding surface can be fragmented or separate from their matrix, and hence composite surfaces are sometimes overcoated with an additional layer of the matrix metal to enhance both mechanical and joining characteristics. However, chemical problems can arise even with such protected composites due to interactions between their constituents. Thus the service and joining temperatures of Al/SiC composites are restricted to less than 450 °C by the need to avoid formation of Al_4C_3 which reacts readily with moist air to cause failure of the fibre–matrix interfaces and mechanical degradation.

4.2 CHARACTERISTICS OF JOINING MATERIALS

4.2.1 Brazes

Brazes are widely available from many sources throughout the industrialized world as thin rods or sheets, and in some cases as powders or pastes. Rods are used when brazing using gas torch and other manual processes, while preformed shapes stamped or cut from sheets to produce geometries, contours and sizes suitable for use with particular types of capillary gaps find application in mechanized brazing processes. Powders and pastes can be used with both manual and mechanized brazing when carefully applied.

Satisfactory brazes wet and spread over the mating surfaces of components in a predictable and controllable manner to fill capillary gaps and form well-shaped fillets without causing detrimental erosion and then solidify to form permanent joints. The property requirements of the joints depend on the proposed application of the brazed components but generally include soundness and strength and often also include corrosion resistance and toughness. A very wide range of brazes suitable for joining particular metallic or ceramic components have been developed and are available commercially. Brazes are also used to join composites, the selection of an appropriate one depending on the identity of the metallic or ceramic matrix of the composite or of any coating on its surface.

Selection of a braze for a particular application must take into account technical factors such as its melting temperature or range and mechanical properties at the proposed service temperature, its chemical physical and mechanical compatibility with the component material and its chemical compatability with the service environment, and its ability to wet and adhere to the component materials. Another severely practical factor to which attention must also be paid is its cost, although the apparent significance of this factor can be deceptive as will be seen later.

Families of braze alloys based on Al, Cu, Ag, Ni and other metals, and in particular on eutectic compositions, have been developed and have found use throughout the industrialized world. The variety of braze compositions that is available commercially is enormous, but the great majority of those referred to in this section have been specified in frequently virtually identical national and international standards.

Al brazes

The compositions of a number of widely available commercial Al braze alloys used to join Al components are summarized in Table 4.10. These alloys are available in a variety of forms and generally require the use of a flux if adequate liquid flow is to be achieved, suitable compositions of

Table 4.10 Some widely used Al braze alloys

Alloy†	*Composition, wt%* Al	Si	Mg	Zn	Cu	*Melting range,* °C
BAlSi-2, 4343	92.5	7.5	–	–	–	577–613
BAlSi-5, 4045	90	10	–	–	–	577–591
BAlSi-4, 4047	88	12	–	–	–	577–582
– , X4003	90	7.5	2.5	–	–	543–599
BAlSi-3, 4145	86	10	–	–	4	521–585
– , 4245	76	10	–	10	4	515–560

† AWS-ASTM classification and UK and US alloy number.

which are described in section 4.2.2. It can be seen from the table that these Al brazes are based on the Al-Si system which has a simple phase diagram comprising a single eutectic composition of Al-12.6Si that melts at 577 °C and two solid phases, Al containing up to 1.6Si and pure Si. Most commercial Al-Si brazes are based on hypoeutectic compositions to which additions of Cu or Zn sometimes are made to decrease the liquidus temperature to as low as 560 °C, albeit by sacrificing some corrosion resistance. For some Al component materials, even lower liquidus temperatures are required, as can be seen by referring to Table 4.2. For joining these materials attention is being paid to the development of Al-Ge and Al-Ge-Si brazes based on the Al-52Ge eutectic composition which melts at 420 °C, but commercial development is still in its early stages. Additionally, the incorporation of a few percent of Mg promotes the flow of the molten Al-Si alloys when vacuum brazing without the use of fluxes and such Al-Si-Mg alloys applied as claddings to cores of, for example, alloy 3003 (Al-1.2Mn), have found extensive use in the manufacture of heat exchangers.

Al and some Al alloys have also been shown to wet at both oxide and non-oxide ceramics at temperatures of 800–1000 °C in low O_2 furnace environments, but no major applications have yet evolved.

Cu brazes

This is the oldest established family of braze alloys and a vast range of useful compositions have been developed and specified in standards. Of particular utility are those based on Cu itself, on Cu-P and Cu-Zn alloys, and examples of these three types are summarized in Table 4.11.

Unalloyed Cu can be used as a braze for Fe, Ni or Ni-Cu components, including stainless steels and similar alloys whose surface films of Cr_2O_3 cease to be a barrier to wetting in low O_2 environments at about 1000 °C.

Table 4.11 Examples of frequently used Cu braze alloys

	Composition, wt%					
Alloy†	Cu	P	Zn	Ag	*Other*	*Melting range, °C*
BCu-1	99.9	–	–	–	0.1	1085–1085
BCu-2	99.85	0.02	–	–	0.13	1085–1085
BCuP-1	94	6	–	–	–	710–890
BCuP-2	92.5	7.5	–	–	–	710–810
BCuP-4	86.25	6.75	–	–	7Sn	650–700
BCuP-5	81	4.5	–	14.5	–	645–800
CZ6	60	–	39.7	–	0.3Si	875–895
CZ6A	60	–	39.4	–	0.3Si, 0.3Sn	875–995
RBCuZn-A	59.25	–	40	–	0.75Sn	888–899
RBCuZn-D	48	–	43	–	9Ni	920–980
CU8	36	–	32	30	3Sn	665–775
AG21	98	–	–	–	2MN	1045–1060

† BCu, BCuP and RBCuZn alloys are AWS classifications,
CU, CZ and AG alloys are BSI classifications.

A range of Cu purities that can be used as brazes have been codified, and Table 4.11 also includes a braze composition containing a small amount of Ni which was added to increase the strength of the solidified joint. The mutual solubility of Fe and liquid Cu is low at brazing temperatures and this results in such a ready flow of pure Cu over Fe components when furnace brazing in inert atmospheres that designers are able to specify very narrow and sometimes even zero or negative, press fit, gaps between components. In contrast, the mutual solubility of Cu in Ni at brazing temperatures is very large so that wide gaps have to be used when brazing Ni components if flow of the liquid is not to be inhibited by interdiffusion. When brazing in air and other oxidizing environments, the surface of the molten Cu needs to be protected by a flux if flow is not to be restrained by oxide films (section 4.2.2), and the use of fluxes can also promote flow by removing oxides from component surfaces.

Another type of Cu braze is based on the Cu-P system which forms a single eutectic composition of Cu-8.3P that melts at 714 °C and when solid consists of a Cu matrix containing up to 1.7 P and a second phase of Cu_3P. As with Al alloys, those used are hypoeutectic variants containing 4 to 8% of P, and the functions of the P additions are both to decrease the melting temperature of the Cu solvent and also to confer a degree of self-fluxing in mildly oxidizing environments by covering the braze surface with a film of liquid phosphate. Additions of P are made also to Cu–Ag alloys to decrease the solidus temperature to 646 °C and both the Ag free and Ag containing variants have found mass application in the joining of Cu pipes. The range of available Cu-P brazes has been extended further recently by use of the rapid solidification technique to produce foils of, for example, Cu-10Ni-8P-4Sn and Cu-10Sn-7P-6Ni, but embrittlement caused by inward diffusion of P significantly inhibits the use of any variant of Cu-P braze to join Ni alloy or Fe or Cu-Ni alloy components containing more than 10% of Ni.

Brazes based on brass have been used for centuries even though the Cu–Zn system contains four brittle intermetallic compounds and no eutectic compositions. Cu-40Zn alloys are the basis for many braze alloys in current use that can be used with a very wide range of component materials, including low and high alloy steels, Cu and Cu alloys and Ni and Ni alloys. However, the relatively poor corrosion resistance of Cu–Zn braze alloys limits the range of service conditions that they can endure. Similarly, the volatility of Zn means that over-heating the braze can cause voids to form in the joints and the affinity of Zn for O_2 also means that flux covers must be used in even mildly oxidizing environments.

The main uses of Cu brazes are for the joining of metal components, but Cu is also used for joining WC inserts onto steel tool holders. Additionally, a number of Cu alloys containing typically 5 to 10% of Ti have been evaluated as active metal brazes for both oxide and non-oxide ceramics, and a Cu-3Si-2Al-2.25Ti composition has been developed commercially for use at 1025–1050 °C. These alloys wet well, but those containing Ti concentrations also tend to form thick and fragile reaction product layers and this mechanical degradation of the joints is enhanced by the stiffness of many braze compositions. The result of this often relatively poor mechanical behaviour is that most commercial development of active braze alloys has focused on brazes based on the ductile Ag-28Cu eutectic.

Ag brazes

Ag alloys are the most numerous, most commercialized family of braze materials. They have been widely used in general engineering for many years and this reflects their excellent performance and ease of use with a very wide range of ferrous, non-ferrous and ceramic component materials. The alloys are characterized by their relatively low melting temperatures, their ability to wet and flow readily to form strong ductile joints with well-formed fillets and their generally attractive colouration.

Pure Ag can be used as a braze but its melting temperature of 960 °C is too high for many potential applications. Thus most Ag brazes are alloys and the great majority of them are based on the Ag-28Cu eutectic composition which melts at 780 °C, and consists of two solid phases, Ag containing up to 8Cu and Cu containing up to 8 Ag. Of technological importance is the fact that the addition of Cu also enhances brazing performance of Ag when it is used to join Fe, Co or Ni and their alloys because Cu interdiffuses with and wets these component materials far better than does Ag. The Ag–28 Cu alloy itself is little used as a braze but it is the basis for scores of more complex compositions that are available commercially and have been codified in standard specifications and recommendations for practice. The compositions and melting ranges of typical commercially available Ag braze alloys are presented in Table 4.12, and it can be seen that there are four dominant modifications of the Ag–28 Cu eutectic: Ag–Cu–Zn, Ag–Cu–Pd and Ag–Cu–Zn–Cd alloys that are used for brazing metal components and Ag–Cu–Ti alloys developed for the brazing of ceramics.

Table 4.12 Compositions of some Ag alloys

	Composition, wt%					
Alloy†	Ag	Cu	Zn	Cd	*Other*	*Melting range,* °C
AG7	72	28	–	–	–	780–780
AG13	60	26	14	–	–	695–730
BAg-5	45	30	25	–	–	677–743
AG5	43	37	20	–	–	780–789
BAg-1a	50	15	16	19	–	620–640
BAg-1	45	15	16	24	–	607–618
AG3	38	20	22	20	–	605–650
BAg-2	35	26	21	18	–	607–702
AG11	34	25	20	21	–	610–670
AG12	30	27	23	20	–	600–690
–	95	–	–	–	5Pd	971–1010
–	90	–	–	–	10Pd	1001–1060
–	80	–	–	–	20Pd	1071–1177
Incusil ABA	59.2	27	–	–	12.5In, 1.25Ti	605–715
Cusin-1 ABA	63.3	34	–	–	1.75Ti, 1Sn	775–806
Silver ABA	92.7	5	–	–	1.25Ti, 1Al	890–912
BAg-19	92.5	7.3	–	–	0.2Li	779–891
BAg-8a	72	37.6	–	–	0.4Li	765–765
BAg-18	60	30	–	–	10Sn	602–718
BAg-7	56	22	17	–	5Sn	618–652
BAg-13	54	40	5	–	1Ni	718–857

† BAg alloys are classified by the AWS,
AG alloys are classified by the BSI,
ABA alloys are trade marks of Wesgo Inc, Belmont, CA, USA.

Additions of Zn and Zn plus Cd further decrease the melting temperature of the Ag–28 Cu eutectic. The presence of Zn also enhances the fluidity of the braze and its ability to wet ferrous components. However, there are also disadvantages, thus the volatility of Zn and its affinity for O_2 can result in harmful fumes that can degrade the appearance and soundness of joints as well as contaminating the surfaces and structures of brazing furnaces, while high concentrations of Zn also diminish the ductility of the solid braze alloys.

Additions of Cd further enhance the beneficial effects to Zn on melting ranges and fluidity characteristics, but extreme care must be taken when using such alloys because prolonged inhalation of CdO fumes can be a serious health hazard. Protection can be gained by using flux covers and well-ventilated work areas, but the growing realization of the potential dangers has diminished the use of these alloys in recent years and has led to their legal banning in some countries.

Ag brazes based on the Ag-Cu eutectic can be and have been used to join ceramic components whose surfaces have been prepared by the 'moly-manganese' process. In this, the ceramic surfaces are painted with a mixture of Mo, or sometimes W, powder, MnO_2, and a glass frit such as $30CaO.40Al_2O_3.30SiO_2$ and then fired at 1400–1500 °C in a damp N_2-$10H_2$ atmosphere. This causes the glass to bond to the ceramic, aided by the MnO_2, and for the metal powder to sinter as a near continuous top layer which is then overcoated with a few microns of Ni, Cu or Au. This technique is well established for Al_2O_3, and has been adapted for use with a number of other ceramics, but achieving reproducible success requires skilful dedication and specialized equipment so such metallizing is usually done by an outside contractor and the principal selection problem facing the joint fabricator is the identity of the metal overcoat. In contrast, ceramic surfaces can be coated directly with metals before brazing in-house if vapour or sputter deposition facilities are available, or indirectly by the even simpler TiH_2 activation process if the components are to be vacuum brazed. This last method involves painting the component surfaces with TiH_2 powder carried in a liquid such as ethylene glycol, $(CH_2OH)_2$, so that the subsequent heating stage of the brazing cycle causes the carrier to evaporate and the powder to decompose to deposit Ti at temperatures of 400–600 °C.

Recent years have also seen the development of active metal brazes that can be used with ceramics without the necessity for prior metallization or coating, and this reflects the growing importance of ceramics as engineering materials and the need to join complex assemblies of ceramics and metal parts. These brazes are based on Cu rich variants of the Ag–28 Cu eutectic to which has been added 1 or 2% of Ti, larger concentrations causing the formation of a region of liquid immiscibility that complicates the flow behaviour of the braze, but even these low concentrations are sufficient to react with the ceramic and cover its surface with a wettable product. Because of the high affinity of Ti for O_2 and N_2, these alloys are used predominantly when vacuum brazing, although in principle they could also be used in low O_2 activity inert gas environments. Alloys containing more than a few percent of Ti tend to be very stiff and hence further modifications have been developed in which In or Sn are also added to increase the chemical activity of Ti and hence the ability of very low concentrations to wet the ceramics. Development of active metal brazes is vigorous at present and suggestions have been made, for example, that additions of Hf rather than Ti should be made to alloys for brazing Si_3N_4 because of the stronger interfaces that can be achieved.

Other additions made to the basic Ag–28 Cu eutectic braze alloy to enhance its ability to join metal components include Sn, Ni, Mn, and Li. Additions of Sn are made primarily to decrease melting temperatures but the other elements also affect wetting behaviour. Thus small Li additions are made to improve the ability of Ag brazes to wet stainless steel and other component materials whose surfaces are covered by tenacious oxide films. Enhanced wetting, of Stellite and other Mo and W alloy components, is also the motivation for adding Mn or Ni, and the presence of Ni additionally improves the corrosion resistance of brazed stainless steel components that are exposed to sea water.

Not all Ag brazes are based on the Ag–28 Cu eutectic and thus Table 4.12 includes examples of alloy compositions in which Pd is added to promote the ability of Ag to alloy with and wet Fe, Ni and Co components, or in which Li or Zn is added to enhance the ability of Ag to make metal–metal contact with components whose surfaces are covered by oxide films.

Finally, it should be noted that Ag and other precious metal brazes are often considered to be at a considerable disadvantage compared to those based on Cu in terms of cost. Ag does of course cost more than Cu, but the cost penalty of selecting a Ag braze is not as alarming as is usually imagined. At the time of writing, Cu costs just over £2000 a ton and Ag is about 60 times more expensive. However, brazing materials are usually in the form of thin rods or sheets whose prices also reflect fabrication costs. Thus the price of brazing foils is about 3 times rather than 60 times greater for Ag, and Ag–28 Cu alloys, than Cu and the price of an individual small braze preform is only a few pence. Similarly, the price of Ag–Cu–Ti active metal braze alloys is even higher than that of the parent Ag–Cu brazes, but the use of such alloys can greatly simplify the joint fabrication process by avoiding the need for metallization processes. Thus selection of a Ag brazing alloy, and even a particularly expensive alloy, may well be more cost effective when account is taken also of its technical performance.

Ni brazes

The melting temperature of Ni is too high for it to be a generally useful braze but effective Ni brazes with acceptable melting temperatures were developed about 50 years ago in response to the need for brazing jet engine components that had to withstand what were then exceptionally high service temperatures.

These brazes were initially empirical developments of wear resistant, hard facing, Ni coating alloys based on the Ni-B, Ni-P and Ni-Si systems. All three systems are complex, with each containing several eutectic compositions and intermetallic compounds. Of particular importance to the development of braze alloys are the three eutectic compositions with the least solute concentration:

- Ni-3.6B which melts at 1093 °C,
- Ni-11P which melts at 880 °C,
- Ni-11.3Si which melts at 1143 °C,

which have matrices of relatively ductile Ni solid solutions.

Table 4.13 lists some of the currently available commercial Ni brazes which have been developed to satisfy particular application requirements and shows that, in addition to P or B and Si, many of the brazes contain substantial additions of Cr to enhance their corrosion resistance and, incidentally, to diminish interdiffusion with Ni, Fe or Co alloys that also contain substantial levels of Cr. The fluidity of the alloys, and particularly the Ni–11P and Ni–14Cr–10P alloys is high, but compositions that are sluggish at use temperatures can be effective fillers

Table 4.13 Some widely used Ni brazes

	*Composition, wt%**						
Alloy†	Ni	B	P	Si	Cr	*Other*	*Melting range,* °C
BNi-3, HTN3	92.4	3.1	–	4.5	–	–	980–1040
BNi-6, HTN6	89	–	11	–	–	–	875–875
BNi-2, HTN2	82.4	3.1	–	4.5	7	3Fe	970–1010
BNi-7, HTN7	75.9	–	10.1	–	14	–	890–890
BNi-1, HTN1	73.9	3.1	–	4.5	14	4.5Fe	970–1045
BNi-5, HTN5	71.9	–	–	10.1	18	–	1080–1135
BNi-8,	65.5	–	–	7	–	23Mn, 4.5Cu	982–1010
–	48.7	2.2	–	2.2	11	36Pd	1004–1051
–	36	–	–	–	–	34Pd, 30Au	1135–1166

† BNi numbers are AWS classifications and HTN numbers are BSI classifications.
* C concentrations are less than 0.1%.

for wide joint gaps. Similarly, B free brazes such as the Ni–19Cr–10Si alloy are used for nuclear applications where neutron irradiation can produce a marked mechanical degradation. A common characteristic of these Ni brazes is that they are rather brittle and contain many intermetallic compounds and this can cause difficulties not only for the user but also for the material supplier when it is necessary to produce thin braze foils. Recent years, however, have seen the development of the rapid solidification technique of foil manufacture and this has mitigated the problem for some standard compositions and permitted the development of modified braze compositions.

In part because of their high-use temperatures, Ni brazes interdiffuse readily with Fe and Ni components and this can have marked effects on the properties of both the joints and components. Unless there is significant interdiffusion, the centres of the solidified joints can be composed of weak and brittle boride, phosphide or silicide layers. Limited interdiffusion encouraged by holding at the brazing temperature, can result in egress of B, P or Si to render the joints more ductile but also unacceptably eroding thin section components and possibly embrittling both thin and thick section components by the formation of brittle compound inclusions in the grain boundaries. Prolonged interdiffusion with thick section components, however, can so disperse the B, P and Si so thoroughly that brittle compounds are no longer present within either the joint or components. The duration of the interdiffusion time needed to achieve such a beneficial microstructure depends not only on the temperature and compositions of the braze and the component but also on the thickness of the joint gap since this determines the size of the reservoir of the embrittling element that must be dispersed. A notable additional effect of the dispersion of B, P and Si is the raising of the temperature at which the joint will remelt so that potential joint service temperatures can be exceptionally high.

Finally, it should be noted that the eutectic based Ni alloys are not the only ones that can be used as brazes. Thus a Ni–40Cu with a liquidus temperature of 1349 °C can be used for the brazing of refractory metals. Similarly, Ni–Ti alloys have been evaluated as brazes for ceramics, but although they can wet well they are too brittle to have yet gained significant commercial use.

Other miscellaneous brazes

While numerous braze metals and alloys have been referred to already, many others have been and are used to join metals and ceramic and some of these are listed in Table 4.14.

The largest group in that table are the alloys based on Au to which has been added Cu or Ni to depress the melting temperature, and further

Table 4.14 Some miscellaneous braze alloys

Alloy†	*Composition, wt%*	*Melting range, °C*
BAu-4	82Au-18Ni	950–950
BAu-2	80Au-20Cu	890–890
BAu-6	79Au-22Ni-8Pd	1007–1046
BCo-1	49.8Co-19Cr-17Ni-8Si-4W-1Fe-0.8B-0.4C	1121–1149
BMg-1	Mg-9Al-2Zn-0.2Mn	443–599
BMg-2	Mg-12Al-5Zn	582–610
–	60Pd-40Ni	1238–1238
–	54Pd-36Ni-10Cr	1232–1260
–	48Ti-47Zr-5Be	893–904
–	45Ti-45Zr-8Ni-2Be	899–899

† BAu, BCo and BMg numbers are AWS-ASTM classifications.

depression can be achieved by the addition of metals such as Zn or Cd or both, but caution must be exercised not to incur health hazards when using alloys containing Cd. While Au brazes have been used by the jewellery trade since time immemorial, current formulations find much wider applications. Thus Au–Cu brazes are used by the electronic industry because of their ability to wet and form ductile joints without excessive interdiffusion with component materials such as Cu, Fe, Ni, Co and refractory metals. Au–Ni brazes can also wet component metals very well but have the additional advantage of retaining high joint strengths even at temperatures such as those encountered in aircraft engines. Au brazes also often contain Ag to decrease their cost without unduly degrading performance, and Pd to enhance wetting behaviour.

Pd itself is used as the solvent metal in some brazes and two such compositions are identified in Table 4.14. However, most of the 'palladium brazes' referred to in the technical literature incorporate Pd only as a solute to promote the wetting behaviour of the pure or alloy Ag, Au, Cu or Ni solvents. Pd brazes are relatively refractory, but even more refractory joining materials with liquidus temperatures higher than 2000 °C have been formulated that are based on Pt–30W and Pt–50Rh compositions.

In contrast, low melting temperature Mg brazes such as the Mg-Al-Zn compositions listed in the table have been developed for the joining of Mg components to minimize problems of galvanic corrosion. The low melting temperatures of these brazes also help to reduce the risk of Mg fires, and it is usual to employ fluxes to promote wetting and prevent combustion.

These brazes can produce joints for use at the relatively high temperature of 150 °C, 0.45 T_M, and are corrosion-resistant if care is taken to avoid flux entrapment and to remove any retained on the joint surfaces.

The table includes a Co based braze which is used for the joining of Co based superalloys. This composition was developed, and has been codified, as an analogue of the Ni–B–Si–Cr brazes referred to in Table 4.13. Finally, Table 4.14 also includes a few Ti brazes which were developed for the joining of graphite and other ceramics. Their compositions make them extremely reactive, and they wet well but their inherent brittleness can cause problems for the joint designer and fabricator.

4.2.2 Fluxes and other ancillary braze materials

A flux is an important constituent of most brazing processes, not only when heating by immersion in a flux bath but also when furnace, torch or induction heating. The function of the flux is to promote joint formation by rendering and maintaining the braze and component surfaces free of oxide barriers during the brazing process. It should be noted however that a braze flux is not intended to clean surfaces by removing grease or corrosion products such as rust. Systematic mechanical preparation and chemical cleaning of the bonding surfaces should be done before assembling the components to braze.

To be effective, the flux must have a lower melting temperature than that of the braze alloy and perform its oxide removal function rapidly at the brazing temperature. To achieve this, the flux must wet the component surfaces yet then must be displaced by the liquid braze alloy to form a metal–metal bond, Figure 4.2.

Many fluxes with proven beneficial effects for the brazing of a wide range of component metals and alloys are available, their compositions and

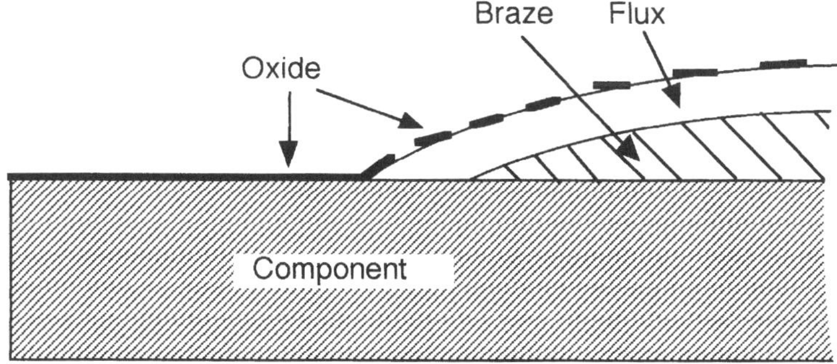

Figure 4.2 Representation of oxide film displacement by a flux and of flux displacement by a braze alloy

characteristics are generally commercial secrets. Thus the joint fabricator would be wise to use that suggested by the supplier of the braze alloy and to follow the recommended procedure with care. Brazing fluxes are normally powder mixtures of compounds such as borates and fluorides, but water is often added by the supplier or user to make pastes that can be applied by brushes or rollers or even sprayable liquids.

Despite the lack of detailed chemical information, some classification of braze fluxes is possible in terms of the component materials, use temperatures and primary constituents. Generally, a flux is regarded as belonging to one of four groups.

- Fluxes for joining Mg alloys that are mixtures of fluorides and chlorides that operate at 480 to 620 °C.
- Fluxes for Al and Al alloys that are generally mixtures of alkali and alkaline earth chlorides and fluorides – sometimes plus AIF_3 or cryolite, $3NaF.AIF_3$ – that have solidus and liquidus temperatures of typically 485 and 580 °C and operate at temperatures of 540–615 °C.
- Fluxes for use with Ag and some Cu brazes to join Cu and Cu alloy, steel and cast iron components are mixtures of alkali borates, fluorides, fluoroborates and unidentified wetting agents that operate at 565 to 870 °C.
- Fluxes for higher temperature brazing of steels and some Cu alloys are mixtures of alkaline borates, boric acid, H_3BO_3, and borax, $Na_2B_4O_7$, that operate at 750 to 1200 °C to promote joining by Au, Cu and Ni brazes. Small additions of SiO_2 and B are sometimes made to these fluxes to achieve high operating temperatures, and the fluoride and fluoroborate contents of these fluxes are kept low.

Removal of chemically stable surface oxides often requires the use of aggressive fluxes. Thus removal of Al_2O_3 from Al surfaces generally requires the flux to penetrate the oxide at flaw sites and then to lift off the oxide film by corroding the metal substrate. This corrosion can continue after the brazing cycle is complete and hence thorough removal of flux after the components have been joined is essential. It is, therefore, noteworthy that Alcan Aluminium Ltd have developed a non-corrosive potassium fluoroaluminate formulation, with the commercial registered name of Nocolok, that melts at 562 °C to dissolve the Al_2O_3 on Al surfaces rather than the substrate metal and hence can be left on the brazed components.

While fluxes can beneficially promote braze flow, the extent of spread must be controllable if joints and fillets are to achieve the desired dimensions and if the liquid metal is to be prevented from spreading on to the jigs that are often used to hold the components in place during the brazing process. While containment of liquid flow can be achieved sometimes by adventitious surface contamination such as oil residues or even finger prints, the more positive step of applying a stop off material is generally taken when containment is important. Stop off materials are fine powders of non-wetting ceramics such as Al_2O_3, MgO, ZrO_2 rare earth oxides or BN suspended in alcohol, water, acetone or other carrier liquids. Selection of a suitable stop off depends not only on its ability to contain braze flow in the particular system being used but also whether it is desired that it should have fast or slow drying characteristics and its ease of application by brush or roller.

Finally, mention should be made of adhesives that can be used to secure braze preforms or powders in place prior to the heating cycle. Since preforms in particular are used when brazing in vacuum or controlled atmosphere furnaces, it is important that the adhesive is fugitive, leaving no char or ash when heated to the brazing temperature. The selection of a suitable adhesive often requires experimentation by the fabricator but volatile organic compounds such as acetone can be useful as can the commercially developed adhesives for use with powdered Ni-B-Si-Cr brazes when joining is to be achieved at high temperatures.

4.2.3 Diffusion bonding interlayers

Both foils and coatings are used as interlayers to promote diffusion bonding of components. The foils are typically 10 to 150 μm thick and can be either pure metals or alloys, while the coatings are usually of pure metals that have been vapour deposited or, more often, sputtered to produce interlayer thicknesses ranging from 0.1 to several microns. The selection of a suitable interlayer material is in some respects easier than choosing a braze because there is no dependence of joint formation on the subtle materials interaction parameter of wetting. In essence, any metallic interlayer material can be expected to bond with any metallic component and many ceramic components if direct contact is achieved between the clean surfaces of the two materials, although the bond may be subsequently degraded mechanically or broken as the bonded assembly cools by the action of stresses induced by mismatched coefficients of thermal expansion. However, the uses of interlayers referred to later are restricted to cases in which permanent high-integrity joints have been produced in the laboratory or on the shop floor.

Again unlike brazes, the metals and alloys used as diffusion bonding interlayers have not been subjected to codification and specification in national or international standards. Nor indeed is there a similar wealth of technical and scientific behavioural data to assist selection of the optimum joining material. Nevertheless, some guidance can be gained from consideration of the principal requirements that diffusion bonding interlayers must satisfy, which can be summarized as:

- a melting temperature in excess of that required for joint formation and, usually, substantially in excess of any that may be encountered in service conditions;
- a low enough yield stress and sufficient ductility to permit deformation to conform to component surface contours when pressure is applied to make the joint;
- a coefficient of expansion similar enough to that or those of the components, and a low enough yield stress and sufficient ductility to avoid cooling stresses that significantly degrade the mechanical properties of the joints;
- a chemistry that promotes the formation of high-integrity and strong interfaces with the components usually by interdiffusion or the formation of thin reaction product layers;
- a chemistry that will not suffer environmental corrosion or induce galvanic corrosion in the proposed service conditions.

The selection of the optimum interlayer material depends not only on its own inherent characteristics but also on those of the components and on the subsequent conditions that may have to be endured by the joint. Nevertheless, for the sake of simplicity and consistency, information relevant to their selection will be grouped according to the chemistry of the interlayer metal or alloy solvent. The first four groups will be presented in the sequence Al → Ag → Cu → Ni, which reflects a progressive change in their refractoriness, expansivity and strength as illustrated in Figure 4.3, while the final group is an empirical assembly of other metals and alloys.

Al and Al alloys

The relatively low melting temperatures and the low yield strength and good ductility of Al and many Al alloys make them attractive candidates as diffusion bonding interlayers, but they also suffer from disadvantages. The coefficient of thermal expansion of interlayers of Al, $23.5 \times 10^{-6}\,K^{-1}$, is far higher than those of component materials they are required to join, but this detrimental characteristic is largely mitigated in practice by the relatively low bonding temperatures that are used and the low yield strength of Al. Of more significance is the fact that the surfaces of the interlayers are coated with oxide films that are difficult to destabilize chemically or disrupt and disperse mechanically to permit metal–metal contact. Further, the affinity of Al for other metals can result in the formation of brittle intermetallic compounds when interdiffusion follows contact and the high electrode potential of Al compared to many other metals means that joints can be corroded in many possible service environments

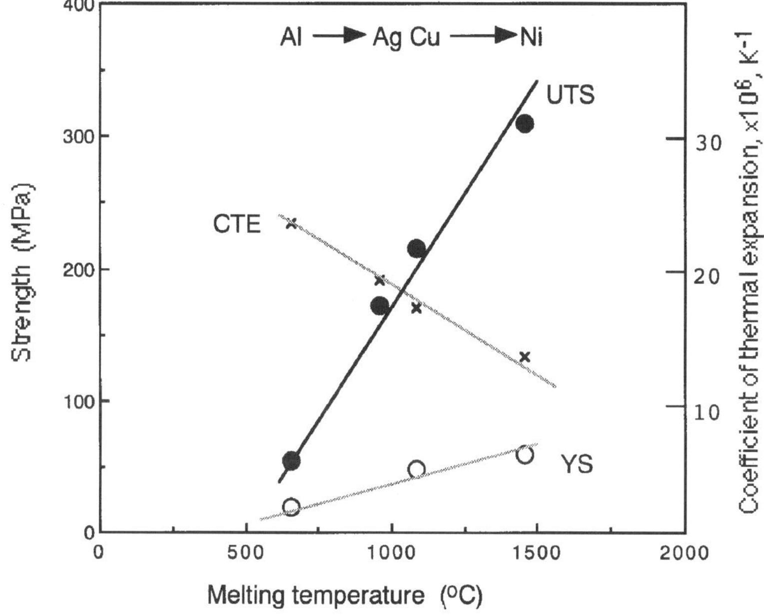

Figure 4.3 Physical and mechanical properties of some pure metal diffusion bonding interlayer materials

These qualifications are largely irrelevant if Al or Al alloy interlayers are selected to join Al components. Thus the low-yield strength of Al foils has been exploited to achieve good confirmation of the surface contours of and ready bonding to stiff Al matrix composites reinforced with ceramic fibres such as SiC, although joining temperatures may have to be kept low and the times short to avoid detrimental interactions at fibre–matrix interfaces. Similarly, the utility of 0.1 mm thick foils of Al–0.6Ga, Al–1.3Li or Al–4.5Mg alloys for bonding the 8090 superplastic alloy at temperatures in the range of 500–540 °C has been evaluated, with the best results being achieved with the Al–Mg alloy because of the ability of its solute to destabilize the oxide films on interlayer and component surfaces.

Despite their tendency to form intermetallic compounds, Al and Al alloy interlayers have also been used successfully to diffusion bond other metals at temperatures providing the impedance of tenacious oxide films is overcome, and both of these effects can be influenced by the detailed chemistry of the interlayers and components and by any treatments given to their surfaces before the onset of the bonding cycle. For example, Al forms embrittling intermetallics with Fe but their growth and hence mechanical significance can be affected by the presence of alloying elements in the interlayers. The presence of up to 2.2 Mg in the Al helps to overcome inhibition of metal–metal contact by surface oxides but the overall result is a degradation of joint quality because the high diffusional mobility of Mg encourages the nucleation and growth of layers of intermetallic compounds at the interlayer/component interfaces. In contrast, the presence of up to 5.6 Si or 3 Cu in the Al enhances joint quality because they impede the nucleation or subsequent growth of intermetallic reaction product layers. Alternatively, beneficial impedance of intermetallic nucleation or growth can be achieved also by coating the Al interlayers or Fe components with Cu, or with Ag, Ni or Zn before bonding.

The bonding of steel components is not the only process to benefit from such chemical control. Thus Ag, Ni or Zn coatings benefit the bonding of Al interlayers to Cu and Cu alloy components by diminishing the formation of reaction product layers of Cu-Al intermetallics during the bonding cycle. Another way of restricting intermetallic nucleation and growth is to limit the bonding temperature and time, thus satisfactory joints have been produced using Al interlayers to bond Mo or Ti components at about 550 °C using times of less than an hour.

Al is a relatively low melting temperature metal but nevertheless it can be used also to bond refractory oxide and non-oxide ceramics such as Al_2O_3, SiO_2, SiC and Si_3N_4 for ambient and modest temperature applications. A very substantial deformation is usually required to disperse the surface oxide from the interlayer and the bonding temperatures approach those needed to melt the interlayers. It is important, however, that the melting temperature is not exceeded because Al does not wet ceramics unless superheat temperatures of several hundred degrees are used. Similarly, Al interlayers can be used to join ceramics to metals and combinations such as SiO_2/Al/steel and Al_2O_3/Al/Ti/Al/Al_2O_3 have found applications when joint designs have allowed differences in coefficients of thermal expansion to be accommodated.

Finally, mention should be made of the deliberate use of oxide films on component surfaces to impede intermetallic formation. Thus a patent exists that describes the usefulness of growing a thick Al_2O_3 film on the surface of Fecralloy before diffusion bonding it to Al_2O_3 components using an Al interlayer.

Ag and Ag alloys

The most numerous and widely used family of brazing materials are the alloys of Ag, but they have been little exploited as interlayers to promote diffusion bonding. There have, however, been demonstrations of the utility of interlayers and coatings of Ag that exploit its ductility and the instability of Ag_2O oxide. Thus Ag sputter coatings have been used to bond to a Ni-20Cu alloy, to low carbon and carbon steels, and notably to bond maraging and stainless steels at a temperature of 400 °C, about only 0.45 T_M for the steels. The room temperature tensile strengths of these latter joints were high, but the ready yielding of the Ag layers determined the somewhat poor creep characteristics of the joints. Ag can also be used to diffusion bond other high-temperature materials such as Ti and Ni–20Cr, but it is noteworthy that the most carefully evaluated possible applications of the benefits to be gained from using Ag coatings have been for Al alloy components whose service temperatures are typically only about 200 °C, 0.4 T_M for Ag.

Ag and Al have significant mutual solubilities, reaching a maximum of 55% for Ag in Al at 560 °C, and Ag can diffuse rapidly into Al from whose surfaces oxide films have been removed, characteristics that are often associated with promotion of diffusion bonding. This expectation has been justified by, for example, the manufacture of joints with room temperature tensile strengths of 60 MPa using an Al–3Mg alloy whose surfaces had been sputter cleaned and coated with 0.1 → 2 μm of Ag before being bonded at 390–450 °C, 0.75–0.82 T_M for Ag. Again, exceptionally strong bonds that required 180 MPa to cause shear failure at room temperature were produced when an Al–6Zn–2.3Mg–1.7Cu alloy clad with a layer of Al-1Zn and then sputter cleaned and coated with 0.1–1 μm of Ag was bonded at 280 °C, 0.62 T_M for Ag, and then heat-treated at 480 °C, 0.85 T_M.

While the use of Ag alloys as interlayers could also facilitate bonding, no major industrial developments have been reported up to now. There is also a shortage of information about the utility of such interlayers for diffusion-bonding ceramics. Interlayers of pure Ag have been used in laboratory studies to bond Al_2O_3 and some other oxide ceramics, but the strengths achieved are not remarkable. However, as with Cu, some enhancement of strength can be achieved by using a mildly oxidizing environment when bonding Al_2O_3.

Cu and Cu alloys

Cu is relatively easy to diffusion bond to itself because of its excellent ductility and the poor stability of its oxides, and both Cu and its alloys have been used as interlayers for joining other metals and ceramics even though their high coefficients of expansion, such as $17 \times 10^{-6}\,K^{-1}$ for Cu, and relatively high-yield stresses can cause problems.

Cu interlayers can be used to form high-integrity diffusion bonds between Fe components, and this satisfactory outcome can be associated with the fact that Cu does not form intermetallic compounds with Fe but has a modest solubility that increases to about 4.5% at the optimum diffusion-bonding temperature of 1000 °C. However, this solubility is affected by the presence of steel alloying elements, being diminished by C and enhanced by elements such as Mn, Ni, and Si, and these effects are reflected in similar changes in bonding behaviour. Coatings can also affect diffusion bonding behaviour and those of Ni, in which Cu and Fe are both completely soluble at 1000 °C, have also been found to be particularly beneficial. The successful bonding of Cu to Pt can also be related to their complete mutual solubility at bonding temperatures of 812 °C, while difficulties can be encountered when joining Cu to Ti with which it forms a family of intermetallic compounds. Physical characteristics also can have significant effects on the ability of Cu interlayers to promote diffusion bonding, thus successful diffusion bonding to refractory metals with low or moderate coefficients of thermal expansion such as Mo, Nb, Ta, V or W can be difficult to achieve but once again the use of a Ni coating can be beneficial in promoting interdiffusion.

Cu interlayers and more usually coatings can also be used to promote the bonding of Al components. Although in equilibrium conditions Cu forms intermetallic compounds with Al, the kinetics of its diffusion into Al from whose surface oxide barriers have been removed by sputter cleaning before being coated are such that near homogeneous structures and strong diffusion bonds can be developed at sub-solidus temperatures of typically 400–500 °C.

Both Cu and Cu alloy interlayers can be used to diffusion bond ceramic components. There have been many demonstrations of the successful bonding of Cu to Al_2O_3 and other oxides, but attempts to diffusion bond to Si_3N_4 have been unsatisfactory, possibly because of the very wide discrepancy in the coefficient of thermal expansion of Cu and the ceramic, $3.2 \times 10^{-6}\,K^{-1}$. When bonding to oxide ceramics such as Al_2O_3, laboratory studies have shown that bond strengths are increased if the Cu contains a small concentration of O or if the normal bonding environment of a vacuum or inert gas is replaced by one that is slightly O_2 active so that there is segregation of Cu–O clusters at the metal–oxide interface. A more conventional way of improving bonding to oxide ceramics is to introduce reactive solutes, so that for example diffusion bonded interfaces between Cu and diamond can be strengthened tenfold by the introduction of low concentrations of Cr or Ti which can react to form carbides.

Ni and Ni alloys

Ni is a ductile high-melting-temperature metal whose oxide can dissociate in good vacua or mildly reducing environments at moderate temperatures such as 1000 °C, 0.73 T_M, and the potential usefulness of Ni and Ni alloy interlayers for producing diffusion-bonded joints with both metal and ceramic components that can endure high service temperatures has attracted particular interest.

Interlayers and coatings of Ni have been used to promote the diffusion bonding of a wide range of metallic component materials and reference has been made above to the use of Ni coatings to promote the bonding of Cu to Mo and W. Ni can also be diffusion bonded to Nb, Pt and Ti, but the use of Ni in the 'activated diffusion bonding' of Ti alloys used to manufacture aircraft engine components is actually a liquid phase process. The most developed and important applications of Ni interlayers to promote diffusion bonding are with steels and Ni base super alloys at temperatures ranging from 900 °C to 1200 °C, 0.68 to 0.85 T_M for Ni. Thus carbon, low- and high-alloy steels such as Fe–0.4C, Fe-1Cr–0.4C, Fe–12.5Mn–1.2C, Fe-4Ni–1.3Cr–0.3Mn–0.45C, Fe–18Cr-9Ni–1.25Mn and Fe-18Ni-12Cr have been successfully diffusion bonded, as have super alloys such as Inconel 600, Nimonic 90 and Udimet 700.

Ni alloys have also been used as diffusion bonding interlayers with metal components, but the favoured solutes have not been the B, P or Si used to develop Ni alloy brazes because they degrade the ductility which is essential for the successful application of an interlayer. Alloys such as Nimonic 75, Ni-20Cr–0.4Ti, are favoured interlayer materials when diffusion-bonding super alloys and high-strength steels because the Cr content can increase the maximum service temperatures of the joints, promote their corrosion resistance and impede egress from the components of their Cr, which often plays a crucial role in enhancing mechanical properties and corrosion resistance. Other solute additives of importance include Co, whose presence benefits the diffusion bonding of component materials such as Udimet 700, and Be whose small atomic radius promotes rapid interdiffusion. Care should be taken, however, when considering the use of any alloy containing Be to become thoroughly acquainted with the dangers of BeO ingestion and the precautions needed to avoid this possibility.

Ni and its alloys have been used extensively in the laboratory and technical development of diffusion bonding processes for ceramics. Pure Ni interlayers have been used to produce strong bonding of oxide ceramics such as Al_2O_3 and ZrO_2 and to composite materials such as WC-6 → 20Co, TiC-15Ni-6Mo, TiC-22Fe-5Cr-3Ni and TiC-29Fe-7Cr-4Ni cermets. However, Ni interlayers can be too aggressive for successful diffusion bonding of engineering ceramics such as SiC or Si_3N_4 because of the low chemical activity of Si in Ni. Thus contact with Ni causes these ceramics to dissociate and for extensive migration of Si into the Ni to occur. This ingress of Si embrittles the interlayer and erodes the ceramic, and can be very fast if it also results in the formation of liquid Ni-Si compositions at the interlayer–ceramic interface.

More generally, Ni alloys are chosen as interlayers for the diffusion bonding of ceramics when they produce slow growing reaction product layers since this characteristic is often associated with mechanical robustness. Thus the strong joints produced when Ni–20Cr alloy interlayers are used to bond Si_3N_4 at 1100 °C are associated with the formation of CrN reaction product layers that bridge the interlayer–ceramic interfaces but are only a few micrometres thick. Similarly, diamonds have been strongly diffusion bonded to Ni–1V and Ni–1.24Ti interlayers following reactions to produce carbide layers that were about 0.2 μm thick at 800 °C after the optimum bonding time of 1 hour, while an Incoloy 909 alloy, Ni–421Fe–12.9Co–4.7Nb–1.5Ti–0.4Si, with a low coefficient of thermal expansion of $8 \times 10^{-6}\,K^{-1}$ was chosen for diffusion bonding to Si_3N_4 and Si_3N_4–TiN and Si_3N_4–TiB_2 composites specifically because it reacted only slowly, producing a reaction product layer less than 1 μm thick even after 4 hours at 1200 °C. Other noteworthy examples of the successful use of Ni alloy interlayers are the bonding of Al_2O_3 by Inconel 600, Ni–16Cr-6Fe, and of SiC by Inconel 718, Ni–19Cr–7Fe–2.5Ti–0.8Nb–0.6Al, and Incoloy 909.

Miscellaneous metals and alloys

Numerous reports of the use of many other metals and alloys in evaluations of diffusion bonding are scattered in the scientific and technical literature, but they are often semi-qualitative and difficult to compare one with another to reveal effects of material characteristics. Nevertheless, some broad observations can be made about the behaviour of certain metals. For example, reactive and relatively refractory Ti can be used to join other reactive or refractory metals such as Hf, Mo, Ta, V and Zr with which it is completely soluble at the high-diffusion bonding temperatures, although problems could be caused by phase transformations as the assemblies cool. However, complete mutual solubility is not a requirement for successful diffusion bonding of high temperature materials, as demonstrated by the joining of Pt to Ti and to Mo with which the mutual solubilities are less than 20%.

Low melting temperature metals including In, Pb, Sn and Zn have also been used as interlayers to bond Al because their easy deformation and flow during diffusion bonding that can rub away the surface film of Al_2O_3 to produce metal–metal contact and bonding. However, their service utility is diminished because of two disadvantages. First, the low melting temperatures of the interlayer materials severely restrict the service temperatures at which the joints can be used, thus Zn is the most refractory of the interlayers but taking $0.5\,T_M$ as its maximum service temperature implies a useful temperature of no more than 74 °C. Second, the low electrode potentials of the interlayer materials compared to Al means that the joints can experience galvanic corrosion in many service environments.

There is a similar mixture of application oriented and generalized or academic studies of the use of miscellaneous interlayers to diffusion bond ceramics. Thus, Australian work has demonstrated the utility of precious metals for bonding oxide ceramic and glass components, Pt interlayers being used to bond Al_2O_3 tubes used in thermocouple sheaths and ZrO_2 tipped O_2 sensors while Au interlayers have been used to bond single crystal windows of Al_2O_3 and to produce vacuum tight seals in Pyrex assemblies while UK studies have used interlayers of Pd and other precious metals to bond laboratory samples of Al_2O_3 in mechanistic studies of the diffusion-bonding process.

Another promising but as yet unexploited development has been the use of Fe and Fe–Cr alloy interlayers to bond Si_3N_4. The strengths achieved with these material combinations were greater than those reported for similar interlayers of Ni and Ni–Cr, presumably due to the higher Si activity in Fe. Other work has shown that Si_3N_4 can be bonded also with interlayers of Co and a wide range of other metals including the refractory metals Mo, Nb, Ta and V that have moderately low coefficients of thermal expansion, as does the ceramic, but care must be taken not to form excessively fragile thick silicide reaction product layers. Further studies of the use of interlayers to diffusion bond Si_3N_4 have succeeded by using the reactive metals Cr and Ti to form thin nitride reaction product layers that bridge the interlayer–ceramic interfaces when optimized bonding conditions are used.

4.2.4 Joining materials for hybrid bonding processes

The selection of joining materials to use in the hybrid processes of concern, transient and partial transient liquid phase bonding and eutectic bonding, is complicated by the fact that the original material undergoes massive interdiffusion and consequently composition and property changes as an inherent aspect of the bonding process. Thus the interlayers or coatings inserted between or applied to the component surfaces to prepare for transient or partial transient liquid phase bonding are subjected to continuing interdiffusion that progressively causes them to form a liquid phase that fills the gap between the components by wetting, to resolidify isothermally to form a permanent joint and, sometimes, to homogenize to such an extent that the concept of a joint seam becomes meaningless. Interdiffusion also controls the progress of eutectic bonding, forming a liquid that fills the gap between the components but continued interdiffusion progressively thickens the liquid layer and resolidification does not occur until the assembly is cooled.

The selection of a suitable bonding material to promote these processes requires knowledge of the interlayer/component phase diagrams and the physical and mechanical characteristics of the alloys that may be formed by interdiffusion as well as the mass transport characteristics themselves. In particular, knowledge is required about the probable migration characteristics of component or interlayer species that can cause embrittlement of the solidified joint. It is also essential that the liquid alloys should wet the components, and this cannot always be assumed to be the case when ceramics have to be joined, and have melting temperatures sufficiently low that no microstructural or mechanical degradation of the components occurs.

Development of these hybrid processes has focused so far on their application to a small number of component materials. Thus applied work on transient liquid phase bonding has related mainly but not exclusively to the joining of high-temperature components so that particular attention has been paid to the interdiffusion and bonding behaviour of Ni and Ti and their alloys. Similarly, the potential partial transient liquid phase bonding has been assessed for Al_2O_3 and Si_3N_4 ceramics.

Probably the best characterized and understood interlayer/component transient liquid phase bonding system is Ag–Cu, but this is not because of some important technical requirement but rather because it has such a simple phase diagram, Figure 3.9, and well-established diffusion parameters that it is almost perfect for testing the validity of the various models that have been developed. Thus studies by several research groups with this system have shown that the two most important – and longest – stages of transient liquid phase bonding are widening of the liquid zone due to dissolution of the component and the subsequent isothermal resolidification.

An application of transient liquid phase bonding that is of more direct technical importance is the work that has been done to develop the technique for joining Ni and Ni alloys, including super alloys and even intermetallics, and steels that often but not always used interlayer or coating materials based on Ni braze compositions. Laboratory studies, for example of the M-007 superalloy, Ni–10.2Co–8Cr–6.2Mo–6Al–4.2Ta, with a Ni–15.5Cr–3.7B interlayer have led to broad confirmation of the conclusions drawn from the Ag–Cu model studies but have also shown that the several sequential stages identified by such studies, such as component dissolution or isothermal resolidification, overlap in time and are sequentially dominant rather than discretely separate.

The use of Ni braze compositions as bonding interlayers for Ni, and Fe, alloy components is attractive because of their relatively low melting temperatures and ability to wet the components, but complications can arise because the B, P and Si melting temperature depressants of the braze alloys can form brittle compounds with Ni and Fe and common constituents of their alloys. It is important, therefore, to consider the rate at which these melting temperature depressants may penetrate and degrade the component materials. Thus the use of an interlayer of Ni-Si-B to effect transient liquid phase bonding of commercially pure Ni and NiAl revealed that the better behaviour was displayed when bonding the intermetallic. Unlike the bonding of Ni, there was no significant mechanically degrading precipitation of borides or excessive dissolution of the component and this was attributed to the two orders of magnitude lower diffusivity of B in NiAl than Ni.

The use of Ni braze alloys to achieve transient liquid phase bonding of steels has been less intensively characterized, but the effects are broadly similar to those observed with Ni alloys. Thus interdiffusion of an Ni–19Cr–10Si interlayer with an Fe–11.5Ni–11.5Ni–2.5Mo alloy and dispersion of the Si deep within the bulk of the steel causes isothermal resolidification by partial homogenization to produce ductile joints free of embrittling seams or reaction product layers of silicides. While the use of such brazes containing B, P or Si as melting temperature depressants has dominated the development of transient liquid-phase bonding techniques for both Ni and Fe alloy components, they are not the only Ni base alloys that can be used to promote effective bonding. Thus Ni–Cu alloys have been used to produce transient liquid phase bonded joints between super alloy UDIMET 700 components that had strengths equal to that of the component material.

Some of the most important applications of transient liquid phase bonding have been in the fabrication of many thousands of Ti alloy aero-engine components, although the process may be referred to by the misnomer of ‘activated diffusion bonding’ or by trade names such as the Rohr process. The bonding of these components has been achieved using coatings of interlayers of Ag or Cu–Ni that were selected because the chemistries of the Ag–Ti, Cu–Ti and Ni–Ti systems and the absence of diffusion impeding oxides on Ti surfaces in the high-temperature low O_2 bonding environments permitted rapid and effective bonding, and a substantial degree of joint homogenization. Thus liquid Ag reacts to form a TiAg intermetallic compound and then a Ti solid solution, containing up to 34% of Ag, that spans the interface between the components. The process is more complicated when Cu or Ni–Cu is used as the bonding agent because both Cu and Ni form low melting temperature eutectic compositions such as Ti–12Cu that melts at 875 °C, Ti–50Cu that melts at 960 °C, Ti–35Ni that melts at 1118 °C and Ti–72Ni that melts at 942 °C. Their formation changes the volume of the liquid phase involved in the bonding process but continuing interdiffusion causes isothermal resolidification and ultimately the formation of Ti solid solutions.

An exception to the generality that transient liquid phase bonding is used for the joining of high service temperature components is the use of Zn and Zn–Cu coatings to promote the joining of superplastic Al alloys. Although sometimes regarded as a form of diffusion bonding, this technique involves the melting of the coating and its interdiffusion with the component before ultimate resolidification, with Zn being chosen as the coating material because of its low melting temperature and solubility in Al of 55% at a bonding temperature of 460 °C.

Partial transient liquid phase bonding has been developed so far essentially for the joining of ceramics to ceramics or metals and hence at least two additional factors must be taken into account when selecting interlayers or coatings. First, it is essential that any liquid formed in the joint gap can wet the ceramic component surfaces, and hence attention has to be given to whether reactive constituents are present in the original coating or interlayer or can migrate into it during the bonding process. Second, the differences between the coefficients of expansion of many ceramics and metals mean that the joints will experience significant strains and hence it is peculiarly important that the joint material in the bonded assembly must be ductile.

Partial transient liquid phase bonding is relatively new and still principally a laboratory development. Nevertheless, the materials used in these, sometimes scoping, experiments illustrate the influence of several important selection factors. Thus, successful joining of Si_3N_4 has been achieved using multiple layer inserts of Ti–Ni–Ti, Ni–Ti–Ni–Ti–Ni, Ti–Ni–Kovar–Ni–Ti, Ti–Ni–AISI304–Ni–Ti and Ti–AuCu–Ni–AuCu–Ti. Thus it is noteworthy that the inserts had cores of ductile materials – Ni, Kovar, an Fe–28Ni–18Co low-expansion alloy or AISI304, an Fe–18Cr–9.5Ni–1.5Mn stainless steel – which could to some degree absorb both Ti not consumed in the wetting reaction and Si released by it without becoming embrittled. Further, all the inserts contained Ti, which can form low melting temperature eutectic compositions with Ni and core alloy constituents, but was present for the specific purpose of promoting the wetting and strong bonding of the ceramic. Ti, however, is not the only reactive metal that can promote the wetting and bonding of Si_3N_4. Brazing studies have shown that Cr can be beneficial and diffusion bonding studies have associated the formation of a thin layer of CrN with strong bonding. Thus diffusion of Cr from an AISI304 core could also have promoted joint formation, and this diffusion effect has been exploited in the successful partial transient liquid phase bonding of Si_3N_4 achieved by using Au coated Ni–20Cr inserts.

The role of reactive wetting in the partial transient liquid phase bonding of oxide ceramics is more difficult to discern since strong joints have been produced with Al_2O_3 using inserts such as Cu–Pt–Cu, Cu–Ni–Cu and Cu–Nb–Cu. Neither Cu or any of the insert core metals wet Al_2O_3.

However, Cu wets the insert core metals very well and hence it must be presumed that it is energetically favourable for the liquid to enter and remain in capillary gaps bounded on one side by Al_2O_3 and on the other by Nb, Ni or Pt. If the liquid does fill the gap between the components it then becomes significant that both solidification of non-wetting sessile drops of Cu, Ni and Pt and diffusion bonding of Cu, Nb, Ni and Pt produce strong interfaces with Al_2O_3.

The selection of materials that can be used to achieve successful eutectic bonding is extremely difficult. Most of the needed phase diagram data are lacking, and while it may be possible to make an informed guess about the likelihood of a particular pair of component materials interdiffusing to produce a eutectic liquid, it is extremely hazardous to suggest whether or not the eutectic compositions will have enough ductility to resist the strains imposed during solidification and cooling. Nevertheless, a number of metal–metal systems have been investigated and a few have shown promise. Thus direct joining of Zircalloy 2 and AISI304 components was successful because Fe, Ni and Cr alloy with Zr to form low melting temperature binary eutectic compositions that melt at temperatures ranging from 1332 to 947 °C and a quasi quaternary Zr–AISI304 composition that melts at 980 °C. The resultant joints were strong and had good corrosion resistance, but they lacked ductility when subjected to rapid or fluctuating stressing and this deficiency has since been overcome by the development of a brazing technique.

Selection of ceramic–metal systems that can be joined by eutectic bonding is also possible in principle, but the range of systems that can be used in practice is very limited due to two main reasons. First, the interdiffusion coefficients of the ceramic and metallic constituents and hence the associated dissolution rates often differ markedly. Second, the presence of substantial concentrations of ceramic constituents usually causes unacceptable brittleness and fragility when the eutectic liquid solidifies. However, joining can be possible if an oxide ceramic component is a passive partner that is merely wetted by liquid derived entirely by interaction of the metallic component with a mildly oxidizing environment. This approach has been developed as the ‘direct bonding’ of Cu to Al_2O_3 by producing a Cu–O composition that melts at 1065 °C, 18 °C below the melting temperature of Cu.

Cu-O is not the only possible direct bonding system, as can be seen by the data assembled in Table 4.15, and patent applications have identified metals such as Fe and Ni as also being capable of being joined by direct bonding. However, most of the metals listed in the table must be rejected from serious consideration. Thus the refractory metals Mo, Nb, Ta and V are embrittled by the presence of even very small amounts of O. Again it is difficult to imagine applications that would require components of Ca, Cs, Nd or Th to be joined to an oxide ceramic.

Table 4.15 Characteristics of some metal–oxygen eutectic systems

Solvent	O content wt%	*Melting temperature* °C	
		Eutectic	*Solvent*
Ca	0.10	839	842
Co	0.23	1451	1495
Cr	1.00	1800	1863
Cs	1.20	−2	28
Cu	0.43	1066	1085
Fe	0.16	1523	1538
Mn	0.40	1220	1248
Mo	4.00	2150	2615
Nb	10.5	1915	2469
Nd	0.20	1010	1021
Ni	0.20	1440	1452
Si	16.0	1360	1414
Ta	6.00	1880	3020
Th	0.01	1735	1755
V	13.5	1642	1910

4.3 EFFECTIVE JOINING MATERIAL/COMPONENT MATERIAL COMBINATIONS

The chemical and physical characteristics of the various families and individual braze alloys and diffusion bonding or transient liquid phase bonding interlayers make them suitable for use with some types of component materials but not with others. Thus there are certain combinations that have been found to be particularly compatible and effective, and some are even enshrined in national standards.

This compatibility effect is most well defined with braze alloys and Table 4.16 provides a broad overview of what is possible by listing some combinations of braze alloys and component materials that are considered to be suitable by British authorities. It should be noted, however, that the inclusion of a particular family of braze alloys in the list does not mean that every member of that family will be suitable. Nor does the listing imply that a braze particular family will be suitable when joining by differing processes, such as torch brazing or vacuum brazing. Although Pd brazes are often mentioned in specifications, they

Table 4.16 Recommended component/braze compositions. After BS 1723

Component materials	*Recommended braze families*
Al alloys	Al
Cast irons	Ag, Cu
Ni alloys	Ag, Au, Cu, Ni
Precious metals	Ag, Au, Cu
Stainless steels	Ag, Au, Cu, Ni
Tool steels	Ag, Au, Cu, Ni
W and Mo	Ag, Au, Cu

are not listed in Table 4.16 because Pd is a solute in Ag and Au brazes rather than the solvent. Further, the listing does not include a recommendation for a braze family that can be used to join Mg alloy components, although this can be done with great care using a Mg braze such as those mentioned in Table 4.14. It is also noteworthy that the table recommends Al brazes as suitable for joining Al alloys but not for steels, Ni base alloys or other major types of metallic component materials, while Ag–Cu brazes are suitable for use with almost every type of component material except Al alloys.

While Table 4.16 provides an overview, there can be significant differences in the behaviour of individual members of a braze and component families and, as elaborated in Chapter 6, restrictions as to which type of brazing process that can be used. For example, Al–Si brazes are unsuitable for use when joining some low melting temperature Al alloy components. Again, Ag braze alloys containing Zn, or Zn and Cd are considered unsuitable for use when vacuum brazing, Ni–Cr–Si–B brazes are suitable for joining low carbon steel except when using resistance heating, and Ni–P and Ni–Cr–P brazes are suitable for joining stainless steel components only in a controlled atmosphere or vacuum furnace. When in any doubt, reference should be made to national and international standards as well as suppliers' literature before making a choice.

Many of the products made by brazing contain component parts of different metals and alloys and hence the braze must be suitable for use with two or more materials. Fortunately, this requirement does not cause many complications. Thus Ag–Cu brazes can be suitable for joining steels, Ni alloys, and Cu alloys, and hence combinations of them. Similarly Ni brazes can be suitable for joining a wide range of both steel and Ni alloy components. However, the need to use Al brazes when joining Al components means that the joining of Al components to others of, for example, steel for which Al brazes are not suitable can cause problems.

The number of interlayer/component material combinations that have been evaluated is less for diffusion bonded, and particularly for transient liquid phase bonded, systems and analyses of behaviour have not yet resulted in the establishment of national or international standards that recommend particular combinations of materials. Nevertheless, Al and Ni and Ni alloy interlayers have been used to diffusion bond many metal and ceramic component materials. While the use of Al diffusion bonding interlayers is restricted to ambient and modest temperature applications, Ni alloy interlayers can and have been used with components that must endure high-temperature service conditions. It is also noteworthy that refractory but ductile Ni alloys such as Ni–20Cr have been chosen as the core material for many developments of transient liquid phase bonding of high-temperature metallic component materials and when partial transient liquid phase bonding ceramics.

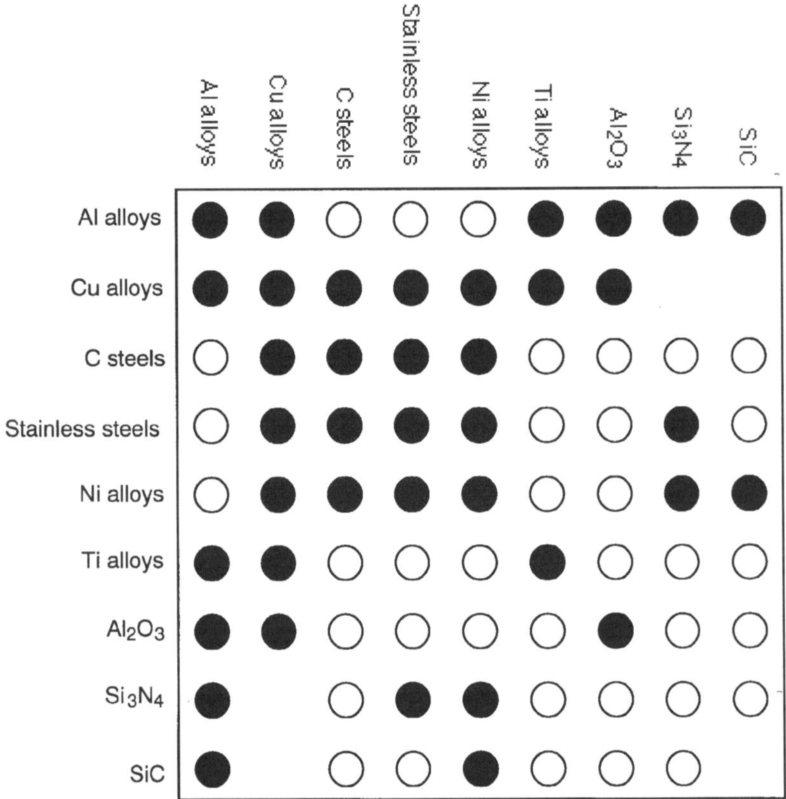

Figure 4.4 Combinations of component materials that can be diffusion bonded. The full symbols identify combinations that can be bonded directly to each other provided the joint design and processing conditions are suitable, and the hollow symbols identify those for which interlayers normally are required

In many and perhaps even most practical cases it is necessary to join components that are of differing compositions and the consequent differing chemical and physical characteristics can be of importance. Thus the joining of Si_3N_4 to Ni base superalloys using Ag-Cu-Ti active metal brazes is complicated by the fact that the Ti is attracted more to the Ni of the metallic component than to the N of the ceramic and hence its ability to form a wettable reaction product layer on the ceramic surface is diminished. Similar effects can occur when bonding using the transient or partial transient liquid phase techniques. For example, Au is insoluble in Mo but Mo and Cu components can be joined successfully because of the ready diffusion of Au into the Cu component. In principle, uneven solubilities or diffusivities also complicate the diffusion bonding of dissimilar components, but in practice the relative slowness of solid state mass transport substantially mitigates the severity of these effects and a vast range of dissimilar materials can be joined using interlayers as shown in Figure 4.4.

In contrast, physical differences between dissimilar component materials, and in particular differences in thermal expansivity, can be of equal importance for joints made by both liquid and solid phase bonding processes. Significant stresses will be generated within a joint if this difference amounts to more than a fraction of a percent as the assembly cools after joining. Thus particularly careful decisions have to be made about joint designs to minimize the strain that will be experienced by the joint, about the use of inserts with intermediate expansivities to decrease the strain gradients within the joint and about the selection of very ductile joining materials that can accommodate substantial strains. Joint design, including the dimensions and placing of interlayers, is discussed in Chapter 5 while the selection of an insert material requires an iterative assessment to ensure an optimum combination of expansivity, rigidity, refractoriness and bonding characteristics. Similarly, the requirement for the joining material to have ductility places a premium on materials such as Al or Ag–Cu alloys and on Al or Ni alloy interlayers when diffusion bonding or transient or partial transient liquid phase bonding are to be used.

FURTHER READING

Further information on braze compositions and properties and diffusion bonding interlayers will be found in:

American Society for Metals, *Metals Handbook*, Vol. 8
American Welding Society, *Brazing Handbook*
British Standards Institution, *BS 1723, Specification for Brazing*
British Standards Institution, *BS 1845, Specification for Filler Metals for Brazing*
Deutsche Industrie Norm, *DIN 8505, Löten (Brazing and Soldering)*

Deutsche Industrie Norm, *DIN 8511, Flussmitell zum Löten metallischer Wekstoffe (Fluxes for metal components)*
Deutsche Industrie Norm, *DIN 8513, Hartlote (Brazes)*
Kazakov, N. F. (1981) *Diffusion Bonding of Materials*, Pergamon Press, Oxford

5 Joint design

A careful selection of a bonding technique, materials and process parameters should result in the production of well-bonded, strong interfaces between the components and the joining materials. However, these interfaces are merely a part of the joint, the good design of which ensures that the chances that these characteristics will have to be displayed are minimized. Thus, good designs generally result in large bonded areas of joints that are subjected to the minimum possible tensile or shear stresses. When identifying such good joint designs, three aspects must be taken into account: the suitability of different types and shapes of joints for use with differing joining processes, the compatibility of the length and particularly the thickness of the joint with the dimensions of the components, and the ease with which the components can be assembled to produce the desired type and size of joint.

5.1 BUTT AND LAP JOINTS

As mentioned in Chapter 1 and sketched in Figure 1.3 there are two main types of joint configuration and each has a major variant. The butt joint is the simplest design with its shape and cross-sectional area being defined by that of the components and it has been used in many laboratory studies. However, this design is not always suitable for applications of real components because it can subject the often relatively weak joining materials to intense tensile stressing. This can arise adventitiously during temperature cycles if the components are of dissimilar materials that have mismatched coefficients of thermal expansion or can be an inherent part of the service requirement such as an ability to resist flexing. The T-joint variant of the butt joint is subject to the same restrictions as the primary butt joint, the bonded area being defined by that of the cross-section of the smaller component except when brazing such thin cross-sections. For these joints, fillet formation can result in a significant increase in the bonded

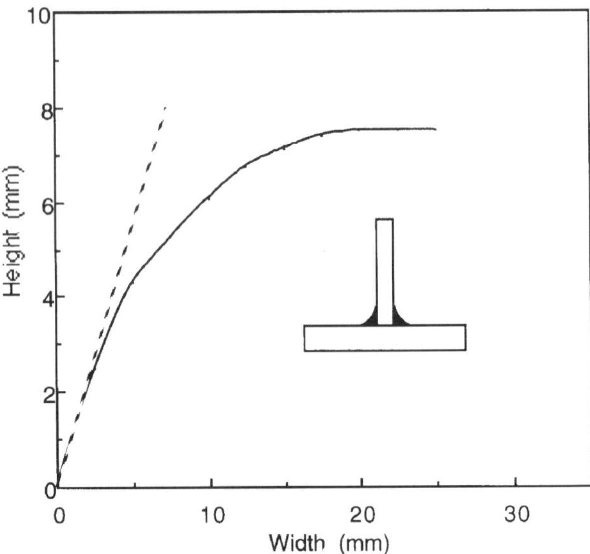

Figure 5.1 Height and width of fillets produced using 1 mm thick sheets of Al to form a T joint. The vertical member was clad with a layer of Al-7Si braze alloy

area but there are limits to this increase for the vertical member imposed by capillary effects that determine the maximum height to which a liquid can rise or at which it can be retained regardless of the volume of liquid available, as illustrated in Figure 5.1.

Lap joints are the second primary type and are also very simple in concept. However, compared to butt joints they have the advantages that their bonded areas do not depend on the cross-sections of the components and external tensile stresses applied to the components are diminished in the joint regions because of the additional cross-sectional area. In the sleeve joint variant of lap joints one component is encased by another. Thus the design is relatively complex but it is attractive for many practical applications. In particular, this design is useful when ceramic components have to be joined to metal structures, or when other dissimilar materials with markedly different expansion characteristics have to be joined. In these cases, the low expansion material is the inner one surrounded by a sleeve of high-expansion material so that cooling from the joining temperatures generates compressive stresses in the joint that enhance its ability to withstand externally applied tensile or shear stresses.

All these types of joint have found application in the workshop as well as the laboratory, although their suitability depends crucially on both the joining technique that is to be used and the stressing conditions the bonded assembly will have to withstand. Thus the common practice of applying an external uniaxial pressure has led to a preference for butt joint designs to produce diffusion bonded assemblies. One example of such a design was referred to in Chapter 2, the Al alloy structure sketched in Figure 2.5, while a second to be described in some detail as a case history in Chapter 8 involves the use of thin Al interlayers to assist the diffusion bonding of a stack of Ti electrode annuli interleaved between Al_2O_3 tubular insulators to produce a kind of club sandwich by applying a substantial uniaxial compressive stress.

Simple butt joint designs are seldom used for brazed assemblies of metal components that have to withstand high-tensile stresses because the cross-sectional area of the joint cannot be much larger than that of the components. Hence joints made with ductile, and generally weak joining materials, will fail by tearing at modest external stresses while those made with strong joining materials are generally also brittle and hence liable to fracture when strained, and particularly when strained in impact. These types of failure are discussed later in Chapter 7 and Appendix D but for the present it is sufficient to note that designs have to be modified if butt joints are needed and have to be brazed. The designs adopted for assemblies of both massive and sheet components normally make use of modifications such as scarf joints with inclined bonding faces, or flanges and sleeves to increase the bonded areas. Some such modifications are shown in Figure 5.2, which also illustrates the use of flanges and buttresses when making T joints using sheet components. These modifications increase the bulk of the joint as well as its area, but this can be avoided by use of the butt-lap design.

Lap joints are much more commonly used when brazing metal components, and the designs frequently exploit the ability of the wetting liquid to flow into capillaries and round bends and corners, enabling braze materials to be located at the entrances of capillary gaps or in recesses within the gaps, as shown in Figure 5.3. As well as simplifying the placing of braze materials, the subsequent flow of the liquid braze and flux along the capillary gap diminishes the chances of gases or flux being trapped within the joint to form unbonded areas. While usually satisfactory as a promoter of joint formation by capillary flow, a possible disadvantage with the lap joint design, and also with the sleeve joint variant, is that the joining of components with markedly differing cross-sections can result in stress concentrations that cause flexure of the thinner member. However, this potential problem can be overcome by tapering the thicker member, as shown in Figure 5.2, or by using a butt-lap design.

The best joint designs generally provide narrow gaps when brazing but this is not always economic, for example the reject rates for components caused by specification of very precise dimensions can be unacceptably high. Again, wide-gap configurations can be encountered when using a brazing technique during the repair or refurbishment of components that have experienced

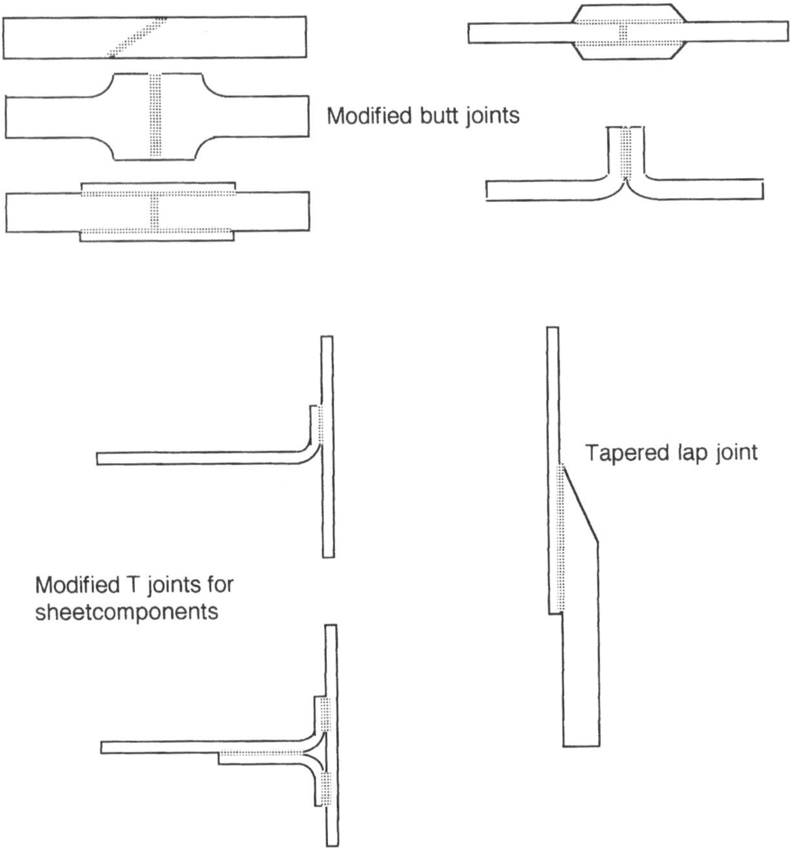

Modified butt joints

Tapered lap joint

Modified T joints for sheetcomponents

Figure 5.2 Some modifications of butt, T and lap joints

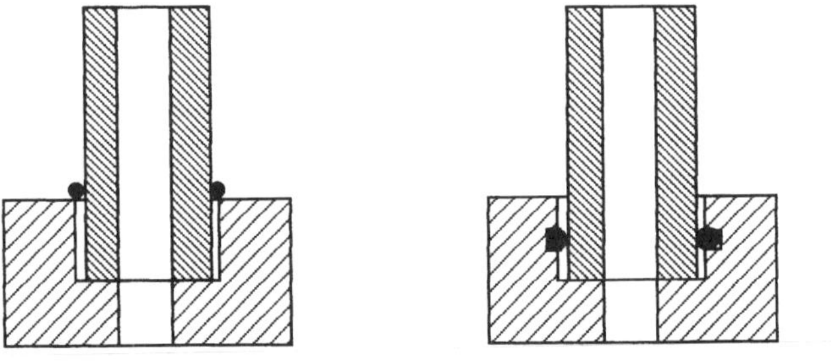

Figure 5.3 Examples of the positioning of braze metal to facilitate joint filling

harsh or excessively long service and require worn areas to be built up or pits and cracks to be filled. In these circumstances, the braze has to act as a padding as well as a bonding agent and simple use of the usually preferred eutectic brazes can be unsatisfactory because the growth of shrinkage voids on solidification of a sizeable pool of molten braze. Some of these difficulties can be mitigated by repair designs in which inserts are used to decrease the gap sizes that have to be filled, but it can be useful also to use brazes with wide melting ranges or to fill the gaps or pits with metal powder, thereby creating a host of small capillaries. For example, strong void free butt joints between Al_2O_3 and Fernico, Fe-29Ni-17Co, or AISI 316 have been made using a 0.5 mm gap by brazing with a Ag-32Cu-10Pd alloy that had a melting range of 817–851 °C. Similarly, the addition of Ni powder when repairing defects in Ni super alloy components using Ni brazes is advantageous because it not only helps to fill pits and cracks but enhances the ductility and toughness of the repair area.

Quite apart from considerations of gap size, the design of brazed joints between unmetallized ceramic components and metallic structures can present special difficulties because of their differing coefficients of thermal expansion. One simple approach to ceramic–metal joint design is to use the ductility of thin metal structures to accommodate the expansivity mismatches. This concept has long been exploited in Housekeeper seals for glass–metal structures and an application to a ceramic–metal system is sketched in Figure 5.4 in which thin metal caps seal a thick ceramic tube to make a vacuum interrupter for use in

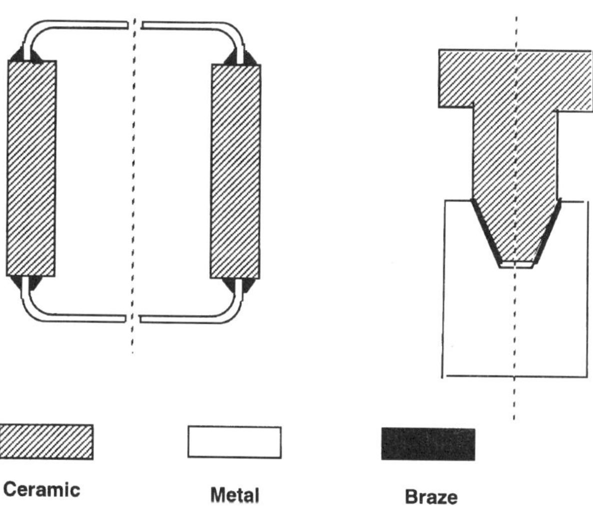

Figure 5.4 Examples of designs for ceramic–metal joints: left, a vacuum interrupter and right, a prototype ceramic/metal shaft

electrical plant and flexing of the metal accommodates the strains imposed by the lower expansivity of the ceramic.

Another solution to the problem of accommodating mismatched expansivities is to use a sleeve joint in which the ceramic is encased within a thin metal member. However, this solution can be impractical when using unmetallized ceramics because the reaction controlled flow of active metal brazes over their surfaces is usually too slow for flow into the capillary gap to result in effective filling. Thus sufficient solid active metal braze has to be placed within the joint gap to ensure that the gap will be completely filled with liquid at the brazing temperature. This requirement is often impractical with a sleeve joint because of the high precision in the machining of the metal and ceramic components necessary to leave a gap of exactly the same size as a tubular braze foil, but one ingenious exception is illustrated in Figure 5.4 in which a prototype turbocharger is produced by bonding a tapered ceramic component to a recessed metal shaft. The active metal braze foil is placed within the recess and the ceramic is then rammed home to squash the foil so that it forms a completely filled annular joint between the two components.

The examples of joint design and comments made so far have not referred to the hybrid processes such as transient liquid phase bonding. This is in part because joint designs specifically tailored for these processes have not yet been developed and because successful hybrid joints have been made using designs intended originally for brazed assemblies. In particular, designs for joining using active metal brazes are suitable because the bonding liquids produced with the three hybrid processes – transient and partial transient liquid phase bonding and eutectic bonding – are produced within the joint gaps by interdiffusion and cannot flow away from their sources.

5.2 JOINT INSERTS AND STRESS REDUCTION

Making a joint by brazing or diffusion bonding using interlayers involves the formation of interfaces between dissimilar materials, and this inevitably results in the generation of residual stresses as the bonded assembly cools. The two materials will have differing coefficients of thermal expansion but their bonding at the interface means that the material with the lower expansivity must be in compression while the other must be in tension. High-integrity interfaces produced by selecting suitable materials and process parameters should not fail because of the action of these residual stresses but their ability to resist external stressing will be degraded. Thus it is important to mitigate the effects of cooling stresses as much as possible and this is mainly a matter of careful designing to minimize the influence of mismatched expansivities.

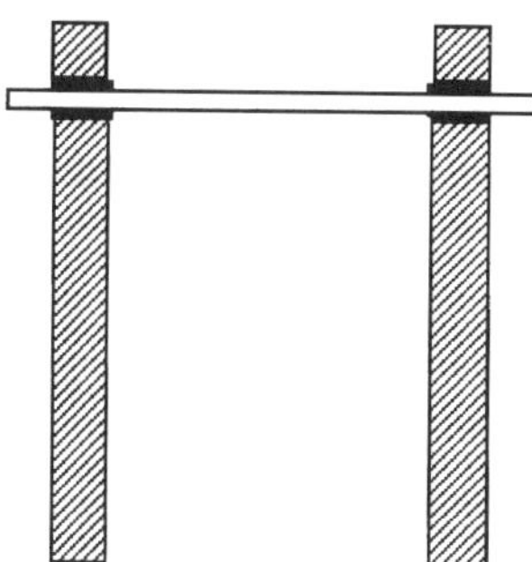

Figure 5.5 The use of redundant end pieces to balance stresses between a ceramic body and metal end caps. The ceramic is indicated by the hatching

While the thermal expansion characteristics of components and joining materials can be markedly different, brazed and diffusion bonded joints are generally so thin and the material used so soft that their ability to generate residual stresses on cooling is trivial compared to that of dissimilar component materials. The size and influence of these contraction stresses can be particularly marked for assemblies of ceramic and metallic components because of their often very different expansivities and the negligible ductility of ceramics. Because of the growing importance of ceramics as engineering materials, considerable effort has been devoted to ways of accommodating this expansivity mismatch and still satisfying the service requirements of ceramic–metal components. One often adopted method of mitigating the effects of mismatched expansivities for bulky ceramic and thin metal members is to use redundant ceramic members to ensure balanced stressing of the metal–ceramic interfaces, as illustrated in Figure 5.5. The approach decreases the likelihood of failure by peeling and is well established in the jewellery trade where it is good practice to enamel both sides of Cu brooches to avoid spalling. Another approach that can be adopted in some applications is to use sleeve joints and much attention has been paid to the use of inserts with intermediate coefficients of thermal expansion. The argument in support of such inserts is that their presence will decrease the strain and stress gradient within the joint and the immediately adjacent component material and hence increase its ability to withstand external stresses. These suggested effects have been confirmed by experimental measurements as well as well-defined modelling.

In practice it has been found that greatest benefit gained from using an insert in the joint between ceramic and metal components is achieved when the insert expansivity is very similar to that of the ceramic. For example, the introduction of a 1 mm thick insert of Mo with a coefficient of thermal expansion of $5.1 \times 10^{-6}\,K^{-1}$ increased the strength from 30 to 370 MPa of a 400 mm^2 square active metal brazed joint between Si_3N_4 and AISI 316 steel, with coefficients of $3.2 \times 10^{-6}\,K^{-1}$ and $15.9 \times 10^{-6}\,K^{-1}$ respectively. This therefore shows the advantage of decreasing the strain gradient across the strong but brittle ceramic–braze interface and concentrating it at a strong and tough steel–braze interface.

The simplest approach to understanding this sort of effect is to assume the ceramic and metal component materials experience only elastic deformation, so the residual stress, σ_R, caused by the mismatching of coefficients of thermal expansion, $\Delta\alpha$, can be derived from the strain produced as the assembly cools from the bonding temperature, T_B, and the elastic modulus, E.

$$\sigma_R = \Delta T_B.\Delta\alpha.E \qquad \{5.1\}$$

In practice, metals are not simple elastic solids but will be deformed plastically once an often low yield stress is exceeded and this deformation allows the metal to conform more closely to the size and shape of the ceramic component thereby decreasing the strain gradient and the residual stress level at the interface. Thus the yield strength of the metal as well as the expansivity and elastic modulus of both the metal and the ceramic should influence the level of residual stresses, and this expectation has been confirmed in practice.

Figure 5.6 plots the residual stress in 25 mm × 15 mm × 1 mm active metal brazed Si_3N_4/metal sandwiches as a function of the mismatch in the

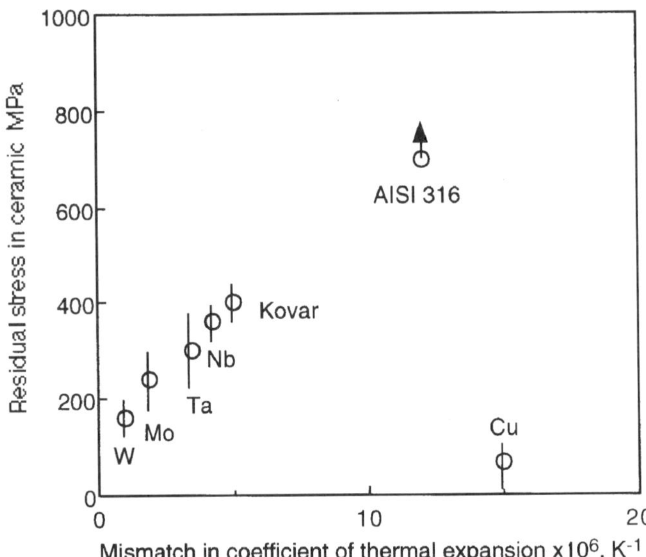

Figure 5.6 The residual stress in brazed Si_3N_4/metal sandwiches plotted as a function of the mismatch in the coefficients of thermal expansion. After Yvon *et al.* (1996)

coefficients of thermal expansion of the two materials. The stress values measured using an X-ray technique increased progressively with the mismatch as the metal component was changed from W to AISI 316 in accordance with an assumption that the cooling stresses produced only elastic strains. However, using Cu as the metal caused a marked decrease in the level of the residual stress despite the fact that it has the highest coefficient of the metals used, and this effect is attributed to its low-yield strength, 54 MPa as compared to more than 180 MPa for the other metals.

The beneficial effect of Cu and other ductile metals as insert materials has found application in a recent development of mass-production automobile components, although the detailed joint designs can be complicated. Thus a Japanese company has produced wear-resistant engine rocker arms by using Ti coated Si_3N_4 tips that are joined to a steel shaft via a Cu interlayer using a Ag–Cu eutectic braze. This same company has also produced the ceramic–metal high temperature automobile turbocharger described as one of the case histories in Chapter 8 in which a finned ceramic wheel was brazed to a metal shaft using a multiple layer insert of W clad with Ni and the entire joint was mechanically encased in a metal sleeve. The beneficial effect of using such an insert was substantial, a 5 mm thick W insert with a 0.2 mm thick Ni cladding decreasing the residual stresses in a brazed Si_3N_4-Nimonic 80A joint from 560 to 40 MPa.

In many developments, the introduction of multiple layer materials has brought a new level of sophistication to the use of inserts. Thus it may be possible to adjust physical characteristics so they satisfy very specific application requirements. Again, the chemistries of the two faces of a multiple layer insert can be chosen so as to permit the use of optimized but differing brazes or diffusion bonding interlayers to effect the joining to the dissimilar component materials. Other novel types of inserts now being evaluated include the use of honeycombs, meshes and corrugated sheet to provide flexibility and hence a measure of toughness.

The development of complex inserts and bonded components requires not only experimentation but analysis and prediction of the stresses generated by both joining and service conditions and this has involved extensive application of the elastic–plastic deformation concept mentioned above. This can reveal complex stress patterns, even for a simple bi-material sandwich such as that shown in Figure 5.7. This sandwich is composed of a stiff ceramic with a low coefficient of thermal expansion and a ductile metal with a high coefficient. The pattern of stresses generated by cooling from the bonding temperature can be derived by first considering the unconstrained contraction of the components, which will be greater for the metal. If the sandwich is constrained so that planar contraction occurs with the dimensions of each component being equal, cooling will generate tensile stresses in the metal as it is stretched and

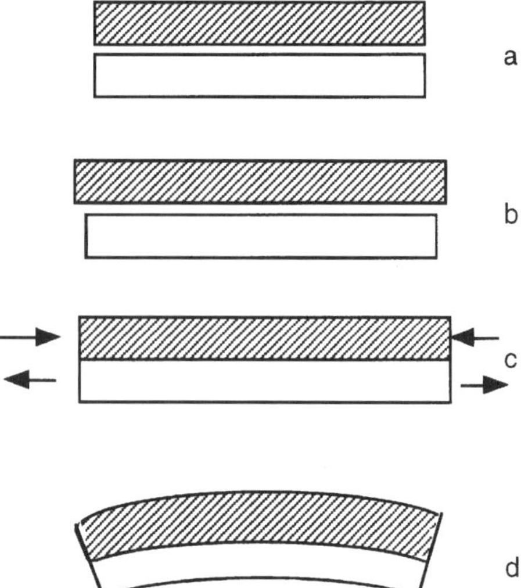

Figure 5.7 The effects of cooling on the stresses in a ceramic–metal sandwich, with the ceramic being identified by hatching. (a) the initial stress free state of separated metal and ceramic components at the bonding temperature; (b) the stress free state of separated metal and ceramic components after having cooled from the bonding temperature; (c) the stressed state of the bonded ceramic and metal components after having cooled when no deformation is allowed; (d) the shape of the bonded ceramic and metal assembly when deformation is allowed during cooling. After Hseu and Evans (1985)

compressive stresses in ceramic as it is squashed and the interface will be a neutral axis where the stress changes from being tensile to compressive. Finally, the components can be allowed to bend so as to balance the moment induced by the asymmetric stress pattern. This has the effect of extending the interface and causing the metal near the interface to deform plastically, thereby relieving some of the tensile stress.

Such analytical models are widely used to predict stress distributions, but modifications have to be made when dealing with structures more complex than a simple sandwich of dissimilar materials. A particularly useful modification is to regard the structure as being composed of a number of small but finite elements with a simple geometry such as a rectangle. The stresses and strains within such a simple element can be calculated fairly readily but are subject to the constraint that the conditions at each wall are shared with the neighbour element. Thus iterative calculations can be performed to construct a stress and strain pattern for the whole structure, as when completing a jigsaw puzzle, and in regions of particular sensitivity such as a joint or near a free surface, the definition of the pattern can be improved by decreasing the size of the element. This finite element technique was developed first to describe the response of essentially two-dimensional structures of purely elastic materials, but the growth of available computing power has permitted further sophistication in recent years to encompass both elastic-plastic behaviour and three-dimensional structures.

The results of such calculations can be used to derive stress patterns and levels within monolithic and bonded components. One such pattern for a ceramic–metal joint is shown in Figure 5.8 to illustrate some of the effects of mismatches in expansivities. The figure shows that the residual stress is not

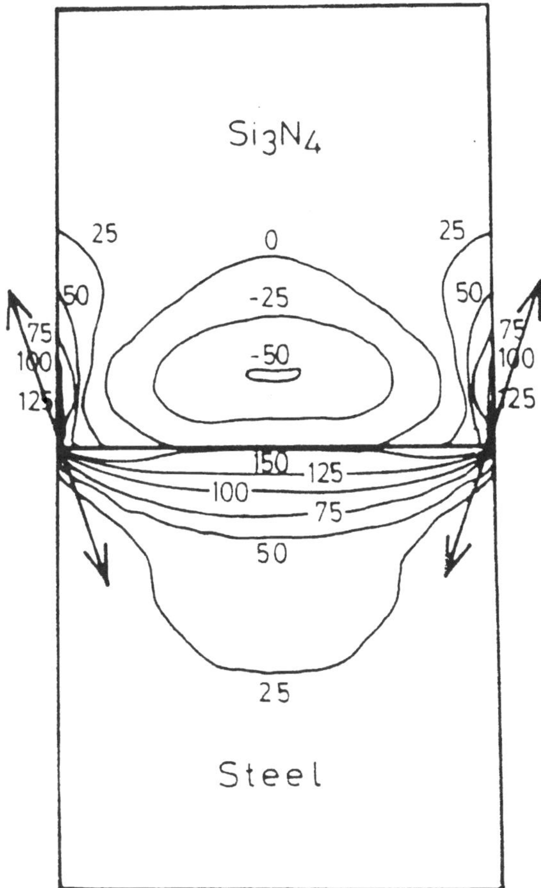

Figure 5.8 Distribution of maximum principal stresses calculated by a finite element method. The joint is of Si_3N_4 bonded to steel and is assumed to have cooled from a bonding temperature of 1000 °C and suffered only elastic deformation. The arrows indicate the position and direction of the maximum tensile stress. Suganuma (1990)

uniform within each component and even varies along their interface. The most detrimental predictions are that tensile stresses will be generated within the ceramic and at the interface, and these stresses will be greatest near where the interface emerges at the free surface. The sites of the maximum predicted stresses are shown in Figure 5.8 and their near perpendicular orientation means that they will be superimposed on any external tensile stress and hence that the measured strength of the joint will decrease.

Finite element modelling also predicts that increasing the bonded area will increase residual stress levels and that the levels will be particularly high at the corners of rectangular joints. These stresses decrease the ability of the joint to resist external stressing and may also cause flaws where some relief has been gained by separation of the components. Further, since it is more difficult to ensure a high standard of surface preparation, for example the surface flatness of components to be diffusion-bonded, there is a greater likelihood that large area joints will have a population of unbonded regions so that the strengths will be not only lower but more scattered.

5.3 JOINT SIZES

Having selected a joint geometry suitable for a particular application and joining process, decisions must be taken about its size. These decisions will affect the complexity of the machining or forming of the components prior to assembly, the readiness and rate of joint formation, the quantity of joining material needed and important service characteristics of the bonded joints such as strength, ductility, corrosion resistance and aesthetic acceptability.

The expected behaviour of lap and sleeve joints can be related to that of model systems developed for chemically inert systems that predict that wetting liquids will flow indefinitely into horizontal capillary gaps and that the penetration of modest distances will be very rapid. Thus the time taken for liquid Cu to penetrate a typical joint length of 1 cm into a gap between the parallel sides of two perfectly wetted components should be 1 second for a 1 μm wide gap and 0.01 second for a 100 μm wide gap. Equilibrium penetration of vertical capillaries is predicted to be greatest for the narrowest gaps since the gravitational resistance will diminish as the gaps are narrowed but the capillary attractions will remain unchanged. Thus the vertical penetration is predicted to increase from 0.33 m to 33 m (!) for Cu flowing into perfectly wetted capillaries when their gap sizes are decreased from 100 to 1 μm, and the times taken to penetrate 1 cm are only a few percent longer than those calculated for horizontal capillaries. The implication of such models, therefore, is that a wide range of gap sizes will result in rapid filling of useful lengths of brazed joints but the most extensive filling will be that of the narrowest gaps.

In practice, few brazed systems are chemically inert and interdiffusion can have profound effects on the influence of gap sizes. Thus erosion of the component material or formation of reaction product layers will alter the gap size while changes in the composition of the liquid will alter its viscosity. The optimum size of joint gap depends on the process, those for vacuum brazing generally being smaller than those needed when using a gas torch and flux, and on the identity of the component and braze material. Further, the optimum braze gap size is often not determined directly by the ease of joint formation but rather by the properties of the solidified joint, and in particular the mechanical properties.

Representative ranges of gap sizes recommended by British and American authorities are presented in Table 5.1 for a number of braze families. It is noteworthy that all the ranges encompass a gap size of 100 μm and this is typical of the values that are specified (in practice at the beginning of a joint optimization process). It is noteworthy therefore that the smallest recommended gap size is for pure Cu used to braze Fe components which it wets very well but with which it is chemically inert. In contrast, the particularly wide gaps suggested for Al brazes relate to the joining of Al alloy components and are due to the ready loss of Si from the braze by diffusion into the component that can result in premature, isothermal, solidification unless the gap is wide enough to act as a reservoir.

The recommended gap sizes are given as ranges, the smaller values of which generally refer to pure metals and eutectic brazes with singular melting temperatures. Brazes with very differing solidus and liquidus temperatures are usually more suitable for joining components when

Table 5.1 Recommended gap sizes for braze families. After BS 1723

Braze	*Gap at brazing temperature* (mm)
Al alloys	0.050 → 0.250
Ag alloys	0.050 → 0.150 (fluxed) 0.025 → 0.125 (controlled atmosphere)
Cu	−0.050* → 0.100 (controlled atmosphere)
Cu-P alloys	0.025 → 0.150 (in air) 0.050 → 0.150 (fluxed)
Cu-Zn alloys	0.050 → 0.150 (fluxed)
Ni	0.000 → 0.125 (controlled atmosphere)
Au	0.025 → 0.125 (controlled atmosphere)

* Produced by using a shrink-on assembly.

narrow gaps cannot be produced, as in some repair operations. As with the case of Al brazes used to join Al alloy components, extensive interactions with the solid that change the fluidity of the liquid braze also diminish the suitability of the lower ends of the recommended ranges of joint gaps. The narrower gaps are also less suitable if the component surfaces are rough because this causes both a variability in the actual size of the gaps and an enhanced extent of interdiffusion by virtue of the increase in the liquid–solid contact area. However, the narrower gaps are generally the more suitable if the brazing process is flux free because it is not necessary to ensure that the gaps are wide enough for the used flux to be swept away from the component surface.

The components to be brazed are assembled at room temperature but the recommended gap sizes quoted in Table 5.1 are for the brazing temperature. When joining components of the same material, the expansion of the gas caused by heating to the brazing temperature is not significant, from say 100 to 103 μm for the gap when a 2 cm diameter brass rod and a brass sleeve with an internal diameter of 2.01 cm is heated to a brazing temperature of 800 °C. However, gaps between dissimilar materials can be markedly affected by the temperature change on heating to induce brazing. Thus replacing the 2 cm rod with one of a low carbon steel will cause the gap to expand from 100 μm to 218 μm on heating, while replacing the sleeve with one of low carbon steel will cause the gap to close, with the separation of the components attempting to change from 100 μm to *minus* 16 μm.

The thickness of joints is not their only important dimension since the failure loads of joints depend also on their bonded areas. Thus using a large overlap length can cause the failure load of a joint to be larger than that of the weaker component even when using a low strength braze. Simple calculation shows that a lap joint made with a braze whose shear strength is σ_S, will fail in the thinner component whose thickness is t^* and tensile strength is σ_T when the overlap distance is L and

$$L > t^*\sigma_T/\sigma_S \qquad \{5.2\}$$

and similarly for a sleeve joint of diameter D′, component failure should occur when

$$L > t^*(2D' - t^*)\sigma_T/2D'\sigma_S \qquad \{5.3\}$$

but in reality longer lengths must be used to include a safety factor, and common practice is to ensure that the overlap length is at least three times the thickness of the thinner component.

The material families listed in Table 5.1 do not include the active metal brazes used to join unmetallized ceramic components. This is because their development and characterization are not yet sufficiently established to be standardized, and they are often used to make butt rather than lap joints. Hence joint gaps are not so rigorously set and, in practice, the manufacture of butt joints with these alloys involves placing one or occasionally more foils of braze alloy between the components and applying a small pressure or dead load to enhance the initial contact, and the foils available from commercial suppliers are typically 50–100 μm thick so that once again gap sizes are usually about or approaching 100 μm.

In contrast to brazed joints, the sizes of diffusion bonded joints that have been or could be used to achieve strong bonding are not constrained. Perfectly smooth, perfectly conforming component surfaces should diffusion bond readily without the need for interlayers providing the fabrication temperature is high enough to permit reasonably rapid interdiffusion. Such temperatures cannot always be used, so interlayers have to be employed to encourage interdiffusion and finite separations of the components, joint gaps, are needed to accommodate them. The sizes of these gaps can vary widely but fall between certain practical limits.

The ideal of perfectly smooth and conforming component surfaces cannot be manufactured but can be approached when diffusion bonding ductile, and particularly superplastic, metal components by flattening microscopic surface asperities and the collapse of macroscopic interfacial voids. The primary role of the interlayer with such components is not to fill a joint gap but the more subtle one of promoting interdiffusion, and this should be achievable by a quite small amount of interlayer material. It is noteworthy, therefore, that coatings with thicknesses of a micron or less have been used successfully to promote the diffusion bonding of superplastic Al alloys.

With less deformable component materials, and most notably with ceramic components, the joints must have a finite thickness to accommodate interlayer foils that have mechanical as well as chemical and physical roles to fulfil. Thus even carefully prepared component surfaces are populated with microscopic topographic features that generate a surface roughness with an average amplitude of typically 1 μm and this implies a peak to valley height difference of about 4 or 5 μm. Particularly when bonding rigid ceramics, it is desirable to avoid direct contact between component surface asperities since this can cause severe, albeit localized stress concentrations that can cause cracking. For such materials, a minimum joint gap of about 10 μm can be expected but the practical lower limit of *interlayer* thickness is larger and typically twice as large, due to the difficulties of achieving effective chemical and mechanical preparation of the interlayer surfaces prior to bonding.

When low-diffusion bonding pressures are used so that there is negligible macroscopic deformation of the interlayer, joints with a wide variety of widths can be made. However, there can be an upper useful limit to the *interlayer* thickness if it is to be subjected to gross macroscopic deformation because of its softness at the bonding temperature. For example, a 100 μm Al interlayer used to diffusion bond Al_2O_3 ASTM test pieces, which are described in Chapter 7, by application of a 50 MPa pressure for 30 minutes at 600 °C was compressed 24%, but a 1 mm interlayer was compressed 83%. The substantial squashing of the 1 mm interlayer required significant movement of the Al_2O_3 test pieces during the bonding process and resulted in extrusion of interlayer material from the joint to produce a flash. Thus the use of thick *interlayers* can cause problems by increasing the difficulty of jigging the assembly to prevent misalignment of the components and introducing an additional operation into the fabrication process to tidy the appearance of the bonded assembly by removing the flash. Nevertheless, grossly deformed interlayers often produce the strongest joints because massive lateral movement enhances the dispersal of oxide films on their surfaces and sometimes it is possible to design the joint and the jigging of the components so that thick but narrow interlayers can be used that deform to fill the joints without producing a flash.

When choosing an optimum joint thickness, attention also has to be paid to its width. Thus increasing the width should increase the pressure needed to achieve the desired deformation of an interlayer because the larger contact area offers more frictional resistance to lateral extension. Theory suggests and experiment confirms that the pressure needed should increase as 0.25 × (width/thickness). Thus further constraints on the choice of an interlayer thickness may be the maximum load attainable with the equipment being used and the compressive strength of jigging devices or even the components themselves.

While there is no particular joint thickness that can be defined as necessary for effective diffusion bonding, strong bonding has been produced using foil interlayers to produce joints with thicknesses ranging from 50 to 500 μm but the frequently preferred sizes typically are about or approach 100 μm. This occurs for reasons not concerned entirely with optimization of the joining process but also the difficulties of handling and the less ready availability of very thin foils of desirable interlayer materials.

Practical considerations of ready availability and handling difficulties are also of significance in explaining the frequent preference for joint thicknesses approaching or about 100 μm when using the partial transient liquid phase bonding technique, although the solubility of the coating and component materials in the core of the interlayer is the prime consideration.

The frequent convergence to a preferred joint gap of about 100 μm whether brazing, diffusion bonding or using a hybrid process is not mere coincidence, but reflects the fact that these techniques produce similar joint structures, a thin seam of soft material sandwiched between relatively massive and rigid components. The ability of the joints to withstand external stressing without deforming or failing diminishes as they thicken, and therefore it is desirable that the thinnest joints be used that can be made conveniently and reproducibly.

5.4 ASSEMBLY

Having selected a type of joint and decided upon its dimensions, it is necessary to ensure that these decisions are realized in practice by firm positioning when the components are assembled prior to bonding and while the assembly progresses through the bonding cycle. One method of achieving these goals is to use external jigging as discussed in Chapter 6. However, it is often cheaper and more convenient to incorporate features in the joint designs that result in a suitable and firm positioning of components that is maintained as they progress through the bonding cycle, and this approach of self-jigging is adopted frequently when components have to brazed.

Many simple techniques can be adopted to produce self-jigging, and some are illustrated in Figure 5.9. The simplest of these is to use the mass, or rather the weight, of components to hold them together by placing the

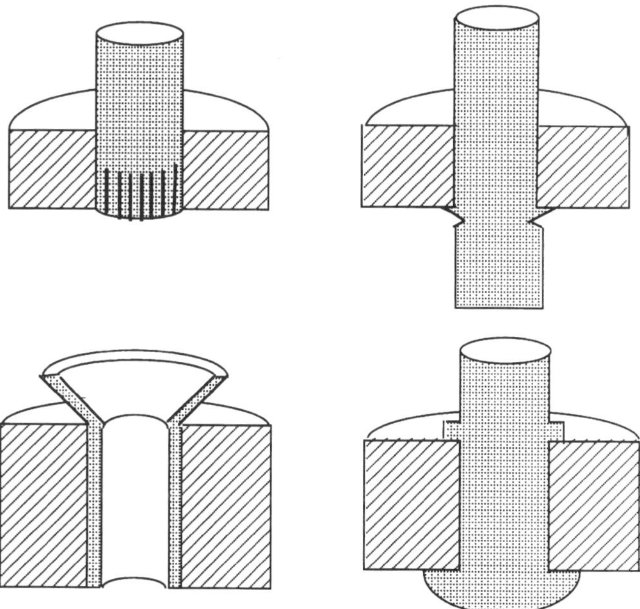

Figure 5.9 Some techniques for locating components prior to bonding. Clockwise from the top left, knurling, punching, riveting, spinning

heavier component on top and this technique is particularly attractive for the assembly of butt joints. Almost equally simple in concept is the use of a tight fit, of an interference fit, to ensure that the components are firmly positioned with respect to each other although it should be remembered that the fit will alter as an assembly of dissimilar material components is heated to the bonding temperature.

Metal components can be deformed after their introduction into an assembly to produce and maintain a tight fit. Thus the surface of a shaft can be knurled, machined into ridges, before being pressed into an opening in the opposite component. Other techniques for producing deformation joints include expanding or swaging, spinning, peening or hammering, riveting and, possibly the simplest of all, punching to produce burrs that lock a metal shaft component into position.

More ambitious techniques for positioning that are sometimes used include screwing and tack welding. In principle, some form of adhesive could also could be used in parts of the component that are not to be bonded but this is rare in practice because of concern about their high-temperature durability. Finally, it should be noted that the jigging to produce brazed joints can be, at least in principle, simpler than those required when diffusion bonding because there is no need for the application of significant pressure. This advantage of being essentially a pressureless process is also a characteristic of the transient liquid phase and partial transient liquid phase bonding techniques.

REFERENCES

Hseu, C. H. and Evans, A. G. (1985) Residual stresses in metal ceramic bonded strips, *Journal of the American Ceramic Society*, **68**, 241–8.

Suganuma, K. (1990) Joining silicon nitride to metals and to itself. In M. G. Nicholas (ed.), *Joining of Ceramics*, Chapman & Hall, London.

Yvon, P., Benoit, M. and Peteves, S. D. (1996) Effects of the metal workpiece properties on the residual stresses in silicon nitride-metal brazed joints, *Journal of Materials Research*, **11**, 3090–8.

FURTHER READING

Anon (1986) *A Finite Element Primer*, National Agency for Finite Element Methods and Standards, UK DTI.

Equipment and process parameters

6

Having selected suitable joining materials and decided upon a joint design, consideration can be given to the choosing of equipment and processing parameters that will ensure effective manufacture of the joints. While choices about materials and joint designs usually have to be made between a few possible options whose number is severely limited by the nature of the components and the proposed service performance of the bonded assembly, those concerned with process parameters are numerous not only because of their variety but also because of the possibility of making incremental adjustments during trial runs to optimize joint quality by fine tuning.

Many aspects of the joining process fall under the heading of this chapter and they will be addressed in an order that reflects broadly the sequence in which they are experienced during the assembly and bonding of components, starting with surface preparation of the components and joining materials and ending with cooling rates.

6.1 SURFACE PREPARATION

6.1.1 Techniques

Surface preparation before bonding is usually necessary for components and joint inserts, and also is frequently necessary for the joining materials. There are three main objectives for preparation techniques. The first of these is the production of surfaces that are free of mobile contaminants such as oil films and surface layers of fragile or thick corrosion products, paint, varnish and other protective coatings that inhibit bonding. Having produced clean surfaces, it is then necessary to ensure that there is macroscopic conformity between the mating surfaces of the components to be bonded. Without this, gap sizes and capillary attraction will vary along the length of a joint to be brazed and may result in incomplete filling by braze alloys, while lack of surface conformity of components to be diffusion bonded may make the achievement of intimate contact impossible when joining either directly or via thin interlayers. Finally, it is important to create microscopic topographies on component and joining material surfaces that encourage bonding.

The removal of rust and other thick corrosion products and the creation of desired macroscopic and microscopic topographies can often be achieved by mechanical treatment of the surfaces. The methods used frequently in practice include machining provided this does not have a significant effect on component sizes and joint gaps, grinding, filing, scratch brushing and grit blasting. Machining and grinding can be used to prepare both metal and ceramic surfaces but there is little or no need to elaborate on their effects beyond emphasizing that they can introduce residual stresses into the near-surface regions and that care needs to be taken to ensure minimal transfer of and contamination by material from the turning or milling tools and grinding wheels, and that lubricants must be removed subsequently by a cleaning process. In practice it can be useful also to avoid producing very smooth surface finishes because brazes can spread better on slightly rough surfaces and populations of microscopic asperities can assist the disruption of surface oxides when diffusion bonding.

When filing, and particularly when scratch brushing, care must be taken not to burnish metal surfaces which is often associated with the creation of a layer of embedded fragments of the surface oxide, and scratch brush wires, because this can degrade wettability when brazing and interdiffusion when either diffusion bonding or brazing. Grit blasting is used mainly to prepare surfaces that are to be brazed and differs from the other mechanical techniques in that the components are treated while in cabinets and hence its progress cannot be monitored so readily by observing the changing appearance of the surfaces. To achieve the best results requires decisions about what blasting pressure and distance should be used, and typical values are a pressure of 300 MPa and a distance of 10 to 20 cm. The blasting grit must be clean and should be angular so that it cuts the surface rather than merely hammering it as would spherical powder. A high blasting gas pressure should be used providing this does not cause distortion of thin components, and other techniques should be adopted if it does so. The grit should be hard and also relatively dense so that fragmented steel, cast iron and Ni alloys are usually recommended rather than ceramics. Some grit is certain to become embedded in metal surfaces and hence another advantage of using metal grit is that these embedded particles will cause less degradation of bonding characteristics than would ceramics. If the component is not to be bonded soon after being grit blasted, embedded steel and cast iron particles may rust and hence it is advantageous to use oxidation resistant grit such as fragments of Ni-17Cr-4Si-4B.

Chemical techniques can be used with metals to produce modest roughening and removal of surface films, such as etching with HCl, HNO_3 or H_2SO_4, washing with acid phosphates or immersing in salt baths but these always must be followed by vigorous rinsing to prevent the build-up of corrosion products, and it should be noted that HNO_3 should not be used with Ag or Cu or alloys containing them. In principle, ion bombardment and other irradiation processes can be used to remove thin oxide films from, and incidentally slightly roughen, the surfaces of metals but such techniques are largely confined to the research laboratory at present and even there are generally done *in situ* during the course of an experiment. Similarly, oxide can be removed from the surface of Cu by heating in a vacuum and other relatively unstable oxide films can be removed by dissolution in their substrate, but such treatments normally form part of the bonding cycle rather than being a separate precursor stage. If reducing atmospheres are used, oxides such as FeO and Cr_2O_3 can be removed by dissociation as well as dissolution and the conditions for this to occur with some oxides of importance are listed in Table 6.1. The conditions needed to achieve such low partial pressures of O_2 by using environments containing H_2 or CO are discussed later.

Heating can also be used to evaporate mobile hydrocarbon contaminates from surfaces, but care must be taken to avoid cracking which leaves an immobile C residue. However, if the surface to be cleaned is an oxide ceramic, hydrocarbons and C contamination can be effectively removed by burning when heating in air at temperatures of about 1000 °C. Most methods of removing mobile contamination, however, employ commercial detergents or organic solvents. Thus alkaline mixtures contain phosphates, carbonates silicates and sometimes hydroxides as well as the

Table 6.1 Dissociation pressures of some oxides at 1150 °C

Oxide	*Log* (O_2 *pressure*, mbar)
NiO	−9.5
Fe_3O_4	−14
Cr_2O_3	−20
SiO_2	−26.5
TiO_2	−28.5
Al_2O_3	−35
CuO	−5.5 at 850 °C
Al_2O_3	−54 at 600 °C

detergents themselves, and can be used in baths, agitated barrels or as sprays to remove machining and grinding lubricants, and also some abrasive debris. Similarly, oils and grease can be removed by petroleum, carbon tetrachloride, ClC_4, and chlorinated hydrocarbons used in baths or as sprays and other chlorinated hydrocarbons such as stabilized trichloroethylene C_2HCl_3 can be used in vapour degreasing processes, but care must always be taken to abide strictly to safety regulations when using such materials.

6.1.2 Monitoring

If surface treatments are to be used, it is essential that their effectiveness be monitored. The ultimate test of this will be whether or not joints are produced that successfully withstand service conditions, but some more immediate form of monitoring is desirable. Unfortunately this is can be difficult to achieve quantitatively.

The removal of gross surface contamination such as paint or rust by mechanical means can be followed by visual observation and graded by comparison with photographs of variously ranked rusty surfaces. Particular attention should be paid to the uniformity of appearance of the treated surfaces since patches of incorrectly prepared surfaces can result in unbonded areas being present within the joint. Although it is difficult to quantify, assessment of roughness by eye or using a reflection meter, or even by gently moving a finger across a surface can provide a somewhat surprisingly reproducible ranking of surface roughness and comparison with roughness standards permits some quantification.

Quantification implies that a number or numbers can be used to describe a characteristic and in the case of roughness these numbers are principally the averages of the amplitude and the wavelength of the height variations. Figure 6.1 shows cross-sections through a very simple rough surface in

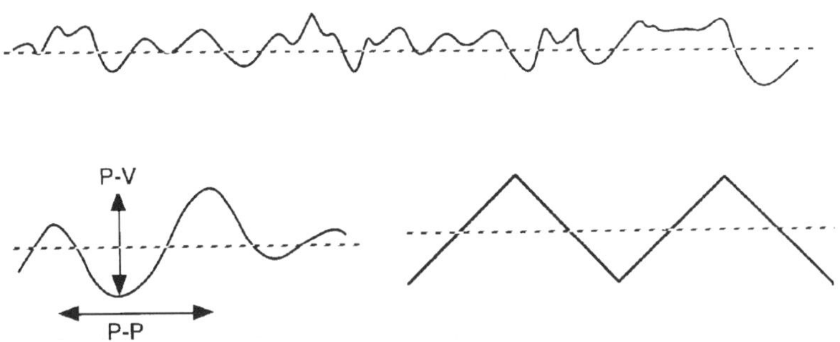

Figure 6.1 Cross-sections through roughened surfaces

which the vertical scale is grossly exaggerated compared with that for the horizontal. The dashed horizontal lines on the sections show the locations of the nominal surface such that the cross-sectional areas of the true surface peaks above, and of the valleys below, are exactly balanced. It is easy to see that the asperities can be characterized in terms of the peak to valley and peak to peak distances P–V and P–P. Although average P–V and P–P values are used sometimes, it is conventional to use the average amplitude, R_a, of the deviation of the surface from its nominal location as a measure of its vertical roughness. As can be seen by considering the very simple profile shown in the bottom right-hand corner of Figure 6.1, the P–V distance must be divided by 4 to obtain the R_a for a surface consisting of angular furrows but the divider for real surfaces is usually somewhat larger. Similarly, the wavelength of a rough surface, λ_a is simply the average P–P distance for features with peaks and valleys on opposite sides of the nominal surface.

Scanning electron microscopy or optical interference microscopy can be used to characterize and quantify the features on a small area of a surface but in practice roughness is commonly measured using surface profilometers, most of which place a stylus in contact with a surface and then traverse it to generate a trace akin to that in Figure 6.1 and values for R_a and λ_a. Just as the styluses of gramophones have been increasingly replaced by the laser sensors of compact disc players, so recent years have seen the development and widening use of devices that use a laser beam to sense height differences so that there is no physical contact with the surface being scanned and some of these laser instruments can also make multiple displaced traverses across a surface to generate a pseudo three-dimensional image. These surface profilometers are laboratory instruments that characterize a small sample placed on a vibration-free table to achieve the maximum precision, but more robust if less precise hand-held instruments are also available that can be used in the workshop and on sites where construction is in progress.

Where surface roughness values are specified they generally cite only an R_a although waviness as well as roughness amplitude can be important when components are to be diffusion bonded without the use of an interlayer. Most widely used preparation techniques generate λ_a values of 10 to 100 μm but it is desirable to avoid the longer wavelengths because the time taken for surface diffusion to increase the bonded area by spheroidizing the elongated voids produced by the initial contact of asperities increases as does their width. The time and cost of preparing surfaces with very low R_a values can be excessive but it is good practice to ensure that R_a is less than about 1 μm when diffusion bonding without using an interlayer. Some relaxation of the R_a requirement can be permitted if an interlayer is used. However, high-integrity joints have seldom been produced using R_a values for component surfaces exceeding 3 μm and those used in practice are often quoted as being less than 0.4 μm for both metal and ceramic components so that a value of about 1 μm can be regarded as typical.

It is known that the spread of well wetting brazes is enhanced by roughening a surface and the use of polished component surfaces should be avoided. However, gross roughness can have a detrimental effect on the mechanical properties of a joint and in the extreme cause unacceptable variations in the sizes of capillary gaps. Optimum R_a values generally lie within the range of 1 to 2 μm with the British recommended practice being that the R_a value of anisotropically roughened surfaces should not exceed 1.6 μm while Americans similarly suggest an R_a value of 0.8 to 2 μm for isotropically roughened surfaces.

The degree of cleaning achieved by surface preparation of the components is even more difficult to characterize despite its importance. The chemistry of surface and adsorbed layers can be defined with a high degree of precision and sophistication by optical and near optical laboratory techniques such as ellipsometry and infrared spectroscopy. Similarly, the beam techniques of electron probe micro-analysis, Auger electron spectroscopy, X-ray photospectroscopy and secondary ion mass spectroscopy are widely used in materials laboratories to provide information about surface chemistry of small samples. All of these techniques require the samples to be contained within an evacuated chamber and valid interpretation of the results needs expert knowledge. Thus application of the techniques is generally impractical in production environments, but results generated by their use in laboratory environments are of considerable value because of the insight they provide into the true state of cleanliness of real surfaces. These are almost invariably covered by thin oxide films even immediately after the removal of previous corrosion products and adsorbed layers of H_2O, N_2, O_2 and organic molecules, but real-time measurements show that desorbtion of gas molecules occurs when the surfaces are heated in vacuum and that in some cases the oxide films disappear as well. However, heating can also facilitate segregation that contaminates surfaces, thus concentrations of even a few parts per million of C can cause sufficient segregation to Ni surfaces to degrade brazing characteristics.

The fabricator needs simple tests to qualify surfaces before they are committed to bonding processes and this need is widely recognized. A number of studies have evaluated the usefulness of measurements of the spreading of sessile drops of H_2O, solvents and even oil and it has been found that this behaviour does indeed relate to the thoroughness of cleaning as well as roughness. The most developed use of such tests is probably the work of the Lockheed Georgia Aerospace Company using the spreading of sessile drops of H_2O to qualify the suitability of surfaces of Al alloy structures for painting, but in that case they were dealing with surfaces that had a virtually unchanging roughness. Similarly, observation of the way H_2O drains from surfaces after rinsing can provide insight, any breaks in the draining front indicating the presence of retained hydrocarbon contamination. In most real cases, however, it is necessary to rely on close adherence to a previously satisfactory cleaning process continuing to yield the desired effects.

6.2 JIGS

The previous chapter described how joint designs can be modified to produce self-jigging assemblies, but this is not always practical and external jigging becomes desirable or even essential. While in some ways a necessary evil because they add to the capital cost and the complexity of the joining process, a positive outcome of their use is that this can allow earlier unloading of cooling assemblies from bonding furnaces and hence increase furnace throughput.

The prime function of an external jig is to hold the components that are to be bonded securely so that the desired joint configurations and sizes are maintained as the assembly is handled, for example while being loaded into a furnace, and while it is being heated, bonded and cooled. The jig should also be relatively easy to assemble and dismantle and to load and unload if joining is to be accomplished in a chamber or bath. There are, however, many other restrictive requirements that must be satisfied as well.

- The jig must not interfere with the bonding process by preventing necessary movement of components when diffusion bonding using an interlayer.
- It must not significantly modify the heat flow by shielding the assembly from heating elements or gas flames.
- It must not be so massive that it becomes a secondary heat source.
- When brazing, it is important that the jig should not cause stressing of components liable to suffer from liquid metal embrittlement by braze constituents.
- The jig should not bond readily to any components or joining materials with which it comes into contact.
- The jig should not become deformed by repeated use.

These objectives and restrictions affect both the design of jigs and the choice of materials used to construct them. Thus it is desirable that contact between the jig and the components should be minimized by using pin points or knife edges and the structure should normally be an open lattice work to minimize both the mass of the jig and its effect on heat flow patterns. The jig materials must be strong at the bonding temperatures and hence refractory metals and ceramics such as Si_3N_4 or Al_2O_3 are often

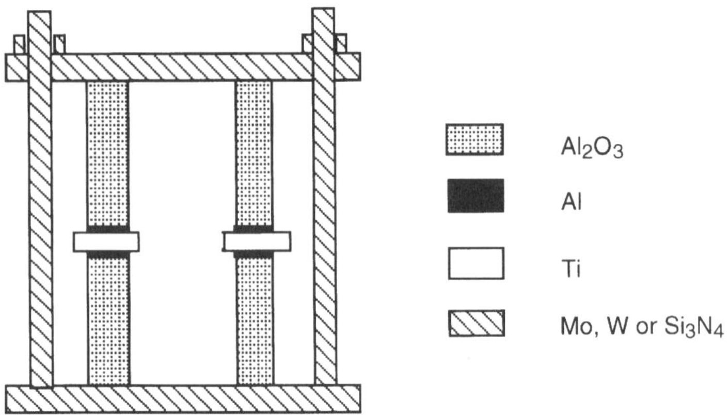

Figure 6.2 Sketch of a ceramic–metal assembly held within a low expansion jig. Note that the figure is not to scale

employed. When a metal is used as the jigging material, it can be helpful to oxidize it, and this is commonly done when using stainless steel, Ni base super alloy or Ti alloy as the jig materials, or to coat with an oxide to diminish the ease with which it can bond to metal components and particularly the possibility of its being wetted and bonded by a braze alloy.

Many jigs are designed so that they exert a compressive stress by using a dead load or taking advantage of differences in coefficients of thermal expansion, thereby increasing the secureness with which the assembled components are held before being bonded. When diffusion bonding, however, the compressive stress can play a much more active role and even be the only stress that is exerted on the components. Thus consider a simple assembly such as that sketched in Figure 6.2 of two 5 cm long Al_2O_3 tubes between which a 3 mm thick Ti cap is to be diffusion bonded by use of Al interlayers with an initial thickness of 0.25 mm. Heating this assembly to a proposed bonding temperature of 600 °C will cause it to expand freely to a length of 10.398 cm, but the use of a Mo clamping frame of end plates bolted together using threaded rods will restrict the expansion to 10.382 cm and if the Al interlayer is the only material to deform this will suffer a 31% compression when using a jig made of Mo, while the compression will be 39% if the jig is made of W and 55% if it is made of Si_3N_4. Unfortunately, the range of low-expansion materials that can be used in compressive jigs is somewhat limited. Thus the low-expansion Invar and Kovar alloys are seldom suitable for use in compressive jigs because they revert to a high expansion mode at quite modest temperatures and also because they are relatively weak materials Similarly, graphite is seldom suitable as a jig material despite its low coefficient of thermal expansion and refractoriness because it oxidizes readily when heated in air, degases when heated in a vacuum and can cause carburization of stainless steels when heated in environments containing H_2. However, large compressions can be achieved also by inserting a high material with a high coefficient of thermal expansion between the components and the end plates of the jig, thus the use of two 0.5 cm thick plates of stainless steel would increase the compression obtained with a Mo jig to 43%.

This approach of using jigs to achieve diffusion bonding is attractive as a means of saving capital cost but also suffers from two disadvantages. First the deformation achieved is very gradual and most of it will occur at temperatures below which diffusion bonding will occur readily. Second, it requires considerable care and skill when tightening the jig bolts to ensure that the pressure is applied uniformly throughout the bonding area. An elegant way of avoiding these difficulties adopted by workers at the UKAEA Culham Laboratory when using Al interlayers to promote diffusion bonding was to use a jig that has a much wider top plate area than the component and to insert a flexible stainless steel pillow, between the top plate and the component assembly. When the bonding temperature was reached, the pillow was inflated with an inert gas to squash the interlayer and promote diffusion bonding. Similarly, Russian authors have reported the use of metal bellows inside the bonding chamber, but no details are known.

6.3 HEAT SOURCES

Brazing and diffusion bonding both require components to be heated to a temperature of 450–1200 °C or occasionally even higher and a wide range of heat sources can be used to achieve this. Many have been mentioned earlier, and each has its own advantages and disadvantages. Each one of the sources mentioned in this section can be used in the production of brazed joints but there are fewer options available when diffusion bonding or when using the hybrid processes. Further, some of the sources are incompatible with the use of specific joining or component materials, so care has to be exercised when choosing a source for a particular application. The characteristics of the principal types of sources can be described as follows.

6.3.1 Gas torches

Gas torches are the most commonly used heat source when brazing metal components. Because they are used to heat in the open workshop or on construction or repair sites, it is essential to use fluxes or self-fluxing braze alloys. Gas torches are the prime heat source for the manual brazing but also find many applications in mechanized brazing processes. When manual brazing it is conventional to use a hand-held torch so that the gas flame concentrates heating near the joint area. Flux is applied as a separate powder by the melting of a coating on a hand-held rod of braze alloy whose tip is moved into the heating zone to touch the joint area. Mechanized brazing often uses more than one gas torch and a rotating table or moving belt to move assemblies into and out of the heating zones, where again heating is usually concentrated near the joint area where both the flux and the brazing alloy have been positioned before the assembly enters a heating zone. When brazing either manually or mechanically it is good practice to position the flames so that the far end of the joint is hotter than its entrance so that the braze is drawn into the joint up a temperature gradient. In contrast to its extensive use when brazing, heating by gas torches is not used for diffusion bonding or the hybrid joining techniques although in principle gas flames could be used as indirect heat sources by being played on the external surfaces of bonding chambers.

The general construction and use of a gas torch were described earlier in the section of Chapter 2 concerned with fusion welding and it will be remembered that the torches generate heat by the reaction of gaseous hydrocarbons or H_2 with O_2. These reactions produce flames that contain several zones in which the proportions of the hydrocarbon or H_2 and O_2 differ, and the hottest part is in the core where none of the reactants is in excess so that reactions such as

$$C_2H_2 + O_2 \rightarrow 2CO + H_2 \qquad \{2.1\}$$

can occur and this particular reaction enables oxy-acetylene torches to generate flame temperatures of up to 3500 °C.

The most used gas mixture when fusion welding is of C_2H_2 and O_2 because it can generate the hottest flames, and these mixtures are also widely used when it is necessary to braze components at high temperatures, as when using Ni braze alloys. Cooler flames more suitable for joining with Ag-Cu and other braze alloys that melt at modest temperatures can be achieved by adjusting the chemistry of the oxy-acetylene torches, by detuning them, but care must be taken to avoid overheating of brazes or components by imprecisely controlled practice. However, lower temperatures can also be achieved by changing the torch nozzles and replacing C_2H_2 with H_2 or other hydrocarbons such as C_4H_{10} (butane), C_3H_8 (propane), or natural gas (a mixture of predominantly CH_4 and about 20% H_2) that are less efficient heat sources but are also cheaper and are less likely to result in overheating. Similarly, O_2 can be replaced by air so that a wide range of gas mixtures are available as heat sources and they can be ranked according to their heat output in the sequence

C_2H_2-O_2
H_2-O_2
C_3H_8-O_2
C_4H_{10}-O_2
Natural gas-O_2
C_2H_2-air
H_2-air
C_3H_8-air
C_4H_{10}-air
Natural gas-air

For example, the temperature reached by the reaction in a neutral butane flame

$$C_4H_{10} + 2O_2 \rightarrow 4CO + 5H_2 \qquad \{6.1\}$$

is less than 2000 °C, and even cooler flames can be generated by mixtures using air. This ranking can be related to the free energy of the reactions, thus the total chemical energy released by the reaction to form CO_2 and H_2O is 15.7 kJ for 1 litre of C_2H_2 and O_2 at NTP and is only 3.4 kJ for a natural gas–air mixture.

Despite the fact that neutral gas mixtures are the most efficient heat sources, it is a common practice to use gas mixtures that are slightly reducing to promote this fluxing action and to protect the component surfaces from additional oxidation while being heated to the brazing temperature. In the case of Al alloys that form particularly tenacious oxide films, this reducing effect can be maximized by the use of H_2-O_2 gas mixture. However, this mixture is not suitable for all components that form tenacious oxides because problems of H_2 embrittlement may arise if it is used with some steel and Ni alloy components.

Hand-held gas torches are the most used source of heat when brazing because they offer the advantages of being simple, low-capital-cost devices that can be used with a very wide range of metal component shapes and sizes. Their heat can be localized so complex structures can be brazed by keeping the torch flames away from sensitive structures such as Al fins on Cu heat exchangers. However, the successful application of hand-held torches requires considerable skill and the resultant slow production rate leads to a high production cost that includes a high labour wage element. These disadvantages can be overcome or at least mitigated by mechanizing the heating process, but only by sacrificing some of the simplicity and flexibility of hand-held torches and by incurring heavier capital costs. While the major use of gas torches is as a direct source of localized heating, they can also be used for indirect and general heating by being used to provide a flame layer above the components, like the grill in a domestic oven, or by being played on the outside of closed brazing chambers or immersion baths. Finally, safety is a major consideration when using gas torches because of the necessary inflammable and sometimes explosive nature of the reactants and because one of the products of the combustion process can be CO.

6.3.2 Induction heaters

The effectiveness of induction heating depends on the effectiveness of electromagnetic field produced by passing a rapidly alternating current through a coil in inducing eddy currents in nearby metallic objects and particularly in components situated within coils. There is no direct contact between the coil and the component and the invariably water cooled induction coil does not get hot itself, but the eddy currents causes heat to be generated in the encircled component and rapid temperature rises to occur. Induction heating can be generated within any electrically conductive material but is especially efficient when used with magnetic metals, and non-conductive ceramic components can be heated indirectly by placing metallic susceptors between them and the coil.

Induction heating uses current frequencies of 5 to 500 KHz and powers of about one to several thousand kilowatts. The higher frequencies, particularly above about 300 KHz, produce eddy currents confined to the near surface regions of components. This 'skin' heating and effect is particularly pronounced with steel components until a temperature of about 800 °C, the Curie point, is reached above which the steel loses its magnetic characteristics.

Conversely, the use of low frequencies such as 10 KHz produces deep penetration of eddy currents and consequent heating. Thus induction heating can be used for either localized or complete heating of components, and sharp definition of localized area, such as the joint, is possible by careful shaping of the coil, and an example in which a coil is placed inside an assembly is described as one of the case histories in Chapter 8.

The separation of the coil from the component surface can have a marked effect on heating efficiency. Narrow separations and close following of the profiles of component surfaces result in rapid heating and thermal efficiencies, but separations of less than about 2 mm can result in short circuiting between the coil and the component. Similarly, the use of wide coil-component separations of up to perhaps 2 cm results in slower heating rates but minimizes the chance of localized overheating which is a particularly important consideration when components have irregular shapes or comprise dissimilar materials with differing responses to induction heating.

Induction coils can be used to provide the heating during both diffusion bonding and brazing. The use of induction coils to diffusion bond requires no special restrictions of the process, but when brazing is restricted in that it is essential to place the braze alloy and any associated flux in position before generating heat, otherwise the hand-held brazing rod will be heated as well as the component. Brazing can be done in the open workshop when fluxes are used but the technique is equally suitable for use with chambers. The induction coil generally surrounds a simple non-metallic chamber such as a sealed tube of quartz or tempered glass that is evacuated or filled with a controlled atmosphere but coils can also be used within relatively large chambers.

The advantages of induction heating are its usefulness when either brazing or diffusion bonding, its low running costs, its potential for producing high heating rates and its capability of being used with both a wide range of metal components and also with ceramics when supplemented by susceptors. However, the capital costs associated with induction heating are relatively high and the design of efficient coils requires both skill and experience. Further, induction heating is a deceptively safe-looking process – care must be taken not to touch the coils when they are active if skin burns are to be avoided and operators must not wear personal jewellery such as rings.

6.3.3 Resistance heaters

These heaters use relatively small electrodes that are brought into contact with the components to be bonded by applying a modest pressure. A voltage is applied to cause a large current to pass between the electrodes, and through the components. Typical parameters are a contact area of 50 to 100 mm^2, a pressure of 5 to 20 MPa, a voltage of 1 to 8 volts and a current of 40 to 100 amperes. Heating is most intense where the electrical resistance is highest, and when high electrical conductivity electrodes of Mo, W, Cu-Cr and other wear resistant alloys with high electrical conductivities are used, the maximum heat is usually generated directly at the interfaces that have to be bonded. However, it is also possible to achieve indirect heating of the components and interfaces to temperatures of up to about 800 °C by using graphite electrodes which become hot when passing the currents and transfer heat into the more electrically conductive metal components.

This form of heating is very suitable for the mass production of simple brazed lap joints in the open workshop using sheets of self-fluxing Cu-P braze alloys placed within the joint gaps or else flux powders and sheets of normal braze alloys, and the process can be mechanized by programming the pressure and power cycles. Resistance heating can also be used to achieve diffusion bonding, but there is often some melting of the material at the initial points of contact so such applications are really hybrids of fusion welding or of brazing when an interlayer is used in diffusion bonding.

Resistance heaters have the advantage of giving very rapid and localized heating and having low running costs. Like induction heating, this source can yield very reproducible heat patterns within the components but its capital cost is much lower. On the other hand, the shapes and sizes of components that can be brazed are more limited and it is not suitable for the joining of high-temperature materials.

6.3.4 Heating elements

These devices raise the temperatures of components to those needed for brazing or diffusion bonding by radiation. They are generally rods, open ended cylinders or spirals of high temperature materials such as Mo, Ta, W, graphite, SiC or $MoSi_2$ but can sometimes be of less refractory materials such as Ni-Cr alloys. A current of typically 200–1000 amperes at a potential of a few volts is passed through the element to raise it to a high temperature and hence to radiate within the visible wavelength range of about 0.4 to 0.8 μm and with an energy proportional to T^4 in accord with the Stefan–Boltzmann relationship. The maximum heating is achieved by designing or placing the elements so that this radiation is concentrated in a hot zone that contains the component and is further enhanced by the use of radiation shields to minimize losses to and through the walls of the chamber. The component temperatures achievable can be very high, more than 2000 °C when using heating elements of graphite or W. If more modest heating will suffice, elements of Mo or Ta can be used to generate temperatures of 1400–2000 °C, and elements of Fe-Cr-Al or Fe-Cr-Co-Al can produce temperatures of up to about 1300 °C while Ni alloys such as those based on Ni-Cr compositions can be used for heating to temperatures of up to about 1100 °C.

The whole component and not just the joint region is heated when using these elements and they are as applicable when diffusion bonding as when brazing. Their use, however, is restricted by the low oxidation resistance of some element materials. Thus Mo, Ta and W can be used only in a vacuum or inert gas environment, while graphite cannot be used in oxidizing environments or even in a vacuum because of problems of degassing. In contrast, heating elements made from alloys that form protective oxide films, such as Al_2O_3 or Cr_2O_3 on Fecralloy, Kanthal or Ni-Cr alloys and can be used in air. Further, heating elements of SiC or $MoSi_2$, which is sometimes called Super Kanthal, can be used in air at temperatures of up to 1500 °C, but it is good practice not to let their temperature fall below about 1000 °C to prevent spalling of their protective SiO_2 layers.

Heating elements can provide moderately rapid heating of components to high temperatures and programming their power input results in very reproducible thermal cycles. The use of low voltages assists safety but care must be taken to avoid short circuits by tools accidentally left in contact with the element, and securing Mo, W, graphite, SiC and $MoSi_2$ elements to their terminals requires great care because of their brittleness or fragility.

6.3.5 Immersion heating

Immersion in a hot bath can be used to provide the heat needed for brazing. In its simplest form, the bath is filled with a passive fluid heat reservoir such as a molten neutral salt, but it is more common nowadays for the bath to play an active role in the brazing process by being filled with brazing flux or a liquid braze alloy covered by brazing flux. The salts, fluxes and braze alloys are themselves heated above their melting temperatures by external heating elements, by gas torches playing on the outside of the bath or by immersed or completely submerged electrodes.

Baths filled with chemically inactive salts can be used at temperatures of up to about 900 °C but are usually restricted to 750 °C or slightly less. Using such baths requires the braze alloy and any associated flux to be positioned at the joint entrances before immersion and their use has diminished in recent years. Salt baths can be used with a wide range of brazing alloys, and also to heat treat, but baths filled with flux are at present effectively restricted to the brazing of Al components at a temperature of about 600 °C. Flux baths can be very large, accommodating even substantial heat exchangers, but since they contain only flux, the components must be either clad with braze alloy or have alloy placed adjacent to the joint entrances before being immersed. Simplest in use when brazing are the baths of flux covered braze alloy. These baths are not limited to a particular type of braze alloy, but considerations of long-term flux stability generally limits their use temperature to a maximum of 900 °C and usually less than 750 °C. Further, these baths tend to be small and restricted to applications such as joining the ends of thin section structures or wire terminations.

Immersion can provide even, reproducible and precise heating of components with a wide range of sizes and shapes. However, inventory and running costs are usually high and components normally require preheating before they are immersed in the bath. Further, flux must be removed after brazing and complete coverage of component surfaces with a braze alloy when using a flux/metal is wasteful.

6.3.6 Other heat sources

Heating elements are not the only source to transfer energy by means of electromagnetic radiation, this is also the transfer process when heating by using electron beams, lasers, infrared lamps or microwave generators. These other heat sources are being introduced into brazing and diffusion bonding, with electron beams and lasers producing localized heating and infrared radiation and microwaves can cause generalized heating. Each type of heat source has its characteristics which bring both advantages and restrictions but it seems probable that their use will increase with time.

The localized character of the heating produced by electron beams and lasers makes these sources most suitable for use when only the joint area needs to be heated. Thus brazing can be accomplished by heating joint regions to cause melting of alloys that have been already placed in position, and a Russian design makes use of three beams spaced equally round the periphery of a circular chamber to provide overlapping and uniform heating for the diffusion bonding of simple butt joints. The electrons are emitted by a W or Ta filament heated to over 2500 °C in a vacuum of typically 10^{-4} to 10^{-5} mbar to generate a current of 50 to 100 milliamperes. The power of electron beam sources is usually 1 to 10 kW with the element being at a potential of 30 to 200 kV and the beam is collimated and focused by electromagnets to give a focal spot about 0.25 to 0.75 mm in diameter. The electron beam can be passed through a thin Al window and hence the components being brazed do not themselves have to be in an evacuated chamber, but this is general practice because the electrons lose energy and the beam becomes scattered by collision with gas molecules at residual gas pressures much above 10^{-3} mbar.

Lasers generate monochromatic coherent light beams that can be focused into small areas to provide intense localized heating. The beams can be generated continuously to achieve thermal penetration of their target or as pulses that are better absorbed by the surface layers of reflective metals. In practice, only two types of laser have yet to find significant application as brazing, or welding, heat sources: a Nd doped solid $Y_3Al_5O_{12}$, NdYAG or neodinium doped yttrium aluminium garnet, in which the electron orbits of the Nd are excited to emit photons with a wavelength of 1.06 μm, and CO_2 plus some H_2 and N_2 at 50 mbar in which stretching of the CO_2 molecules causes emission of photons with a wavelength of 10.6 μm. The NdYAG sources have a power of up to about 1 kW but those using CO_2 can generate ten times this output. The laser energy is absorbed by the surface of the component as the beam is traversed and hence the flow of heat into the joint will depend on the laser power and the processing speed. In practice the heat intake is usually quite low, but sufficient to cause melting of the braze alloy because it occurs over a very small area. Laser heating can be done in air and therefore can be used to produce joints requiring fluxing of the braze alloys.

Infrared heating uses high-intensity quartz lamps with powers of 5 kW or more to provide heat, using arrays of lamps and reflectors to generate light with a wavelength of about 1 μm that is focused on the components.

The heating effect depends on the thermal characteristics of the components and is inversely proportional to the square of the distance between the lamps and the components but temperatures of up to 1700 °C can be reached and hence this permits joining with virtually any brazing alloy.

Microwave heating is similar to that by infrared lamps in that again the energy generated by a magnetron or variable klystron oscillator is radiated and the whole component or merely the part incorporating the bond zone can be heated. However, the wavelengths are much longer, between 1 mm and 30 cm, and their absorption by the target causes resonance of specific chemical bonds. This resonance causes heat generation due to dielectric loses in ceramics, but in practice many of the sources being evaluated for the joining of ceramics use a frequency of 2450 MHz, and wavelength of 12.2 cm, which is the same as that used in domestic microwave ovens to cook by causing resonance of bonds in H_2O.

Microwave heating can be used to generate very high temperatures, for example Si_3N_4 has been heated to 1650 °C using 2450 MHz microwaves although there is evidence that high temperature sintering of this ceramic and also Al_2O_3 is better achieved using a frequency of 28 GHz. The energy available within the heating zone is maximized if reflection from a wall opposite to the source produces a resonant pattern of microwaves pattern. Thus furnaces employing sources capable of generating microwaves at varying frequencies can be fitted with a moveable opposite wall whose position can be adjusted to make the most effective use of the available radiation. It is not necessary for the microwave source to be within the bonding chamber but can be separated from it by a window of mica or other transparent material, so that the heating of Si_3N_4 was performed in a N_2 atmosphere to prevent dissociation of the ceramic. Since microwave heating due to resonance of chemical bonds is very specific, the gas around a component is not heated and hence the temperature of the surface of the component rises more slowly than does that of the bulk, the cake cooks from the inside. Further, microwave heating can achieve very fast heating rates but care has to be taken to avoid thermal overshoot due to the use of too large power sources or changes in dielectric characteristics with temperature.

The main advantages of these other heating sources is that they are clean, can be used with small components and by choosing one or another, localized or general heating can be achieved. The localized nature of the heating achieved by electron beam and laser sources makes them most suitable for brazing, while the general heating possible with infrared and microwave sources makes them as suitable for diffusion bonding and hybrid joining processes as for brazing. The disadvantages of these techniques include high capital costs, and especially so for electron beam heating, the small sizes of the heating zones produced by laser and electron beam heating and the extra special care that must be taken to protect operators from direct exposure to the sources.

6.4 FURNACE CHAMBERS

Most brazing and diffusion bonding processes have to be performed in a furnace chamber because of the nature of the heat source or the materials that are being used. In practice, there are two main types of furnace chamber, each with many individual variants that have particular characteristics, advantages and disadvantages. One major grouping is that of continuous furnaces, the use of which involves placing the component on some conveying system which carries it through various locations in the furnace so that it experiences a complete bonding cycle before being carried out of the furnace and unloaded.

The second grouping encompasses batch furnaces in which the component or batch of components to be bonded are loaded into and are often sealed within a furnace chamber where they stay immobile while the power input to the furnace is varied to create a bonding cycle. Most existing and new equipments will be some sort of batch furnace chamber because this is the most versatile design, permitting fitting of most kinds of heat source and the use of both vacuum and controlled atmospheres as the bonding environment.

6.4.1 Continuous furnaces

Continuous furnaces are essentially tunnels containing various regions through which components travel to experience a brazing cycle that lasts from between a few minutes to a few hours, and they are widely used for the mass production brazing of small and usually self-jigging components such as Cu brazed low C steel automobile parts. The furnaces are relatively large and high capital cost items that can take a number of forms. Their hot zones can be produced usually by electrical heating elements, or by gas, or oil, burners but occasionally by induction coils or infrared heaters and the environments experienced by the components can be air or more usually neutral or reducing atmospheres. One example of such a furnace is shown schematically in Figure 6.3, and this particular furnace is of a common sort in that it is designed for controlled atmosphere brazing and uses heating elements to produce the temperature cycle needed for bonding. Moving the components through the furnace, which is usually achieved by placing them in a basket or on a pallet resting on a wire mesh or chain belt or else on rollers that can be driven, causes them to experience a processing cycle with three distinct stages. The initial stage is the introduction of the components into a controlled atmosphere in which they are heated to just below the temperature required to melt the braze alloy. The intermediate stage heats the components to a higher temperature so that the braze melts and flows to form a joint. The final stage is simply the forced or free cooling of the brazed components in a controlled atmosphere.

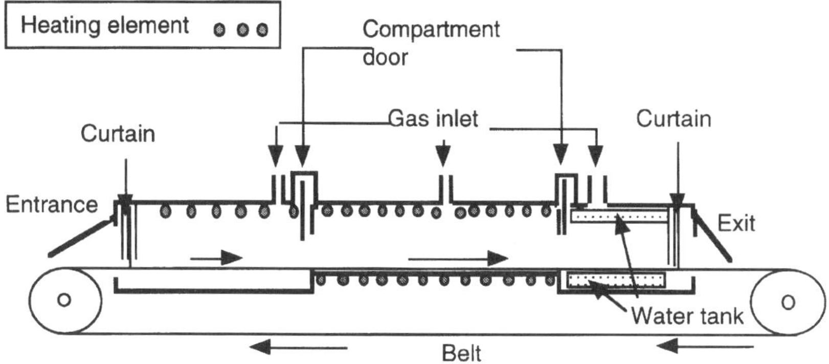

Figure 6.3 Schematic arrangement of a continuous conveyor belt furnace that uses a controlled atmosphere

Clearly using a continuous furnace requires the braze alloy and any flux to be placed in position before the component is loaded. Other restrictions are that particular care must be taken to prevent workshop air from passing through the easy entrances into the furnaces. The usual method of preventing this ingress is to use the controlled atmosphere at a slight positive pressure. However, when using atmospheres that contain or produce H_2 or CO it is necessary also to ensure that they are safely burnt, converted to H_2O or CO_2 before passing out of the furnace into the workplace. Thus it is a common practice to use walls of flame in the entrance and exit regions to ensure burning of controlled atmosphere gases and impede entry of workshop air.

6.4.2 Batch furnaces

The batch furnaces that are used to braze or diffusion bond can be exceedingly varied in form and are used to join assemblies that can be small and simple or large and complex. Their output is generally slower than that of continuous furnaces, but they can be used to produce a multitude of small and simple self-jigged assemblies or more complex and often larger structures that require external jig. As with continuous furnaces, braze alloys and fluxes, and also diffusion bonding interlayers, must be placed in position before loading of the furnace chamber commences.

Batch furnaces can be very flexible in design and effective when used in a wide range of applications. Thus, the furnace chambers can be open or sealed, with those of the sealed type being loaded through a front opening door or top lid or by raising the chamber to load from the bottom. Any of the heating sources mentioned in the previous section may be used except gas torches, and even this qualification needs elaboration because gas and oil burners can be used. The internal environments within the chamber may be air, a controlled gaseous atmosphere or vacuum.

The simplest form of batch furnace chamber is a gas or electrically heated open ended muffle furnace in which is placed the assembly to be bonded, in practice brazed at a modest temperature. Loading can be when the furnace is cold or at the brazing temperature so heating rates can be slow or fast and obviously the braze alloy and any associated flux must be positioned before the assembly enters the furnace. The presence of air within the chamber makes it particularly necessary for the flux to be able to spread well on component surfaces so that they are protected from oxidation when heated to the brazing temperature.

A design development of this furnace is to use an end-sealed quartz tube as a chamber to encase a small assembly being brazed using induction heating. Joining is achieved after the tube has been evacuated or evacuated and back filled with a controlled atmosphere. Another similarly simple concept, albeit often physically substantial in realization, is the retort furnace which has a chamber shaped like a bell, or an inverted bucket, that is raised to allow loading from below. After being loaded the chamber is lowered and sealed to a base plate by an impermeable gasket or O-ring before being evacuated or, again, evacuated and backfilled with a controlled atmosphere. Heat is supplied to this type of furnace chamber gas by external heating elements or gas or oil burners, and the maximum-use temperature of about 1000 °C for this type of furnace chamber is dictated by the ability of the chamber walls to withstand hot corrosion. Somewhat higher temperatures, perhaps up to 1150 °C, can be achieved by using a double-walled retort in which heating elements are located between the walls in a space that is usually evacuated. Additionally, refractory bricks or fibrous insulation may be placed between the heating elements and the outer wall to decrease wasteful heat losses.

Most double-walled batch furnace chambers provide a front opening door or top lid and are so constructed to provide a cold wall capability by filling the interwall gap with flowing water which may in fact be warmed to prevent condensation when the furnace is open and exposed to normal, undried, workshop air. The chambers can be fitted with induction coils or more usually with internal heating elements that are surrounded by packs of radiation shields whose function is to reflect heat into the central hot zone which would otherwise escape to the walls of the furnace chamber. Other packs of radiation protect the top and bottom surfaces of the chamber so that even a modest flow of water is sufficient to keep walls cold and rubber O-rings can be used to make vacuum seals. In practice, the chambers are evacuated to pressures of 10^{-3} to 10^{-6} mbar, and may be backfilled with high purity Ar or He. These environments permit the use of refractory metals such as W and Ta to construct both the heating elements and the radiation shields so that temperatures in excess of 2000 °C can be reached. The hot zones of furnace chambers intended for very high temperature use are often as small as a few millilitres, but chambers operating at modest temperatures can be large so that substantial aero-engine components are brazed at 950–1150 °C and Al heat exchangers with volumes of up to ten cubic metres are brazed at about 600 °C.

There are many variations of cold wall batch furnaces. Thus the double-wall construction can be replaced by a single wall to the outside of which is brazed a closely packed Cu coil through which passes the cooling water. Again, the furnace can be fitted with sealed ancillary loading and unloading chambers that can be evacuated or evacuated and backfilled with an inert gas. Passage between the ancillaries and the main joining chamber is effected by the intermittent opening of doors and this allows a higher throughput to be achieved so that the main chamber is not directly exposed to workshop air and hence does not require prolonged re-evacuation. Similarly, the heat source can be an electron beam rather than a heating element providing the vacuum quality is good enough not to cause the beam to lose energy and become scattered by interaction with residual gas molecules. The mean free path between any residual molecules should be greater than the distance between the source of the electron beam and its target, and taking this as 10 to 100 cm implies a necessary evacuation to less than 10^{-3} to 10^{-4} mbar.

The variation of most significance for the readers of this book is probably the use of evacuable and cooled batch furnace chambers to achieve diffusion bonding. Most diffusion bonding chamber designs incorporate the water cooling and radiation shields characteristic of evacuable batch furnace chambers used for brazing but differ because of the incorporation of a hydraulically operated ram that passes through the top or bottom plate of the chamber. Vacuum tightness even when moving the ram is maintained by using preferably at least two rubber O-rings to seal the metal shaft of the ram where it passes into the chamber. The rams themselves often incorporate extensions of a high-temperature Ni super alloy or a ceramic such as graphite or Si_3N_4 that can endure the bonding pressures and temperatures without being deformed. A large laboratory version of such a chamber is shown in Figure 6.4 along with its associated hydraulic pressure, temperature and vacuum control and monitoring equipment This particular equipment was manufactured by Thermal Technologies Inc. of Santa Rosa, California and can operate using a hot zone with a diameter of 80 mm and a height of 150 mm at a bonding temperature of up to 1800 °C when the chamber is evacuated or backfilled with an inert gas. The maximum load that can be imposed is 200 kN and this is used to cause compressive stresses of 100 MPa or more on the samples that are being bonded between graphite ram extensions. Pro-

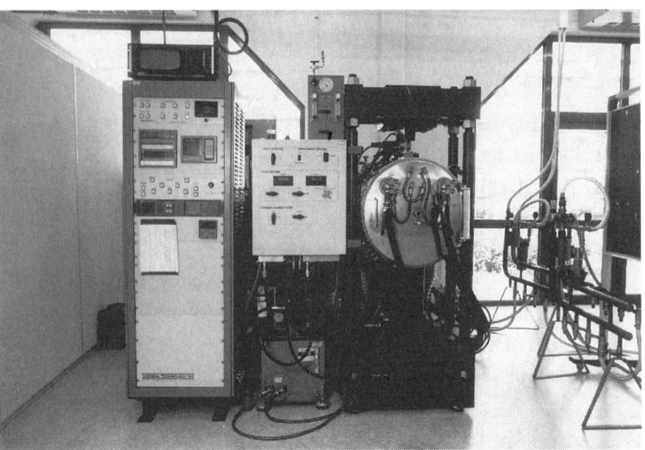

Figure 6.4 Large laboratory diffusion bonding furnace, (top), overall view showing the control equipment and (bottom) view with an open chamber door. (Courtesy of Dr S. D. Peteves)

duction chambers that operate at modest temperatures can be much larger, so for example hot zones of 600 mm diameter and 1000 mm high have been used for diffusion bonding with Al interlayers.

In some designs of diffusion bonding furnace chambers, the ram is fully within the chamber and is moved by external hydraulic systems that compress a bellows, but a much more important variation is the use of hot isostatic pressing, HIP, chambers. These are thick walled sealed devices that are filled with Ar, N_2 or a mixture of gases pressurized by external hydraulic equipment and heated by internal refractory metal or graphite elements. Laboratory HIP chambers can be used to achieve pressures of up to 200 MPa and temperatures of about 2000 °C in Ar and somewhat less in N_2 within a working zone diameter of perhaps 70 mm and height of about 100 mm. Production facilities generally achieve more modest pressures and temperatures but they can be very large, typically up to 100 MPa and 1400 °C but with a working zone of up to 1000 mm in diameter and 2000 mm in height. Not surprisingly, the cycle time of large production units is long, perhaps a day.

Because the pressure in such devices is isostatic, the assembly joint gap must be protected from ingress of the pressurizing gas if subsequent complete closure and bonding is to be achieved. In practice the whole assembly is encapsulated to prevent ingress, usually in a cheap disposable material like mild steel or glass but occasionally it is possible to design so that the outer part of the assembly itself is a capsule. Clearly, the size of production HIP facilities can allow the diffusion bonding of very large metal–metal or metal–ceramic assemblies, but another exciting possibility is their use for simultaneous production of a large number of smaller bonded components. Diffusion bonding can often be combined with heat treatment of the components, but the necessary use of encapsulation when bonding by the HIP process also allows diffusion bonding to be combined with powder consolidation to form assembly members *in situ* as well as closure of casting voids.

Finally, note should be taken also of microwave furnace chambers that can be used to diffusion bond ceramics without the need for metal interlayers. The chamber itself is a relatively simple box with a polymer window through which the microwaves pass from an external generator. Within the chamber, adjustable partition walls are used to collimate and control the proportion of the microwaves impinging on the components to be diffusion bonded by movement of externally operated rams. Outside the chamber is not only the microwave generator but control equipment to monitor the temperature of the components and prevent thermal overshoot.

6.5 BONDING ENVIRONMENTS

The selection and control of the gaseous environment is a crucial aspect for successful and reproducible brazing and diffusion bonding. The environment affects the surface chemistry of both the components to be joined and the joining materials themselves and hence exerts a fundamental influence on the process since joining is the conversion of surfaces to interfaces.

The gaseous environment used to effect joining is seldom air because it generally causes the formation of oxide films that act as barriers to the direct contact between the component materials that is needed to create an interface. There are, however, some exceptions such as when joining with self-fluxing Cu-P braze alloys which depend on oxidation to form the phosphate flux or when diffusion bonding wrought iron by hammering in a smithy. The use of hand-held gas torches to braze in open workshops or construction and repair sites is another apparent and much more widely occurring exception, but in fact the assembly of components to be joined is not accomplished in air but rather in an artificial micro-environment. Thus braze melting and flow into the joint is achieved using heat from a chemically reducing inner flame while preheating and often fluxing of the components ahead of the braze flow uses the cooler outer flame.

When using a hand-held gas torch, the operator has an immediate guide as to whether the joining process conditions, including the gaseous environment, are adequate by observing the melting and flow of the braze alloy. However, this direct observation of the bonding process is not often possible when furnace brazing and is impossible when diffusion bonding. It is, therefore, for these processes that especially careful thought must be given when specifying the environment and means to achieve and control it. The three main types of gaseous environment used to furnace braze or diffusion bond are vacuum, inert gases or gas mixtures such as Ar or He and reducing gases such as H_2 or CO.

6.5.1 Vacuum

Perhaps the biggest single barrier to the successful achievement of brazed or diffusion bonded joints is the presence of oxide films on the surfaces of the braze alloys, diffusion bonding interlayers and metal components. These oxides form initially by reacting with the O_2 present in air, but they can thicken substantially when heated if O_2 is available. Thus the most significant fact about the use of an evacuated furnace chamber is that it has a low partial pressure of O_2.

Evacuation of most industrial vacuum furnace chambers is effected by the use of an oil filled diffusion pump backed by a rotary pump, as sketched in Figure 6.5. The rotary pump is a mechanical device that sweeps away gas using a paddle wheel effect and oil seals to prevent backflow while the diffusion pump traps gas molecules within a shower of condensing droplets of oil. The essential design features of the system sketched in the figure are the use of a rotary pump to achieve the initial evacuation of the furnace chamber to a pressure of typically 10^{-1} to 10^{-2} mbar and to then back a diffusion pump that can be brought into operation to decrease the furnace chamber pressure further from 10^{-1} to 10^{-4} to 10^{-6} mbar. The location of the valves to shut off any connection between the diffusion pump and the chamber during the initial pump down is important if damage is to be avoided by cracking of the pump oil. The pipes between the various components of the system and the chamber

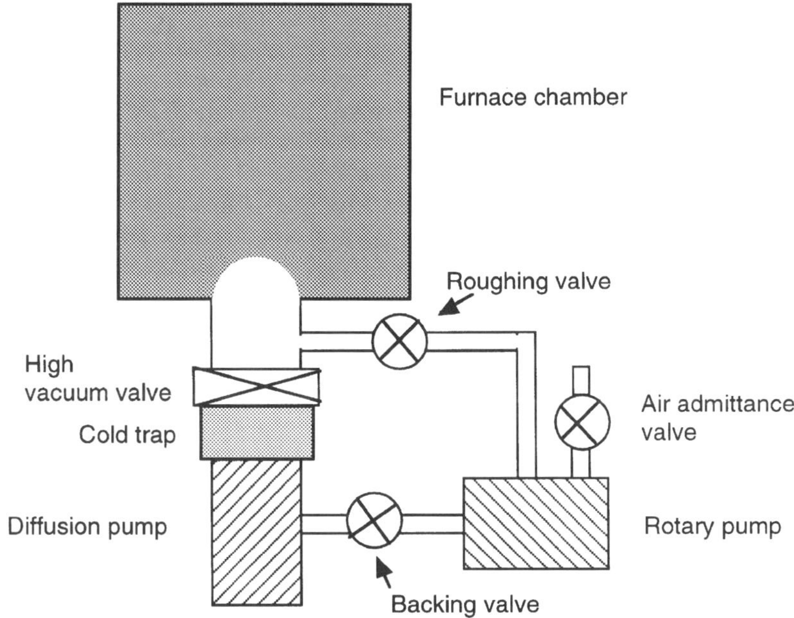

Figure 6.5 Sketch layout of a typical vacuum furnace pumping system

should be as wide and as short as possible. Preferably, they should be as wide as the inlets to the diffusion and rotary pumps so that viscous drag has an effect on pumping speeds and pressure gradients along the pipes are minimized. The locations of vacuum gauges are also important, and it should be noted that at least one gauge should be immediately adjacent to or actually fitted to the furnace chamber. Fitting a gauge away from the chamber may be more convenient but it can result in a falsely low pressure reading.

The efficiency of adsorption of gases on cool surfaces and their removal by a pumping system is not the same for all gases, and hence the composition of the residual gases is not that of air. Thus mass spectroscopy of the residual gases present in the furnace shown in Figure 6.4 after evacuation to a pressure of 10^{-6} mbar revealed a composition that included 2.5×10^{-7} mbar of CO, 2.5×10^{-8} mbar of CO_2, 9×10^{-8} mbar of H_2O, 8×10^{-8} mbar of N_2 and 1×10^{-9} mbar of O_2 plus small traces of other unidentified gases, possibly including hydrocarbons. In this particular case the very low level of O_2 in the residual gas and the dominance of CO is associated with the presence of graphite furniture in the furnace, while an O_2 level of about 1% of the total residual gas content is typical of graphite free all metal furnaces. While these decreases in O_2 levels benefit joining processes, the presence of hydrocarbons in the residual gases can contaminate component surfaces to prevent or at least degrade wetting and bonding behaviour. These gases come from the rotary and particularly from the diffusion pump, but their ingress into the chamber can be diminished considerably by fitting a cold trap, often filled with liquid N_2 to give a temperature of $-200\,^{\circ}C$, between the diffusion pump and the furnace chamber as shown in Figure 6.5. If a very clean system is required, an oil-free turbo-molecular pump can be used to produce a vacuum of typically 10^{-6} mbar at bonding temperatures or special systems can be designed that incorporate sputter ion pumps to evacuate the chamber to 10^{-10} mbar or less. However, such equipment is at present more characteristic of the research laboratory than the shop floor.

Pressures of 10^{-4} to 10^{-6} mbar and partial pressures of perhaps 10^{-6} to 10^{-8} mbar of O_2 are too high to cause dissociation of the oxide films on the surfaces of components except of Ag or Cu and the main benefit of such low pressures is that they decrease the arrival rate of O_2 molecules at component surfaces. As shown in Figure 6.6 the predicted arrival rate and hence the time taken to form a monolayer is directly proportional to the gas pressure. It can be seen that a monolayer of O_2 should take several minutes to form at a bonding temperature of $1000\,^{\circ}C$ if the vacuum quality is good, and such low rates allow the oxide films on Ti to be dissolved faster than they are reformed. Relatively low O_2 arrival and oxide reformation rates also assist the removal of oxide films by other processes, such

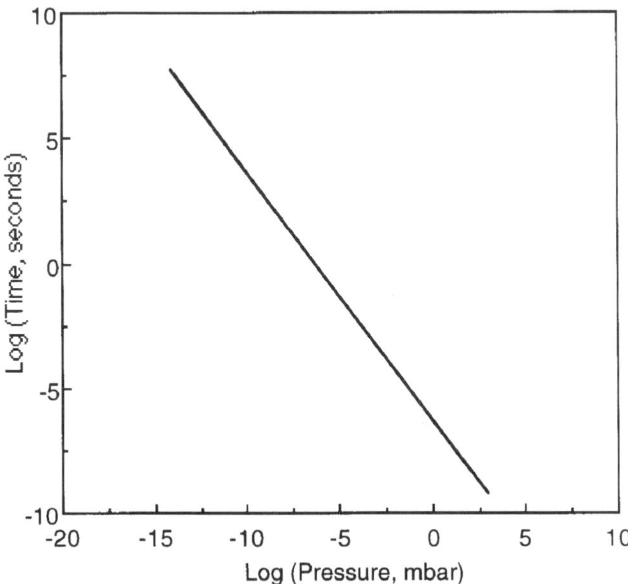

Figure 6.6 The predicted effect of pressure on the time to form a monolayer of adsorbed gases at $1000\,^{\circ}C$

as chemical interaction with their substrates, undermining by the braze alloy and vapourization of substrate constituents. In principle, removal by fluxing will also be assisted but in practice fluxes are very rarely used when vacuum brazing.

Vacuum chambers have to be pumped continuously while they are in use because they are never quite leak tight and because constructional and other materials can evolve gas when exposed to vacuum at a high temperature. The seals on vacuum chambers are normally of rubber because it can conform easily to seal quite large gaps. However, gas can still permeate through the rubber and diffuse along the rubber/door interfaces. Higher quality but still not perfect sealing can be achieved using metal gaskets, usually of Cu, but this practice is expensive because the gaskets cannot be reused with complete confidence and even one-time use requires complex modification of the surfaces to be sealed so that they cut into the faces of the gaskets. Since total sealing cannot be achieved, it is important to know the actual leak rate so that its possible impact on the joining process can be assessed. A first indication of whether the chamber is sufficiently leak-tight can be gained by watching the pressure change after closing the valve between an evacuated chamber and the diffusion pump. More specifically, the location of leaks can be identified by attaching a mass spectrometer gas analyser to the chamber and squirting He at various areas on the outside of an evacuated chamber. If the leak rate is unacceptably high, seals should be examined and replaced. It is not good practice to achieve a 'better' vacuum with a leaky system by using a more powerful pump because this will merely suck more air into the chamber.

The metals used to construct the vacuum furnace chamber and its furniture are usually covered not only by an oxide film but by adsorbed gases, typically several monolayers of weakly bonded hydrocarbons or H_2O or a strongly bonded monolayer of O_2, H_2, N_2 or Ar. When exposed to a vacuum, these gases will desorb, but the rate at which this occurs depends on the metal, its surface treatment and the length of time for which the system has been pumped. Thus the room temperature desorption rates from the surfaces of W, Cu and Fe that had been pickled and degreased with acetone have been found to be 0.045, 0.55 and 0.95 $ml.m^{-2}.h^{-1}$ respectively after pumping for two hours. Cu that had not been pickled evolved gas at four times as fast, and Cu that had been neither pickled or degreased evolved gas more than ten times as fast. If pumping is continued, the rate of desorption gradually decreases, but the evolution rate can be enhanced dramatically by heating.

While much of the gas desorbed will have little effect on the bonding process and can be removed quite easily by the pumping system, its evolution will slow the evacuation time and it is good practice to scrupulously clean and degrease a new chamber or furniture and to bake the equipment, perhaps by doing a dummy run at a modest temperature without water cooling, to accelerate the desorption of gas. Thereafter, the chamber should be kept evacuated always except when being loaded or unloaded and it is good practice to backfill the chamber with Ar when about to unload rather than allowing workshop air to be adsorbed on the chamber surfaces. This air may well be damp and hence allow many monolayers of H_2O to be adsorbed and at least transiently increase the oxidizing potential of the residual gas during evacuation of the chamber.

Procedures such as these should ensure that vacua of 10^{-4} to 10^{-6} mbar are achieved within the bonding chambers. These are the levels generally used when brazing or diffusion bonding in a vacuum furnace, but in fact such high-quality environments are not always needed to achieve bonding. Thus the vacuum conditions that have been found to be necessary in practice to achieve effective brazing can be more than 1 mbar, as summarized in Table 6.2. In fact, brazing can be achieved with some of these systems in even poorer vacua than is suggested in the table but other aspects of the end product may be unacceptable. Thus Cu can be used to braze stainless steel in a vacuum of 10^{-2} mbar but the component surface assumes an olive-green hue that is usually unacceptable to the end user. Similarly, the brazing of Ti by Cu in a vacuum of only 10^{-1} mbar degrades the end-product quality by allowing excessive diffusion of O_2 into the Ti, and the brazing of Al components with Al-Si-Mg alloys in poorer vacua than suggested in the table results in excessive formation of MgO that impairs the appearance of the component surface and ultimately can damage the furnace. The effect of vacuum on the quality of diffusion-bonded components is similar. Thus the most perfect diffusion-bonded interfaces between dissimilar materials, without any voids and with very close matching of the atomic lattices, have been produced between Nb and

Table 6.2 Examples of the vacua required for successful brazing

Vacuum	*Braze alloy type*	*Component material*
$>$ 1 mbar	Cu-P, Ag	Cu
$\sim$ 1 mbar	Cu, Ag	Cu, low C steel
$< 10^{-3}$ mbar	Cu, Ag	Cu, low C steel, low alloy steel
$< 10^{-3}$ mbar	Ni, Au, Al-Si, Ag, Cu	High alloy steel, Ni alloys, Al Ti, Zr, W
$< 10^{-4}$ mbar	Ag-Cu-Ti	Ceramics

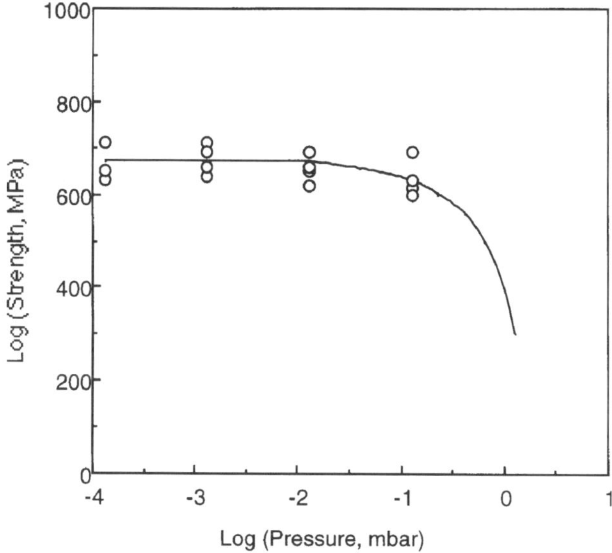

Figure 6.7 The effect of vacuum quality on the strength of diffusion bonded Fe-0.45 C. After N. F. Kazakov (1981). The bonding temperature was 1000 °C, the pressure was 20 MPa and the bonding time was 5 minutes

Al_2O_3 produced using an ultra high vacuum of 10^{-10} mbar, but Russian work has shown that the strengths of diffusion bonded C steel joints are not degraded until the vacuum is poorer than about 10^{-1} mbar, as shown in Figure 6.7. In practice, good technical vacua of 10^{-4} to 10^{-6} mbar are commonly employed when using evacuated chambers to avoid discoloration of components and the danger of residual gas being trapped as bubbles at joint interfaces.

6.5.2 Reducing environments

Reducing atmospheres are used to promote brazing by removing the surface films of oxides that prevent metal–metal contact. In principle, such atmospheres could also be used to promote diffusion bonding using metal interlayers, but this approach has not yet been adopted in a substantial manner.

The classic reducing atmosphere is H_2 which, for example, reacts to remove the oxide from Fe by forming H_2O

$$FeO + H_2 \rightarrow Fe + H_2O \qquad \{6{\cdot}3\}$$

and this reaction will continue until a chemical equilibrium is reached, until the ratio $(a\text{Fe}.pH_2O)/(a\text{FeO}.pH_2)$ reaches a value determined by the

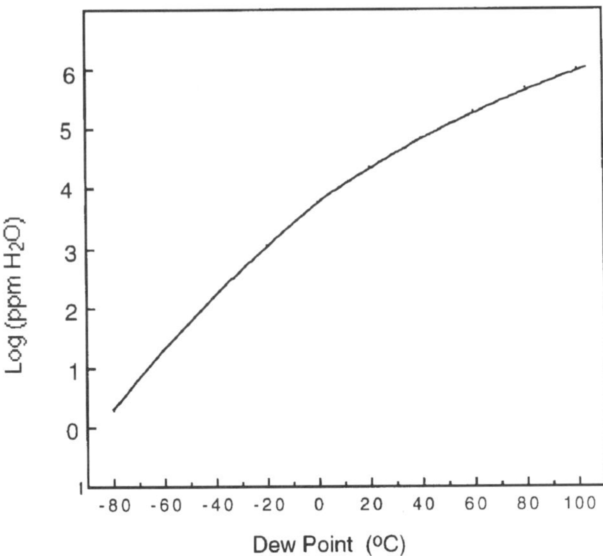

Figure 6.8 The effect of temperature on the H_2O concentration of saturated gas mixture

free energies of formation of H_2O and FeO which are such that the reaction will continue at room temperature, 27 °C, until the ratio reaches a value of 0.0064, and since the solids involved in this reaction have unit activities, this will also be that of the ratio of the gas pressures and concentrations. However, a gas atmosphere at a pressure of 1 bar that contains 0.64% of H_2O will start to form liquid droplets, a dew, when its temperature is decreased to 10 °C as shown in Figure 6.8 and the measurement of such a Dew Point or condensation temperature is the means of assessing the H_2O/H_2 ratio and hence the reducing capability of hydrogenous gas mixtures. In this context, it should be noted that even pure H_2 has a Dew Point, of about −60 °C which is equivalent to an H_2O concentration of 12 parts per million, because it will react with any oxide or adsorbed O_2 on the metal surfaces of gas bottles and pipe work.

The measurement of a Dew Point is used widely in industry to monitor whether hydrogenous gas compositions are sufficiently reducing to remove surface oxides, and the Dew Points of 1 bar of gas that are in equilibrium with a number of oxides at different temperatures are shown in Figure 6.9. Reduction of the surface oxide will occur if the Dew Point of a gaseous atmosphere is lower than the values plotted in the figure and it can be seen that these Dew Points vary considerably for metals of interest as constituents of components or joining materials.

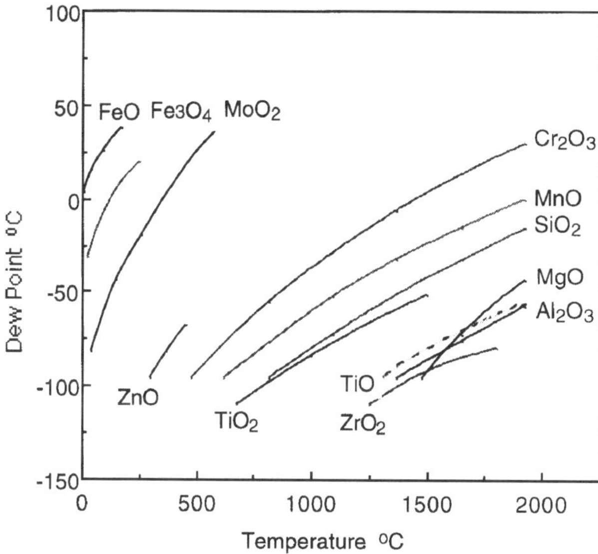

Figure 6.9 The Dew Point temperatures at which H_2O/H_2 mixtures are in equilibrium with various oxides at various bonding temperatures

Since the data shown in Figure 6.9 result from thermodynamic calculations, they provide no indication of the rate at which a surface oxide will be removed, but the rate that is achieved in practice in a closed chamber can be assessed by continuously monitoring the Dew Point. Similarly, the rate of oxide removal in a chamber using a continuous gas flow can be monitored by measuring the Dew Point at the gas outlet as well as inlet. If the total gas pressure is less than 1 bar in a closed chamber if, for example a N_2-H_2 mixture is used rather than pure H_2, then reduction will still occur at the same H_2O/H_2 ratio but the Dew Point temperature will be lower because the partial pressure of H_2O is less. Similarly, if the oxide being reduced is not that of a pure metal M but of an alloy constituent, the ratio $(a\underline{M}.pH_2O)/(aMO.pH_2)$ is unaffected but $a\underline{M}$ will be less than unity and hence the H_2O/H_2 ratio can be larger and conversion of MO to M can be achieved at a higher Dew Point. However, the introduction of small concentrations of very oxygen active solutes can have a marked effect on the Dew Point needed to ensure complete oxide removal from the surfaces of metal components. Thus the Dew Point needed to ensure oxide removal from the surface of a stainless steel at 1150 °C is decreased from about −25 °C to −45 °C by the introduction of 0.5 Al.

It is usually desirable to achieve the lowest practical and economic H_2O content of H_2 atmospheres, but there are exceptions such as the need to use damp H_2 or N_2–H_2 mixtures as environments for the moly-manganese process of ceramic metallization. The optimum degree of dampness for this process varies from application to application because H_2O may be picked up by desorption within the furnace and could be generated by other deliberate or adventitious reducing reactions that are not part of the metallization process, but it is often equivalent to a Dew Point for the inlet gas of close to room temperature. It is common practice to dampen the H_2 by bubbling at least part of the inlet gas stream through H_2O, and good practice to monitor the Dew Point of the outlet gases and use that information to control the proportion of inlet gas that passes through the bubbling bath.

As already indicated, pure H_2 is not the only reducing atmosphere. It is quite common practice to use it as a part of a N_2–H_2 mixture produced either by cryogenically pure gases to produce a mixture containing up to 30% of H_2 with Dew Point of nearly −70 °C or by dissociating NH_3 to produce a mixture with a H_2 concentration of 75% and Dew Point of −55 °C. Apart from these hydrogenous mixtures, even more extensive use is made of mixtures using the reducing capability of CO produced by the partial combustion of hydrocarbons such as CH_4, C_3H_8 and C_4H_{10} in generators before they are admitted to the furnace chamber. There are two main categories of such atmospheres, exothermic mixtures in which the air/gas ratios are 5 to 10 and considerable heat is released and endothermic mixtures with lower air/gas ratios of typically 2. Using a low air/gas ratio results in the production of principally CO and the release of H_2, but high ratios also cause CO_2 and H_2O to be produced by reactions such as

$$2CH_4 + O_2 + 4N_2 \rightarrow 2CO + 4H_2 + 4N_2 \quad \{6.3\}$$

$$2CH_4 + 3O_2 + 12N_2 \rightarrow CO + CO_2 + 2H_2 + 2H_2O + 12N_2 \quad \{6.4\}$$

Some of the gas compositions that can be produced are summarized in Table 6.3 along with their Dew Points. Which composition should be chosen for application depends in the first instance on the chemical

Table 6.3 Some typical burnt gas furnace atmospheres

	Composition				
Gas type	CO	CO_2	H_2	N_2	*Dew point* (°C)
Rich exothermic	10	7	13	70	20
Lean exothermic	3	13	3	81	20
Rich endothermic	23	0	40	37	−40
Lean endothermic	20	0	35	45	−40

stability of the oxides that must be reduced by H_2 and CO and hence the extent to which reduction to form H_2O occurs and reactions such as

$$FeO + CO \rightarrow Fe + CO_2 \qquad \{6.5\}$$

proceed. Thus the reducing capability of these gas mixtures can be measured not only by their Dew Point but also by their CO/CO_2 ratio. For example CO/CO_2 mixtures free of H_2 will reduce the oxides on Cu and Ni at 1000 K, 727 °C, even if their CO concentration is only 0.1%, while the oxide on Fe will be reduced only if the CO concentration is at least 55% and those on Cr and Zn require CO concentrations of more than 99%. Examination of the data summarized in Table 6.3 shows that endothermic mixtures will be more effective oxide removal agents than exothermic mixtures because of their lower Dew Points and higher CO concentrations. However, such mixtures are inherently expensive because of the greater hydrocarbon consumption during their production.

While the use of reducing gases is an effective means of removing the surface films of oxides that impeded joining processes, some of the other chemical characteristics of hydrogenous and partially burnt hydrocarbon atmospheres can cause problems. Thus contact with H_2 can cause embrittlement, particularly of high-yield strength steels and Ni alloys, Fe-Si and Co alloys, while mixtures containing N_2 can produce scaling of brazes or components containing Ti, Cr and other strong nitride formers. Similarly, the varying C potential of gas mixtures derived from hydrocarbons can result in decarburization of high C steels or conversely carburization of low C steels.

Even more important than these technical effects on the joining process or the components are considerations of safety. Great care must be taken when using either hydrogenous or hydrocarbon derived mixtures because their mixing with air can result in ignition of H_2 at temperatures of 575 °C and of CO at 650 °C and above. Mixtures of air containing 4 to 75% of H_2 are explosive while CO can be toxic at high concentrations. The safe limit for CO in air is only 0.01% and special care has to be taken not to exceed this limit because its presence is not detectable by human senses. Thus, furnace chambers must be thoroughly purged with an inert gas or evacuated before hydrogenous or hydrocarbon derived mixtures are introduced and waste gases should be burnt completely and continuously or vented to air outside the workshop.

6.5.3 Inert environments

Non-reactive environments can be produced also by using inert gases such as Ar or He at a pressure of generally about 1 bar. The prime attraction of these gases to the user is of course their virtually complete chemical inertness, but there are physical differences between the gases that could affect their efficiency as joining environments. Thus the specific heat capacity and thermal conductivity of He are much larger than those of Ar but which gas is used in a particular application is largely dependent on availability and cost factors, and He is used more in North America than Europe.

The main advantage of such gases is that they are indeed chemically inert. However, no real material can ever be completely pure and hence many uses of Ar and other inert gases require that they are passed through some purification process being introduced into the furnace chamber. Pressurized gas bottles containing Ar with a purity of up to 99.996% are readily available commercially and if H_2O is the only impurity then the gas has a Dew Point of about −50 °C. In practice, the impurity level of the gas that enters the furnace chamber may have been increased due to desorption of O_2 and H_2O from the walls of the pipework between the bottle and the chamber inlet. Fortunately, passing the gas through commercial purification trains immediately before it enters the furnace chamber can decrease the levels of both H_2O and O_2 to about 1 ppm or less, a Dew Point of −80 °C, while N_2 levels can be decreased by using simple workshop facilities.

These purification processes make use of both physical and chemical effects. Thus H_2O can be removed by passing the gas through a 'molecular sieve' of a zeolite such as the $Na_{12}(AlO_2)_{12}(SiO_2)_{12}$ structure which consists of open cages and tunnels onto whose surfaces gases can be adsorbed and cause hydration by the attachment of up to 26.7 molecules of H_2O. Similarly O_2 can be removed by using a molecular sieve whose surfaces have been coated with a very thin layer of Fe that reacts to form FeO. Both types of molecular sieve have long lifetimes because they can be reactivated by heating in a dry environment to drive off H_2O or in H_2 to change FeO back to Fe. Another commonly used approach is to equilibrate the gas with a drying agent such as those listed in Table 6.4. Other less commonly used techniques for decreasing H_2O levels include passing the gas through concentrated H_2SO_4, to decrease the level to about 3×10^{-3} ppm or cold trapping with 'dry ice', solid CO_2 to decrease the levels to about 1 ppm. A common laboratory practice is to remove O_2,

Table 6.4 The effects of drying agents on gas dampness

Agent	H_2O *content, ppm*	*Dew point* (°C)
$CaCl_2$	1650	−17
Silica gel	66	−50
$MgClO_4$	5	−70
BaO	1	−82
P_2O_5	0.1	−97

H_2O and also N_2 by passing the gas through a heated tube packed with Ti or Zr swarf. In fact any very reactive metal may suffice for such permanent chemical fixation so that in special circumstances a pot of liquid Na can be used to remove O_2 and H_2O while N_2 can be removed as well by a pot of liquid Li or even U powder.

Apart from their chemical inertness, the use of Ar or He environments is attractive because the gases are not explosive, inflammable or toxic. Nor does their use require a capital investment in a high-quality vacuum pumping system, although it is a common practice to evacuate batch furnace chambers before backfilling them with the gas. The physical presence of a significant gas pressure in the chamber also to some extent suppresses volatilization and dissociation of joining and component materials and hence extends the range of materials and conditions that can be used to achieve bonding. Thus joints can be brazed with the braze alloys containing Zn or Cd that could not be used in vacuum and the temperatures that can be used to diffusion bond Si_3N_4 can be increased.

There are, however, drawbacks to the use of inert gas environments. Thus they are of little effectiveness in removing oxide barriers already on the surfaces of metal components or joining materials and hence it is crucially important to use thorough oxide removal processes before loading the components into the bonding equipment. Further, although an impurity level of 1 ppm of H_2O is impressive, this is merely equal to the total gas content of a barely acceptable vacuum of 10^{-3} mbar. Similarly, it has been commented already that using gaseous environments can be a cause of problems when diffusion bonding due to entrapment at the joint interfaces. The presence of such gas pockets will prevent direct contact between the surfaces to be bonded and the application of a bonding pressure, of typically 10 to 1000 bars, will compress the trapped gas and hence increase the partial pressure and chemical activity of trace constituents such as H_2O, O_2 and N_2. Further, applications such as the active metal brazing of ceramics with alloys containing Ti will require commercial grades of gas to be purified to achieve very low H_2O, O_2 and N_2 levels to render it even more inert before it is introduced into the furnace chamber.

Such treated gas will be lost when the chamber is opened unless it is first re-evacuated and the used gas stored before being re-purified and introduced once more.

6.5.4 Oxidizing environments

Although the presence of O_2 in a gaseous environment is usually detrimental, there are situations in which the use of slightly oxidizing environments can be beneficial. The only substantial application of this effect to be developed so far is the direct bonding of Cu to Al_2O_3 at the Cu-Cu_2O eutectic temperature of 1065 °C, for which success requires the gaseous environment to contain at least 1.5×10^{-3} mbar of O_2. However, laboratory and other studies have shown that this is not the only potential application. Thus laboratory studies have shown that the presence of a small amount of O, 0.025%, in solid Cu strengthens the joints achieved by diffusion bonding to Al_2O_3. Similarly, beneficial effects of O_2 on the wetting of TiO_2 have been found in laboratory studies using Ag and Au, while the patent describing the direct bonding technique for Cu-Al_2O_3 joining also quotes the benefit of using environments containing 10^{-6} mbar of O_2 when joining Cu to Cu and of 10^{-4} mbar of O_2 when joining Ni to Ni or Fe to Fe.

When using slightly oxidizing environments, it is often convenient to achieve the desired composition by mixing two streams of inert gas, one of which contains an amount of O_2 While experience will often determine the valve settings and gas pressures that produce the desired results, it is prudent to monitor the actual O_2 content of the inlet gas using a solid state sensor based on the effect of O_2 activity on the electrical conductivity of ZrO_2.

6.6 Bonding temperatures, times and pressures

Previous process parameters have been discussed one at a time because they are totally independent of each other but the process parameters to be discussed in this sub-section are presented as a set because they are strongly inter-related. Thus the selection of both the bonding temperature and bonding time directly affect the extent of interdiffusion that occurs during brazing, transient liquid phase bonding (TLPB) or partial transient liquid phase bonding (PTLPB) and diffusion bonding. Further, in the case of diffusion bonding, variations in the bonding temperature also affect the yield strengths of the component and interlayer materials and hence the external pressure that must be applied to produce intimate contact by causing the surfaces to conform and disrupting and dispersing oxide films that may be coating the bonding surfaces.

These process parameters also differ from those discussed earlier in that their importance varies markedly with the nature of the bonding process so that, while an external pressure is essential when diffusion bonding, its brazing or joining by the TLPB or PTLPB processes may not be necessary or merely be required to hold the assembly of components in place before they are bonded. This varying importance of the parameters for different joining processes therefore makes it convenient to consider their role for each of the main joining processes in turn.

6.6.1 Brazing

Temperature is the most important process parameter that needs to be selected when brazing both metal and ceramic components. When joining metal components, it is essential that the braze alloy is fluid enough to flow into capillary gaps or form fillets. Thus it should be sufficient merely to exceed the melting temperatures of pure metals and eutectic compositions to achieve satisfactory flow because of their modest viscosities.

In practice, slightly higher minimum brazing temperatures are generally used to prevent the total loss of fluidity because of small and perhaps unavoidable temperature variations, decreases, within an assembly of components or from one production run to another. Thus for Cu, it is common practice to use a minimum super heat of about 10 °C. However, braze suppliers and technical organizations recommend not only minimum but maximum brazing temperatures that are determined by factors such as their volatility when used in a vacuum furnace, about 1150 °C for Cu, or the extent and rapidity with which they erode component metals and alloys or interdiffuse to form fragile reaction products, or so change the liquid composition that there is a serious loss of fluidity. Such recommendations are made not only for pure metals and simple eutectic compositions but also for the wide range of non-eutectic braze alloys that are commercially available, examples of which are given in Table 6.5. It can be seen by comparison of Tables 6.5, 3.1 and 3.2 that once more the minimum temperature at which an alloy can be used to braze is generally only slightly higher than its liquidus and this is illustrated in Figure 6.10 which compares the liquidus and the range of recommended use temperatures for several families of braze metals and alloys.

For the alloys that generated the data plotted in the figure, the minimum brazing temperature is on average only 6 °C higher than the liquidus while the maximum is a superheat of 95 °C. There are, however, significant differences between the various families of braze alloys. Thus, the fact that the braze alloy does not have to be completely liquid to be able to flow is exploited in the use of Al brazes such as the Al-7.5Si alloy which has a use range of 577 to 612 °C that is also its solidus–liquidus range. Such exceptionally low use temperatures when employing Al brazes is attractive because their usual application is to join Al alloy components that have solidus temperatures little above the liquidus temperatures of the brazes. This practice is effective for these particular braze/component systems because of their chemical similarity and the ready diffusion of the Si melting temperature depressant but in general the use of brazing temperatures lower than the liquidus should be avoided because of the danger of liquation causing the liquid phase to spread and flow away to leave behind a rubble of material with a high solidus temperature.

While the recommended maximum use temperatures for Al brazes are low, those for other families are substantially higher than their liquidus temperatures. For the brazes used to generate Figure 6.10, these superheats were on average 55 °C for the Cu brazes, 85 °C for the Au brazes, 120 °C for the Ag brazes and 140 °C for the Ni brazes. These differences have arisen largely because the braze alloy compositions, and particularly those of Ag and Ni brazes, have been adjusted to satisfy many requirements other than simply being fluid and able to wet component materials. Thus the frequent requirement for corrosion resistance had an important influence on the development of both Ag and Ni brazes, the desire to

Table 6.5 Brazing temperatures for some widely used alloys

Alloy		*Brazing temperature range* (°C)
Type	*Composition*	
BAlSi-3	Al-10Si-4Cu	522–585
BAlSi-2	Al-7.5Si	577–612
BAlSi-4	Al-12Si	577–582
–	Al-7.5Si-Mg	577–598
BAg1	Ag-24Cd-16Zn-15Cu	618–760
BAg1a	Ag-19Cd-16Zn-15Cu	635–760
BAg2	Ag-26Cu-21Zn-18Cd	701–843
BAg5	Ag-30Cu-25Zn	743–843
BAg7	Ag-22Cu-17Zn-5Sn	651–760
BAg18	Ag-30Cu-10Sn	718–843
BAg8a	Ag-28Cu-0.2Li	766–871
–	Ag-28Cu	779–889
BAg19	Ag-7.3Cu-0.2Li	876–982
BAu2	Au-20Cu	890–1010
BAu4	Au-18Ni	949–1004
BAu6	Au-22Ni-8Pd	1046–1121
BCu1	Cu-0.10rem	1093–1149
BCu2	Cu-0.15rem	1093–1149
RBZn-A	Cu-40Zn-1Sn	910–954
RBZn-D	Cu-41Zn-10Ni	938–982
BNi6	Ni-11P	927–1093
BNi7	Ni-10.1P-14Cr	927–1093
BNi8	Ni-23Mn-7Si-4.5Cu	1010–1093
BNi2	Ni-7Cr-4.5Si-3.1B-3Fe	1010–1177
BNi3	Ni-4.5Si-3.1B-3Fe	1010–1177
BNi1	Ni-14Cr-4.5Fe-4.5Si-3.1B	1066–1204
BNi5	Ni-10.1Si-18Cr	1149–1204

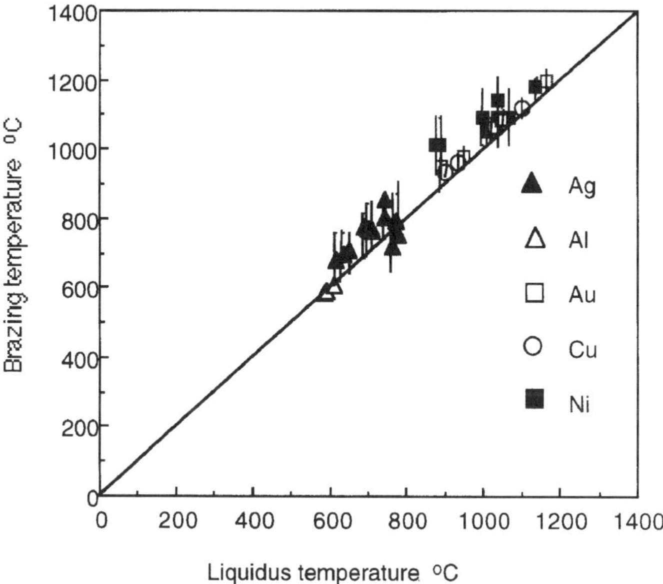

Figure 6.10 The liquidus and recommended average use temperatures of families of braze alloys. The vertical bars indicate the ranges of recommended use temperatures

match the colour of jewellery metals affected the formulation of Ag brazes and optimization of the beneficial effects of interdiffusion in homogenizing the composition and enhancing the ductility of joints between stainless steel and Ni alloy components affected the specification of Ni brazes.

One result of the differing ranges of use temperature and of the varying degrees of super heat needed to initiate effective brazing is that the rigour with which temperature needs to be controlled varies from application to application. Thus very close control is needed when brazing with Al alloys, certainly no poorer than $\pm 10\,°C$ and thus the temperature stability achievable when using a flux bath as a heat source is attractive. In contrast even $\pm 30\,°C$ would suffice for some of the Ag and Ni alloy brazes but in practice closer controls are desirable because of the effects that variable temperatures can have on the microstructure and hence mechanical properties of the components being joined.

Controlling the temperature when furnace brazing requires monitoring the output of the heat source and whenever possible also the actual temperature of the components being joined and this latter process may require the use of more than one thermocouple when complex assemblies are being brazed. When direct measurement of the component temperature is not possible, other techniques must be adopted to provide guidance for process control. Thus adjustment of the gas supply to torches, and the positioning of hand-held torches, can be guided by observation of the melting and flow of the flux and braze alloy. Similarly, setting a high furnace temperature and then varying the speed of a conveyor belt can be used to alter the maximum temperature achieved by component assemblies.

Varying the speed of a conveyor belt will also affect the time that components spend above the minimum temperature for braze flow. Prolonging the time spent in this temperature regime and decreasing the heating rate once the braze is liquid should increase the extent of braze flow. Theory suggests that the time taken for a liquid metal to flow the few millimetres that are all that are usually required to form a brazed joint will be a few seconds at most for chemically inert systems and experimental observations have demonstrated that a few minutes are sufficient even when flow is impeded by oxide films. Thus short dwells at the brazing temperature may be sufficient and indeed times of a few seconds are all that are generally needed when brazing using localized heating with a gas torch or an electron beam, and even shorter times can suffice when resistance brazing. Further, the total duration of the brazing cycle, including the times taken to heat and cool, can be very short when such localized heat sources are used or when rapid heating sources such as induction heaters or immersion baths are used.

In contrast, times of typically up to 30 minutes or more at the brazing temperature and intermediate dwells during heating to the brazing temperature are often introduced when furnace brazing as shown by the examples given in Table 6.6. These prolonged brazing cycles have little or

Table 6.6 Some successful brazing cycle durations quoted in the ASM Metals Handbook

Component and braze materials	*Furnace type and environment**	*Time (minutes)*		*Brazing temperature* (°C)
		total	*brazing*	
Cast iron, BAg-1	C, E	144	15	700
Stainless steel, BAu4	B, H_2	240	8	1020
Stainless steel, BCu1	B, H_2	180	25	1125
Cu, BAg1a	C, H_2	9	1	680
Al, Al-Si-Mg	B, vacuum	> 16	1	590

* B Batch furnace, C Conveyor furnace, E Exothermic gas

nothing to do with the actual joint filling process but reflect the need to achieve a uniform and controlled temperature throughout the assembly to avoid warping and, sometimes, to promote interdiffusion during or after brazing. These dwells can often constitute most of the total time needed for the braze cycle and in particular very extended post-brazing dwells can be needed to allow sufficient interdiffusion between constituents of Ni brazes and the component material to achieve a homogeneous joint microstructure free of embrittling phases.

The most commonly used Ni brazes contain Si or Si and B which depress the melting temperature but also degrade solidified joints by forming seams of brittle intermetallic compounds. This embrittlement problem is particularly acute with brazes containing B because of its negligible solubility in Ni. Both B and Si can be removed from the joints by diffusing into the component material and the extent of this interdiffusion will depend on the diffusion coefficient and hence the temperature, the time at temperature and the size of the B reservoir, or the width of the joint gap, that must be depleted. Formation of ductile joints free of seams of intermetallic compounds, therefore, can be promoted by using narrow joint gaps, high brazing temperatures and long brazing times and it is important to be able to specify a time that will be sufficient to ensure that a joint of a certain size will become single phase when brazing at a particular temperature. This information can be derived with confidence only from experimental studies and the observed interplay of the three variables can be illustrated conveniently by use of TETIG (TEmperature-TIme-Gap) diagrams such as that shown in Figure 6.11. It is noteworthy that the thicker gaps require many hours at temperature before they become free of embrittling seams and in practice it is often convenient to affect this microstructural change by post-fabrication heat treatment rather than occupying the brazing furnace for such prolonged times.

The final parameter that needs to be considered when contemplating the brazing of an assembly of components is whether a pressure needs to be applied and if so at what level. It is impractical to apply a pressure to induce entry of a non-wetting liquid into a capillary gap when brazing but external pressures can be of assistance in holding the components in position while they are heated and joined. These pressures are usually applied by means of jigs or dead weights and are typically less than 0.1 MPa and are very rarely more than 1 MPa. In fact the actual level of the pressure is unimportant when brazing metal components provided it does not cause them to deform or the joint gaps to be closed.

These and the previous comments on brazing process parameters relate to the joining of metal components. The same parameters are of concern when joining ceramics with active metal braze alloys, but their importance and values differ. Thus the temperature used when joining ceramic components must be sufficiently high not only to ensure that the active

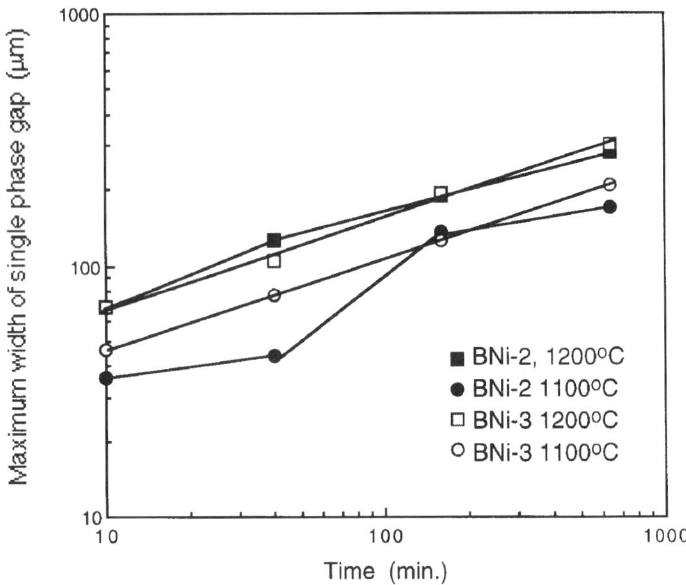

Figure 6.11 The TETIG diagram for the brazing of stainless steel with BNi-2 (Ni-7Cr-4.5Si-3.1B-3Fe) and BNi-3 (Ni-4.5Si-3.1B-3Fe) alloys. After Johnson (1987)

braze alloy melts and but also that it reacts rapidly enough with the ceramic surface to produce a wettable reaction product layer over which it will flow in an acceptable time. The recommended minimum brazing temperature for an Ag, Au and Cu base active metal braze is usually close to and occasionally even lower than the liquidus, Table 6.7, but sometimes substantially higher temperatures have to be used. Thus the chemical stability and physical tenacity of the surface oxide on Al prevents it from freely wetting even in a good vacuum when using superheats of less than about 200 °C. Similarly, the need to produce a 1 μm thick continuous layer of reaction product to achieve satisfactory bonding when brazing SiC with Ag-28Cu-2Ti lead to the use of a brazing temperature of 950 °C, a superheat of more than 100 °C (Boadi *et al.*, 1987) while laboratory work has shown that a Sn-1Ti solder alloy requires a superheat of 300 °C before it will wet Al_2O_3 or C at about 600 °C.

The practice of placing the active metal braze alloys within a joint gap prevents the sluggishness of their reaction controlled flow rates from affecting the duration of brazing times. In practice ceramics are generally brazed in vacuum or controlled atmosphere furnaces and it is customary to dwell at the bonding temperature for 15 to 30 minutes to ensure that all parts of the components that need to be joined have achieved the desired

Table 6.7 Recommended use temperatures for some commercial active metal braze alloys

Braze name and composition*		*Temperatures* (°C)	
		Liquidus	*Use*
Incusil-ABA	Ag-27Cu-12.5In-1.25Ti	715	700–750
Cusin-1 ABA	Ag-34Cu-1.75Ti-1Sn	806	810–840
Cusil-ABA	Ag-35Cu-1.75Ti	815	830–850
Silver-ABA	Ag-5Cu-1.25Ti-1Al	912	900–950
Gold-ABA	Au-3Ni-0.6Ti	1030	1025–1050
Nioro-ABA	Au-15Ni-0.75Mo-1.75V	960	1000–1050
Copper-ABA(L)	Cu-2Al-3Si-2.25Ti	1024	1025–1050

* Trade marks of Wesgo Metals, Belmont, CA, USA

temperature. These dwell times usually are sufficient also to ensure the growth of a reaction product layer with a thickness of typically 1 or 2 μm that is continuous and flaw free so that the resultant interface between the ceramic and the solidified braze is acceptably strong. If such optimum thicknesses and structures are not produced, it is often simpler to alter the temperature slightly because the large activation energies for diffusion in the solid reaction products makes their growth constants far more sensitive to changes in temperature than time: for example, a typical activation energy of about 500 kJ/g mole implies that raising the brazing temperature from 850 to 875 °C is equivalent to increasing the time at temperature from 10 to 30 minutes. However, the total duration of a brazing cycle is usually far longer than these dwell times to ensure that the assembly of ceramic components is heated uniformly and hence does not experience differential thermal expansion strains and stresses as it is heated to the brazing temperature, and the temperature cycle illustrated in Figure 6.12 is typical of those used in practice.

The practice of placing foils of active metal braze alloys within the joint gaps also affects the role played by externally applied pressures when joining assemblies of ceramic rather than metal components. Thus it cannot be guaranteed that the foils will have the same configuration as the joint, in particular they may be slightly curved while the joint is flat, and hence the use of pressure of perhaps up to about 1 MPa can be of use in promoting the effective filling of the gap before the temperature is raised to melt the braze alloy. Again, when brazing with Al or Al alloys, the

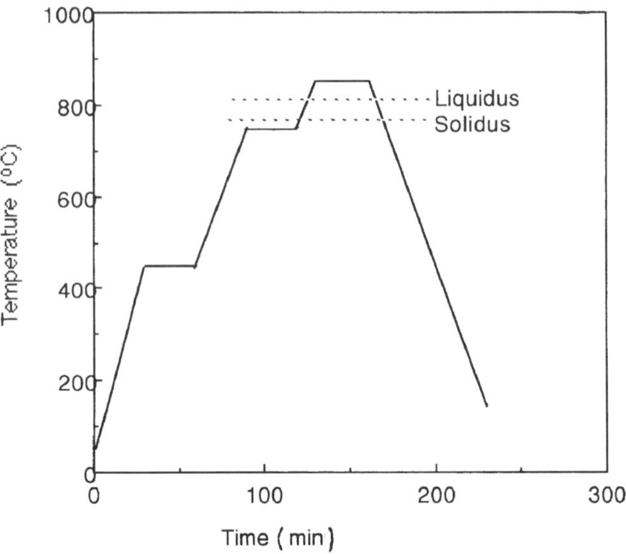

Figure 6.12 Time–temperature cycle used when brazing small Si_3N_4 blocks with a Ag-35Cu-2Ti alloy

application of a small external pressure can be of benefit if it is sufficient to cause asperities on the ceramic surfaces to puncture the oxide film on the braze surface and so permit it to come into direct contact with the ceramic at a lower temperature than would otherwise be possible.

6.6.2 Diffusion bonding

Of the three process parameters being considered in this section, the selection of a bonding temperature is usually the most important decision when specifying the conditions in which to diffusion bond components. The choice of a suitable temperature will depend on the materials involved and also whether the components are being bonded together directly or via an interlayer. Considering first the direct bonding of metal components, the minimum useful bonding temperature will be determined principally by the deformation characteristics of their surface asperities and when super-plastic forming/diffusion bonding also by of those of the bulk material. However, attention may also have to be paid to a maximum useful bonding temperature defined by the onset of grain growth or microstructural coarsening which results in degradation of mechanical properties.

Direct bonding of metal components usually requires a temperature in excess of $0.5\,T_M$ to ensure that diffusion occurs at a useful rate. In practice, the temperatures to produce strong bonding are often between 0.6 and 0.75 T_M as can be seen by examining Table 6.8 which summarizes the process parameters that have been found to produce well bonded and strong joints for a number of metal–metal and alloy–alloy systems. Among the pure metals, the behaviour of Al is notable since the recommended bonded temperature of 540 °C is equivalent to 0.87 T_M and reflects the chemical stability and physical tenacity of the film Al_2O_3 present on Al surfaces that acts as a barrier to prevent bonding. However, bonding can be achieved at much lower temperatures if that barrier is removed, thus Suga *et al.* (1988) bonded Al at room temperature, 0.32 T_M, in an ultra-high vacuum facility by using sputtering to remove surface layers of the oxide and the metal immediately before bonding.

The pressures needed to achieve strong bonding between metal components are usually only a few MPa but their actual values have to be defined by experimental work. The optimum value for a particular application depends on the bonding temperature that has been chosen since this affects the stress needed to cause the surface asperities to yield either virtually instantaneously or progressively by creep, and hence the extent to which the applied pressure creates intimate contact between the components by squashing their surface asperities.

Table 6.8 Conditions used to achieve strong diffusion bonding of some metals and alloys

	Conditions			
Material	*Temp.* (°C)	*Pressure* (MPa)	*Time* (min.)	*Environment*
Al	540	3	20	Vacuum
Al-0.9Si-0.6Mg-0.5Mn	550	4	25	Vacuum
Al-6Cu-0.4Zr-0.3Mg (Super plastic alloy)	510	4	240	Air
Cu	700	7	120	Vacuum
Fe	850	15	10	Vacuum
Fe-0.4C	1000	7	10	Vacuum
Fe-1Cr-0.4C	1000	10	60	Vacuum
Ni	1000	15	10	Ar
Ti	930	7	180	Vacuum
Ti-4Al-6V	950	0.7	120	Ar
Ti-4Al-4Mo-2Sn-0.5Si	950	2	120	Vacuum

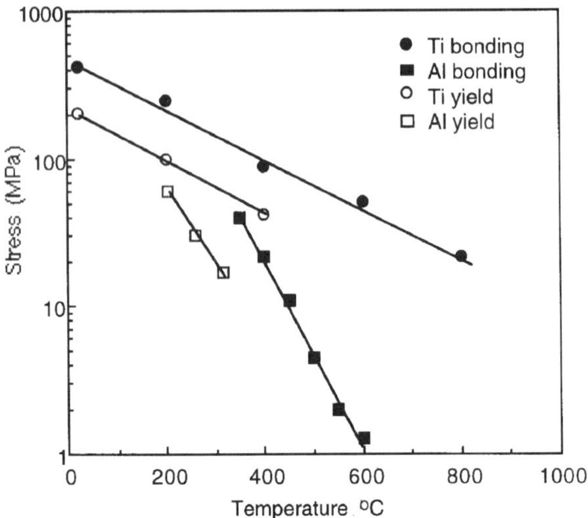

Figure 6.13 The effect of temperature on the compressive stresses needed to produce strong bonding of commercial purity Ti and 2xxx series Al alloys. Also included are the tensile yield strengths

Figure 6.13 illustrates the effect of temperature on experimentally derived optimum values of the bonding pressure for Ti and a group of Al alloys. The figure shows that changing the temperature has a marked effect on the pressure needed to achieve strong bonding and this effect can be described by the empirical relationships

$$\log Pext = -GT + H \qquad \{6.6\}$$

$$P = \exp H.2.303.\exp(-GT) = U.\exp(1/GT) \qquad \{6.7\}$$

where G, H and U are constants. The data plotted in the figure show that the pressures needed to bond Ti are larger than those for the Al alloys and that the effect of temperature on the bonding stress values as defined by the value of G is more marked for the Al alloys than for Ti. These differences are due to the greater refractoriness of Ti which affects its deformation characteristics in both compression and tension. Thus Figure 6.13 also includes values for the tensile yield stress, defined by the 0.2% proof stress, of both Ti and the Al alloys and it is noteworthy that those for Ti are again higher than those for the Al alloys and less affected by temperature. Comparison of the two sets of data shows that the optimum compressive bonding stresses are about double those for tensile yielding so that literature values of yield stresses such as those quoted in Smithells can be used to provide guidance as to the approximate value of the optimum bonding pressure. This relation to the yield stress also accounts for the frequent similarity of the optimum bonding stresses at the preferred temperatures of about 0.7 T_M since the effect of temperature on yield stress values can be normalized for many simple metals in terms of T/T_M.

Subjecting a component to an unchanging external stress or load at high temperatures will cause it to deform continuously, to creep, and hence progressively to flatten and close interfacial voids when diffusion bonding. Closure of interfacial voids and the growth of bonded areas is also promoted by diffusion controlled mass transport driven by the energy decrease achieved when surfaces are converted into interfaces. Thus prolonging the duration of a diffusion bonding process should enhance the integrity of the joint, but prolonged dwells at high temperatures in specialized equipment will increase the fabrication costs and can cause degradation of the component materials by coarsening their microstructures. It is important, therefore, to select a bonding time that is not excessive but nevertheless is sufficient to produce the desired joint properties.

The growth of interfacial bonding proceeds at ever slower rates as the duration of the bonding process is prolonged because the resistance of the asperities to further yielding and surface energy driving force for diffusional processes diminishes as the bonded area increases. This effect was illustrated in Figure 3.21 for the diffusion bonding of Fe components at 700 °C by plots of experimental measurements and theoretical predictions. The actual extent of bonding achieved after a certain time depended on material and process factors that can interact in a complex manner. Decreasing the surface roughness from many micrometres to about one micrometre usually has relatively little effect on the extent of bonding, but further decreases to substantially less than a micrometre can be beneficial. Increasing the temperature or applied pressure results in the more rapid growth of the bond area as indicated by other laboratory data reported by Derby for Fe that is assembled in Table 6.9 and similar studies using Ti-6Al-4V components showed the time required for good bonding to be

Table 6.9 Effects of time, temperature and pressure on the bonding of Fe samples

Time (min.)	*Temperature* (°C)	*Pressure* (MPa)	*Bond area* (%)
10	850	7	71
60	850	7	79
10	700	7	54
60	700	7	62
10	700	8	55
10	700	17	63
60	700	8	63
60	700	17	70

achieved at 900 °C increased from 20 to 80 minutes as the applied pressure was decreased from 400 to 100 MPa.

The sensitivity of the time to the effects of several other parameters results in a very wide range of times having been used to produce strongly diffusion bonded components. Even a cursory glance at the technical literature reveals successful bonding having been achieved using times ranging from less than a minute to a day. Most bonding cycles, however, employ times of between 15 minutes and about 4 hours, if only to ensure a total cycle time that fits comfortably within a working shift or day.

The same considerations regarding the selection of the bonding temperature, pressure and time apply when joining dissimilar material combinations but the selection is heavily influenced by the characteristics of the less refractory material. Thus the diffusion bonding of Al to stainless steel in vacuum is best accomplished at 550 °C using an applied pressure of 5 MPa maintained for 10 minutes, conditions similar to those reported in Table 6.8 for the effective bonding of Al and some Al alloys to themselves. However, there is always a possibility of interdiffusion causing mechanical or chemical degradation in the interface zone when dissimilar materials are joined and this can impose an additional restriction on the selection of bonding times. For example, the time used to join Al to stainless steel has to be moderately brief if the growth of thick and fragile aluminide reaction product layers is to be avoided. Less dramatic but also of concern is the possibility of causing the production of substantial zones where mechanical properties are degraded because of unwanted carburization or decarburization when joining steels with differing C activities.

Similar considerations must be taken into account when using interlayers to promote the diffusion bonding of metal and particularly of ceramic components but fortunately the selection of an interlayer material can ease some of the restrictions on the choice of processing parameters. Thus the use of interlayers of Ni or Ni-Cr alloys to join Ni base super alloys diminishes the chemical driving force for potentially detrimental interdiffusion while the usual thinness of such interlayers limits the volume of component material that can be affected by any interdiffusion that does in fact occur. The temperatures employed to achieve bonding when using an interlayer are determined primarily by the characteristics of the interlayer material except for systems in which interdiffusion can result in the formation of eutectic compositions with low melting temperatures. However, the pressures required are often substantially higher than those needed when bonding blocks of interlayer material because the thinness and geometry of the interlayers restrains the yielding and flow of the interlayer material, so that pressures of 100 MPa or more have been used when employing Ni interlayers to promote the joining of high alloy steels at temperatures of 1100 to 1150 °C. As when diffusion bonding components of dissimilar materials, the times required to achieve the best possible bonding may be restricted by the need to avoid degrading composition changes.

A particularly important application of interlayers when diffusion bonding is to achieve ceramic–ceramic and ceramic–metal joining. The same considerations have to be taken into account when selecting suitable bonding temperatures and pressures as when joining metal components and often very similar conclusions are reached. Thus the bonding temperatures are restricted by both the melting of the interlayers themselves and the possibility that they will interdiffuse with the component materials to form low melting temperature compositions. An example of this latter restriction is the effect of Ni-Si eutectic compositions that can form and melt at temperatures of 1255 and 1143 °C when Ni interlayers are used to promote the bonding of SiC or Si_3N_4 ceramics. Similarly, the pressures needed for effective bonding are again a function of the interlayer thickness and geometry as well as their more fundamental yield characteristics.

Some of the sets of conditions that have been used to achieve effective bonding of ceramics in vacuum using foils that are about 100 μm thick are summarized in Table 6.10. It can be seen that the conditions selected can

Table 6.10 Conditions used to achieve strong diffusion bonding of ceramics using metal interlayers

System	*Temp.* (°C)	*Pressure* (MPa)	*Time* (min.)
Al/Al_2O_3	600	50	30
Al/ZrO_2	540	2	20
$Al\text{-}4.5Mg/Al_2O_3$	600	50	90
AA-7075*/Al_2O_3	360	6	6000
$Au\text{-}Al_2O_3$	1000	10	240
Cu/Al_2O_3	1000	1	600
Cu/Al_2O_3	1000	4	240
Cu/Al_2O_3	1000	6	120
Cu-0.27Cr/Diamond	800	36	80
Ni/Al_2O_3	1390	4	120
Ni/Al_2O_3	1100	10	120
Ni/ZrO_2	1100	10	60
Ni/Diamond	800	36	180
$Ni\text{-}Si_3N_4$	1000	10	60
$Ni20Cr/Al_2O_3$	1100	27	120
$Ni20Cr/Si_3N_4$	1200	100	60
$Ni20Cr/ZrO_2$	1100	10	120
Ni-1.25Ti/Diamond	800	58	40
$Pd\text{-}ZrO_2$	1100	10	60
V/Si_3N_4	1050	20	90

* AA-7075 = Al-5.6Zn-2.5Mg-1.6Cu-0.23Cr

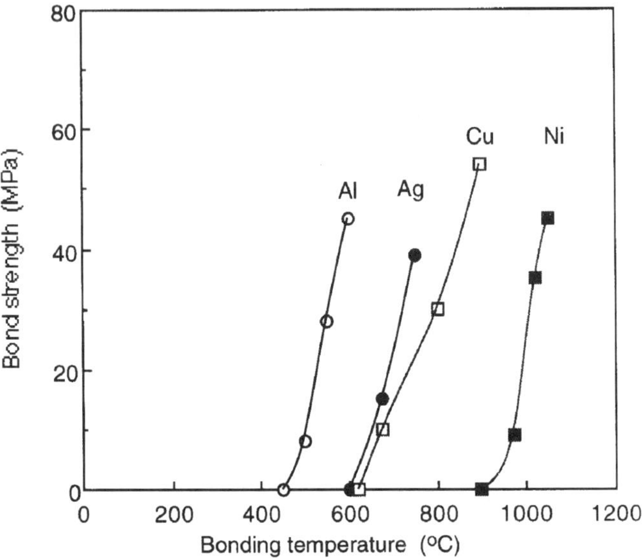

Figure 6.14 The effect of fabrication temperature on the tensile strength of Al_2O_3 samples bonded with metal interlayers. Bonding was achieved in an evacuated chamber by applying pressure of 50 MPa for 30 minutes

be quite varied but some general conclusions can be drawn. Thus the temperatures used to achieve bonding are related to the melting temperature of the metal interlayer rather than some characteristic of the ceramic but are often quite high with values for Al, Cu and Ni used to bond Al_2O_3 being greater than O.90 T_M. However, the temperatures needed to initiate bonding can be much lower and, as illustrated in Figure 6.14 for the bonding of Al_2O_3, are often about 0.7 T_M.

Table 6.10 also shows that the pressures used to achieve bonding are often quite high and the bonding times are long, reflecting the constraint imposed on deformation of the thin foils by the rigidity of the ceramic components and the frequent association of bonding with the formation of a reaction product layer whose refractoriness slows its growth to an optimum thickness. If this optimum thickness has been achieved during the bonding process, subsequent post-fabrication annealing or use at high service temperatures will cause the reaction product to become too thick and hence degrade the mechanical properties of the joint, but sometimes post-fabrication annealing can be used for controlled optimization of mechanical properties, as illustrated by Figure 6.15. In that particular case, bonding was associated with the formation of a layer of $MgAl_2O_4$ spinel at the metal ceramic interface and the data illustrate the effect of changing the heat treatment temperature from 600

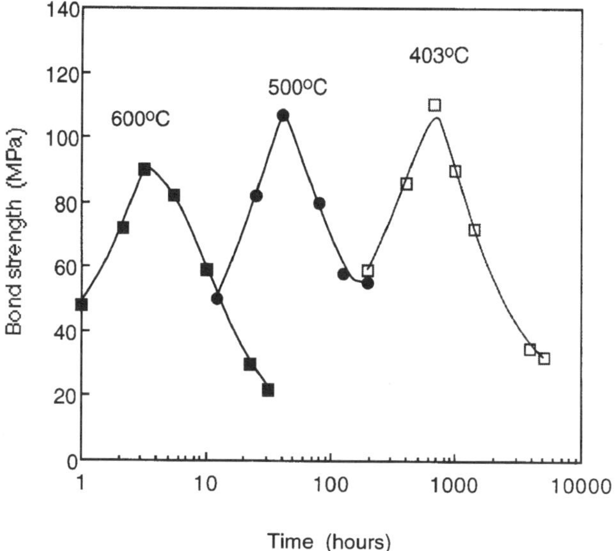

Figure 6.15 The effect of post-fabrication heat treatment on the tensile strengths of Al_2O_3 samples bonded with Al-4.5 Mg

to 403 °C on the rate of growth of the reaction product layer to an optimum thickness.

The tabulated data also provide evidence that there is not one set of optimum conditions but rather an envelope of sets in which the adjustment of one process parameter allows a variation to be made in the value of another. Thus the data for the diffusion bonding of Al_2O_3 at 1000 °C using Cu interlayers reveal a trade off between the applied pressure and the bonding time. While most of the process parameter values listed in Table 6.10 are determined by the properties of the foils, comparison of the data for bonding Al_2O_3 and ZrO_2 with Al and Ni foils suggest that ZrO_2 may be inherently easier to bond.

It can be seen that defining the optimum process parameters for the diffusion bonding of ceramics using metal interlayers is a particularly difficult task. Some general guidance can be gained by reference to previous experience but it is usually necessary to resort to proving trials to arrive at a final selection of values.

6.6.3 Hybrid processes

As might be expected, the influence of the bonding temperature, pressure and time on the quality of joints produced by the hybrid processes varies from one to the other and the sets of process parameters are often similar to mixtures of those found best when brazing or diffusion bonding. Thus the minimum temperature that can be used to join a metal to a ceramic or a dissimilar metal by the direct bonding or eutectic bonding technique is in both cases the melting temperature of the eutectic formed by metal–oxygen or metal–metal interactions, but the maximum useful temperatures can differ substantially. When direct bonding, the maximum temperature that can be used is determined by thermodynamic factors and in particular by the usually small liquidus depression achieved by interaction with the gaseous environment: thus in the case of the most exploited system, the bonding of Cu to Al_2O_3, the range is only from the 1066 °C melting temperature of the Cu-Cu_2O eutectic and 1085 °C melting temperature of Cu itself. In contrast, the maximum temperature that can be used when joining by the eutectic bonding process is less than that of either metal and is determined by the kinetics of the interdiffusion and dissolution processes and a frequent desire to restrict the width of the liquid region.

The direct bonding and eutectic bonding techniques can be used with relatively few systems and provide little room for manoeuvre when selecting process parameters, but transient liquid phase bonding, TLPB, and partial transient liquid phase bonding, PTLPB, offer more scope for possible variations when deciding the conditions to be employed for a particular application. The operation of these techniques involves complex materials interactions and the selection of optimized processing parameters depends on considerations of both thermodynamic and kinetic factors such as phase diagrams and rates of interdiffusion and dissolution.

When using the TLPB technique to effect joining of metal components, the selection of an externally applied pressure is the least important of the decisions that have to be taken about processing parameters since its purpose is merely to ensure that there is good contact between the interlayers and the components to be joined and to hold the components together so that the desired joint sizes and geometries are maintained while the assembly is heated to and dwells at the bonding temperature. In contrast, the selection of a bonding temperature is of crucial importance and the actual values that have been shown to be successful are often modest superheats above the melting temperature of the interlayer or the eutectic formed by interdiffusion of the interlayer and the components. Thus the temperatures used for joining by TLPB resemble those employed when brazing, but the times for creation of optimized joint structures by allowing interdiffusion to continue until the liquid within the joint gap experiences complete isothermal resolidification are much longer than those needed for brazing and are often even substantially in excess of those employed when diffusion bonding. Increasing the temperature will increase the rate of interdiffusion between a TLPB interlayer and the components being joined, but does not always result in faster isothermal resolidification because the temperature increase also usually increases the extent to which the interlayer composition must be changed before resolidification occurs. These effects are illustrated in Table 6.11 for a number of specific systems by the temperatures and times found necessary to achieve complete isothermal resolidification.

Some of the data in Table 6.11 also reveal effects of interlayer composition on TLPB that may be useful to consider when selecting process parameters for other systems. Thus it is noteworthy that changing the interlayer from Cu to Ag-20Cu dramatically decreased the recommended fabrication time when joining Ag because the substitute interlayer composition is closer to the Ag-7Cu that must be achieved at the joint centreline for isothermal resolidification to be completed. The table also shows that isothermal resolidification of an interlayer of Ni-P is far slower than that of one of Ni-Cr-B, reflecting the differing mobilities of the two melting temperature depressants.

Similar processes of interdiffusion, dissolution and isothermal resolidification affect the selection of the parameters to use when making ceramic–ceramic joints by PTLPB, but there are also additional complications. Thus only the surface coating rather than the whole interlayer melts and its isothermal resolidification depends primarily on the unidirectional diffusion of low melting temperature constituents into the core. At the same time, some

Table 6.11 Temperatures and times required to complete isothermal resolidification of joints when bonding by the TLPB technique

*Materials** *Component/interlayer*	*Process conditions*	*Comments*
Ag/Cu, 80 μm	820 °C, 200 h	Ag-28Cu T_E is 780 °C
Ag/Ag-20Cu, 75 μm	820 °C, 10 h	
M007/MBF-80, 30 μm	1125 °C, 2 min.	MBF-80 melts at 1065–1075 °C
M007/MBF-80, 30 μm	1150 °C, 80 min.	
Ni/Ni-11P, 30 μm	1100 °C, 51 h	Ni-11P T_E is 875 °C
Ni/Ni-11P, 30 μm	1250 °C, 15 h	
6061-Al_2O_3**/Cu, 10 μm	580 °C, 20 min.	Al-33Cu T_E is 548 °C
AISI 316/BNi-5, 80 μm	1050 °C***, 6 h	BNi-5 melts at 1080–1135 °C

* M007 Ni-10Co-7Cr-6Mo-6Al-4Ta-1.3Hf
MBF-80 Ni-15.5Cr-3.7B
6061 Al-10Mg-0.6Si-0.25Cr-0.2Cu
AISI 316 Fe-17.5Cr-12.5Ni-2.5Mo-1Mn
BNi5 Ni-18Cr-10Si

** The interlayer was between 6061 and Al_2O_3 components.

*** Preceded by a dwell at 1150 °C for 15 minutes.

Table 6.12 Process parameters found to result in the strong bonding of some ceramics by PTLPB

Materials		*Process conditions*		
Component	*Interlayer core/coating*	*Temp.* (°C)	*Time* (h)	*Pressure* (MPa)
Si_3N_4	Nb/10 μm Ni	950	3	20
Si_3N_4	Ni20Cr/2.5 μm Au	1000	4	0.5
Al_2O_3	Ni/3 μm Cu	1150	6	5
Al_2O_3	Pt/3 μm Cu	1150	6	5

interaction of the liquid with the ceramic components is necessary to ensure bonding and this often results in the formation of a reaction product layer and sometimes in the migration into the liquid of temperature depressants or embrittling constituents such as Si. These additional considerations make the prediction of suitable bonding times and temperatures for joining by PTLPB particularly difficult, but some conditions that have been found to be successful in practice are summarized in Table 6.12.

The few data assembled in the table present a number of noteworthy features. Perhaps the most notable is that sometimes substantial pressures were used when making the joints even though pressure fulfils no fundamental function in the PTLPB process. In fact it is probable that these pressures are of no real significance because no explanation was given for the use a 20 MPa pressure when bonding Si_3N_4 with Ni coated Nb, and the 0.5 and 5 MPa pressures used with the other systems were merely the lowest possible that could be applied to hold the components with the equipment used to make the joints. More significantly, the temperatures used to make joints by PTLPB can be related to those needed to melt the coating metals, or eutectics they form with the cores of the interlayer. However, the times taken to achieve a strong bond are far shorter than those listed in Table 6.11 to achieve complete isothermal resolidification, and this is due primarily to the thinness of the coatings and hence indirectly of the liquid layers that need to be resolidified.

REFERENCES

American Society for Metals (1972) *Metals Handbook*, **8**
Boadi, J. K., Yano, T. and Iseki, T. (1987) Brazing of pressureless-sintered SiC using Ag-Cu-Ti alloy, *Journal of Materials Science*, **22**, 2431–4.

Johnson, R. (1987) High temperature nickel braze filler metals. In *High Technology Joining*, British Association for Brazing and Soldering, Abington, Cambridge.

Kazakov, N. F. (1981) *Diffusion bonding of metals*, Pergamon Press, Oxford.

Suga, T., Miyazawa, K and Yamagata, Y. (1988) Direct bonding of ceramics and metals by means of a surface activation method in ultrahigh vacuum. In M. Doyama, S. Somiya and R. P. H. Chang (eds), *Proceedings of the MRS International Meeting on Advanced Materials*, **8**, Metal-ceramic joints, 257–63.

7 Joint properties and behaviour

All the decisions about what technique, joint design, material and process parameters to use discussed in the preceding chapters had as their objective the reliable production of acceptable joints by economic means. In this context, acceptable means joints whose structures and mechanical, physical and chemical characteristics enable them to satisfy the performance criteria set for in-service behaviour. These previous chapters have contained passing qualitative references to joint characteristics when considering what factors influence the optimization of joining processes, with joints being described as 'of high integrity', 'sound', 'strong', 'robust' or 'tough'. In practice, joint characteristics must be quantified to identify acceptance limits and to provide reassurance to users that the components should be able to endure service conditions and lifetimes without failing or being degraded to an unacceptable level. The expectations of users are of course application oriented and hence seldom related to fundamental joint characteristics. Thus a joint in a garden tool will have to resist very varied mechanical and other abuse that cannot be defined easily in terms of stress levels or corrosion behaviour. Similarly, the joint in an item of jewellery may be visible and hence must have and retain an appearance and colour whose acceptability can be assessed only by the ultimate customer. Nevertheless, fundamental characteristics measured by well-established laboratory procedures are of importance because they provide understanding of the effects of design, materials and process parameters that can be used to make more efficient decisions in future applications.

It is necessary to consider for each and every application what definable joint characteristics are required, and this may involve several rounds of discussion with the customer before quantifiable targets can be set. This chapter discusses what targets might be set for the structural, mechanical, physical and chemical characteristics of joints and how these values might be measured before giving examples of actual values.

7.1 MACROSCOPIC STRUCTURE

7.1.1 Characteristics

The desirable macroscopic structure of a joint is one of the aspects of joint design considered in Chapter 5, and essentially two types of such structure exist. In the first of these, the joint is made directly between the components so that their mating surfaces become the joint interface as when diffusion bonding. The other and more common case is that of a joint produced using some foreign material to achieve bonding and this results in a joint that has not only a distinctively different composition to that of the components but also idiosyncratic mechanical and physical properties.

The first type of joint is usually regarded as the ideal because it is free of the chemical, physical and mechanical discontinuities caused by the presence of a joining material that is distinctively different to the components. Even when the component materials are dissimilar, this ideality may be approached to some extent if sufficient interdiffusion occurs to replace the material discontinuity originally present at the interface between the components with a chemical gradient between the two components so that there is an associated continuum of mechanical and physical properties.

More usually encountered, however, are joint structures produced using foreign materials to effect bonding. In the case of brazed joints, the resultant macroscopic structure can be that of a fillet or, more commonly, a seam between the components formed by the braze alloy flowing into a joint gap. This seam structure is typical also of diffusion bonded joints produced using metal foils and of joints created by some of the hybrid joining processes that use coatings on component surfaces to facilitate bonding. In practice, some joints produced with a seam structure can be processed to make their chemistry similar to that of the components so that it is no longer a simple matter to identify the boundaries of the seam. The main example of such processing is the use of prolonged heat treatment times to remove embrittling species such as Si or B from Ni brazed joints between Ni alloy components, to dissipate the Ag interlayer of diffusion bonded Al alloy components or to homogenize transient or partial transient liquid phase bonded joints. Such treatments that blur the identity or structure of the joint are, however, very seldom applied to components displaying braze fillets.

To fulfil their intended functions, joints must be free of voids, cracks and inclusions and the adoption of the good practices described in Chapter 6 should ensure that they are well filled and sound. If defects occur, such as the trapping of gas bubbles or islands of flux within the seam of a brazed joint, this may be due to faulty design, material or processing decisions rather than being an inherent characteristic of the system. Thus inappropriate design and jigging can prevent the liquid braze, and the liquid flux, from flowing freely through the joint to the exit to produce an observable fillet, or 'witness' and unwise selection of the braze alloy and heating rate can result in liquation. If ceramics are being joined by an active metal braze alloy, it is necessary to place the alloy within the joint gap because of its very slow rate of liquid flow but this practice can result in the formation of shrinkage porosity unless care is taken to ensure that the liquid is compressed by a small external load. Again, the use of excessive braze alloy to produce a large fillet can result in shrinkage porosity being caused by thermal gradients within the joint so that the free surface solidifies before the interior. It is worthwhile noting further that there is seldom any mechanical or other advantage to be gained from the deliberate production of large fillets.

7.1.2 Assessment

Non-destructive techniques can be used to characterize and assess the quality of the macroscopic structure of joints and in particular the extent to which the joint gaps are filled and free from porosity or cracking that degrades mechanical properties. The range of techniques that are suitable for the assessment of joints between metallic components is quite wide but is restricted for the assessment of joints between ceramic components or dissimilar ceramic and metal components. The use of non-destructive assessment techniques is attractive and adopted in practice but it is also often prudent to supplement them by destructive assessment of a limited proportion of the bonded assemblies of components.

The simplest non-destructive assessment technique is visual examination, and reference has been made already to the proof provided by a witness that a braze has flowed along the entire length of a capillary joint. Careful examination of a witness, however, can provide additional information about joint structure and quality. The presence of cracks or shrinkage porosity on the surface of the witness provides at least grounds for suspecting that they may also be present within the joint, while the contours of the witness can provide guidance as to the contact angle between the braze alloy and the component and hence indirectly to the likelihood of poor filling of the joint. To maximize the usefulness of such observations, it is often helpful to make comparisons with photographs of the witness surfaces of similar joints that are known to suffer from various types of defect.

When it is difficult to distinguish between small crevices on a witness surface and cracks that penetrate a significant distance into the joint, the use of a dye penetrant can be of assistance. This is a liquid carrier such as water or an organic solvent containing a dye or fluorescent material that is brought into contact with the edge of a joint and allowed to penetrate. Excess liquid is then wiped off and the joint edge is inspected for liquid oozing out of cracks, a process that can be encouraged by dusting the surface with a fine powder that exerts a capillary attraction. In principle, this technique can be used also to identify the presence of surface cracks in bonded ceramic components close to their joints that may have been caused by mismatched coefficients of thermal expansion, but care has to be taken not to be confused by the effects caused by the small percentages of porosity that are present in many commercial ceramic materials. Another possible limitation of this technique is that the dye penetrant must be removed if the joint surface is to be built up further by techniques such as vapour deposition or additional brazing, and such removal can be difficult to accomplish.

Cracks that penetrate right through the joint will cause its tightness and such defects can be identified in tubular structures by both leak and pressure testing. If open ends or ports in the structure are closed by blanking plates and O-ring seals, the interior of the component can be filled with gas at a pressure of more than 1 bar and the presence of a leak is made manifest by a fall in pressure after the component is isolated. The location of the leak can be detected by bubbles forming at the surface when the component is immersed in a bath of water or the outside surface of the joint region is brushed with soapy water. Alternatively, a closed-off component can be evacuated and leaks identified by using a mass spectrometer to monitor the ingress of a gas such as He that is squirted at various suspect regions using a fine valved nozzle. Mass spectrograph devices are very sensitive and commercially available devices can detect leaks smaller than 10^{-14} $l.s^{-1}$ when subject to a pressure difference of 1 bar, a rate normally described as 10^{-11} $torr.l.s^{-1}$. It should be noted, however, that liquid can temporarily seal a small leak and hence a very careful drying procedure must follow observation of bubbles if He detection is to be used subsequently.

The other main techniques for non-destructive assessment of macroscopic joint structures involve more specialized equipment but two are well established and a third is assuming growing importance. Ultrasonic and radiographic inspection techniques are widely used means of detecting discontinuities such as voids and cracks in brazed and diffusion bonded joints provided that they are not buried deep within thick structures. The ultrasonic vibration must penetrate the component to reach the joint and does so at a velocity that depends on the elastic properties and density of the component material. However, this penetration is accompanied by attenuation of the vibrational amplitude at a rate which also depends on material properties. Ultimately the attenuated wave reaches the free surface of a crack or void or merely the other side of the component and this causes a proportion to be reflected while the rest continues. Thus examination of a reflected ultrasonic signal can reveal additional features in areas where there were cracks or voids, and using a source operating at, typically, 5–25 MHz enables flaws in brazes or diffusion bonded interlayers with diameters of about 1 to 0.1 mm to be detected. Such defects can be detected also by radiography using transmitted beams of X-rays whose attenuation can be measured by an absorption coefficient that depends on material characteristics, and hence enables the presence of voids and to a lesser extent of cracks to be detected by examination of images produced by the transmitted beam. Defect images revealed by radiography provide positive indication of the presence of flaws but care should be taken not to assume that the absence of such images is proof that the joint has a perfect structure.

A third but less well-established or widely used technique is thermography, which is particularly suitable for use with thin-walled structures such as the diffusion bonded/super plastic-formed sandwich structure sketched in Figure 2.5. The technique is based on the fact that the presence of flaws and inhomogeneities in the structure of a component can result in the formation of hot spots at its surface. Thus a component structure heated by infrared lamps will exhibit surface temperature variations that can be revealed by using thermal sensitive materials such as phosphors. The hottest spots will be produced where the component is thickest and large areas of voids and cracks will show up as cool dark regions. More sophisticated applications of the technique expose one face of a component to a flash of infrared radiation while monitoring the temperature of the opposing face or scanning it using an infrared detector.

The information about the macro-structure of the joints provided by these non-destructive techniques is indirect, and depends on the flaw having an effect on some physical characteristic of the joint. Usually of more importance is the effect that flaws will have on the mechanical properties of joints, and therefore the proof testing of a small percentage of the products is sometimes used to provide additional confidence in the results of the non-destructive testing. Proof testing involves stressing the component to a level typically 10 or 20% higher than it is expected to encounter in service. If the component survives without apparent degradation being revealed when assessed again using non-destructive techniques, it is considered fit for service. However, reservations can be felt about this approach because it is not certain that a once-stressed component will be able to survive a second exposure to a high stress once in service. If the component should break when stressed, examination of the fracture surface for evidence of inclusions, voids and other flaws can assist identification of the cause of failure.

A totally destructive but direct method of establishing whether or not a joint is free of voids, cracks and other macroscopic flaws is to prepare it for metallographic examination. In essence this involves the cross-sectioning of a component or test sample in a plane normal to the joint, mounting in some holding material such as Bakelite and grinding and polishing the cut surface to a fine finish before examining its structure for evidence of flaws using optical or scanning electron microscopes. Those unfamiliar with the technique will find more information in Appendix C and should refer to authoritative texts such as those by Greaves and Wrighton (1967) and the American Society for Metals (1973).

7.2 MICROSCOPIC STRUCTURE AND CHEMISTRIES

7.2.1 Characteristics

Even if a joint has a satisfactory macroscopic structure, it may still perform poorly in service due to its detailed microscopic structure or microchemistry, both of which can have a profound influence on mechanical and other properties. In the special case where the components are of the same material and are being joined by diffusion bonding without the use of an interlayer, the ideal microstructure is that of a continuation of the grain structure of each component. The original location of the interface should be unmarked by the presence of markers such as microscopic voids or remnants of surface films of oxide or even of small variations in microchemistry such as high solute concentrations of O or of the surface oxide cation.

In the more common case of joints that take the form of a macroscopic seam between the components, the structure and chemistry is usually far from ideal and their performance can be affected by the presence of porosity, inclusions and unbonded interface regions even less than a micrometre in diameter which will not have been revealed by examination of joint macrostructures. In addition to such flaws, the chemistry and hence the identity and often also form of the phases present in a joint can determine its mechanical behaviour, and the presence of brittle phases and in particular a continuous central layer of a brittle phase is especially detrimental. Thus the ideal microstructure of a seam joint is that of a single-phase material whose chemistry and mechanical properties such as micro-hardness are closely similar to those of the components that have been joined. Additionally the joint material must be in intimate contact with the components, entering every minute crevice and cranny present on their bonding surfaces. This intimacy often involves or results in the formation of a layer of reaction product at the joint/component interface and it is important that this layer is free of cracks and in practice usually has a thickness of no more than one or two micrometres.

7.2.2 Assessment

For both seamless and seamed joints, assessment of microscopic aspects provides not only a basis for understanding the service behaviour of joints but also for appreciating the processes involved in joint formation. It is important, therefore, to gain the fullest practicable information from techniques for characterizing the microstructures and microchemistries of joints at the microscopic scale, and of observing the kinetics of joint formation in real time.

The most widely used and usually adequate technique for characterizing microstructures and also gaining insight into the associated microchemistry is the examination of metallographic structures, the preparation of which is described in Appendix C. Examination using an optical microscope is usually facilitated by etching the polished surfaces to reveal the shapes and sizes of the grain and phase structure of the joint. This technique can also provide information about the microchemistry of the joint by identifying the particular phases present through their differing sensitivity to etching, especially when this produces distinctively coloured surface films that reflect the phase chemistries.

Beam devices that operate in evacuated chambers, and particularly electron beam devices such as a scanning electron microscope, are now widely used to provide detailed microstructural and microchemical information about samples or sometimes even small components. Microstructural observations at magnifications of 100 000 × or more can be made using scanning electron microscopes which have the advantage of possessing a greater depth of resolution than their optical counterparts. Additionally, semi-quantitative chemical information can be derived from analysis of the X-rays generated by interaction of the electron beam that can penetrate metal surfaces to the depth of a few microns. This information can be presented either as a graph of intensity variations of a particular spectral line generated by a specific element during a single scan across the sample or as a two-dimensional map of the intensity variations revealed by rastering. Similarly, precise quantitative information can be obtained by electron probe micro-analysis, EPMA, of the spectra of X-rays generated by a finely focused electron beam, with spot sizes as small as 1 μm diameter and sensitivities of 100 ppm or better. The compositions of the outer atomic layers of surfaces can be obtained by even more sophisticated techniques that analyse the spectra of electrons generated by surfaces bombarded with X-rays, XPS, or electrons, AES, or of the secondary ions produced by bombardment with heavy ions, SIMS.

Some of these techniques, and in particular scanning electron microscopy and EPMA, can be used to examine the fracture surfaces of joints that have been stressed to destruction, or simply have not bonded, and hence to identify the structural features and microchemistry associated with mechanical failure and the inhibition of bonding. Additionally, sometimes real-time observations of the melting and flow of braze alloys, and phase changes and growth of reaction products occurring at bonded interfaces can be made using hot stages fitted into optical or scanning electron microscopes as illustrated in Figure 7.1.

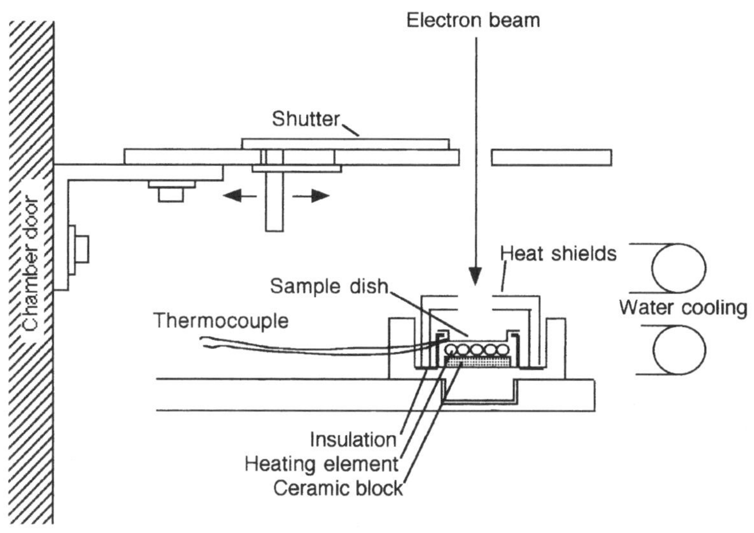

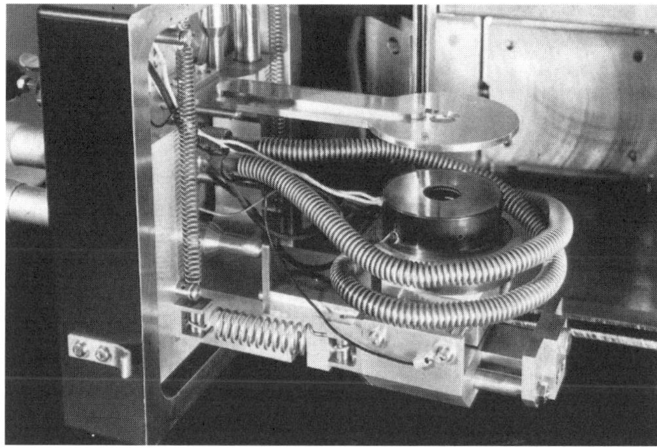

Figure 7.1 Sketch and appearance of a water cooled hot stage fitted on to the door of a scanning electron microscope. This particular stage can be used to produce video images of braze melting and flow in vacuum at temperatures of up to about 900 °C. (Courtesy Dr J. C. Ambrose)

7.2.3 Observations

It is impractical to comment even superficially on each of the multitudinous material combinations that are used in brazing and diffusion bonding, but some of the frequently observed features can be illustrated by reference to the microscopic characteristics of a few systems that cast light on the development of joint structures or can be used to refine the selection of materials and process parameters.

Elimination of interfaces

Microscopic examination shows that conversion of separate samples of Cu or of Ti into one macroscopic unity by diffusion bonding involves the dissolution of their surface oxides and the growth of grains and migration of grain boundaries to eliminate all evidence of the original location of the interface. Thus anything which impedes grain boundary movement, such as residues of the oxide films originally on the metal surfaces, will inhibit the achievement of an ideal structure.

In practice, the components to be joined are rarely of exactly the same material and joining by brazing or by transient or partial transient liquid phase bonding necessarily involves the introduction of dissimilar material into the joint. Even when diffusion bonding, it is a common practice to use interlayers to promote the bonding process between components of both identical and dissimilar materials. Thus interdiffusion across an interlayer/component interface usually occurs during joining processes, and this can change the local chemistry and ultimately result in the complete homogenization of the component/joining material couple and disappearance of the interlayer as a distinct phase. Such a disappearance was observed by Ceccone *et al.* (1995) and is illustrated in Figure 7.2 for a pair of Ni-20Cr sheets joined by coating their mating surfaces with 2.5 μm of Au and heating under a small pressure for 4 hours at 1000 °C to effect transient liquid phase bonding. This treatment caused the Au to disappear as a separate phase and decreased its concentration at the original centre line of the joint to little more than that deep within the Ni-20Cr, a clear demonstration that the joining process was successful. However, this experimental observation also provides some validation for the modelling of the bonding process.

The phase stabilities and diffusion coefficients for the Ni-Cr-Au system are not known, but well-established data are available for the Ni–Au system that show there is complete mutual solubility of Au and Ni at 810–955 °C and that a liquid forms with an initial composition of Au–15Ni when the temperature rises above 955 °C. The diffusion coefficient of Au in Ni at 1000 °C is $9 \times 10^{-15}\, m^2.s^{-1}$ and, depending on which specific model of the transient liquid phase bonding process is used, this leads to

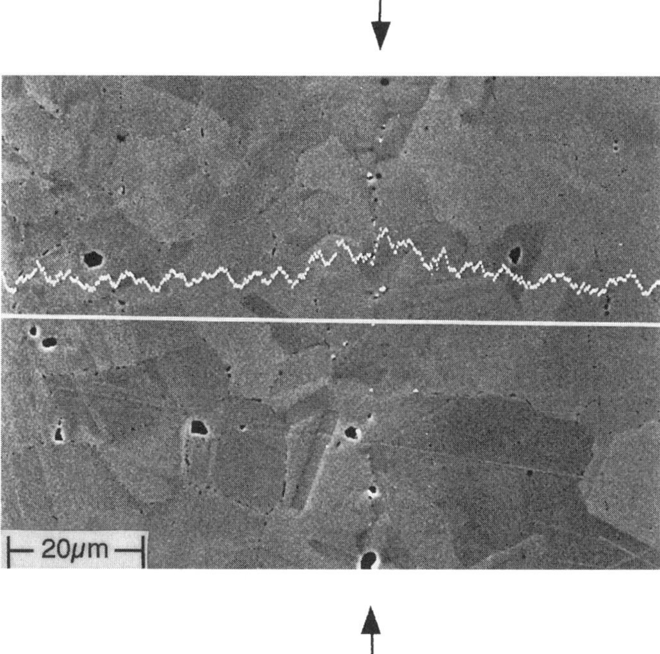

Figure 7.2 Scanning electron micrograph of a near homogeneous cross-section through a pair of sheets of commercial quality Ni-20Cr coated with Au and joined by transient liquid phase bonding. The arrows mark the centre of the original metal–metal interface and the flat white line shows the location of the line scan. (Courtesy Dr S. D. Peteves)

predicted times of 2.9 to 4.9 hours to produce isothermal resolidification of the original interlayer material as an alloy with a composition of Ni-15Au, in reasonable agreement with the experimental observation.

Formation of reaction product layers at interfaces

Quite often, but not always, interdiffusion between dissimilar component materials or between a joining material and the component results in the formation of a new phase that appears as a reaction product layer at the component/component or joint/component interface. Thus a diffusion bonding study by Morretto *et al.* (1993) involved microstructural examination of interfaces produced by using Ni-20Cr interlayers to join Si_3N_4 and a bonding chamber that had been backfilled with 1 bar of Ar that contained 50 ppm of N_2 and, therefore, was sufficiently N active to prevent dissociation of the ceramic. The equilibrium phase diagram of the Ni-Cr-Si-N system is not known but even that recently determined for

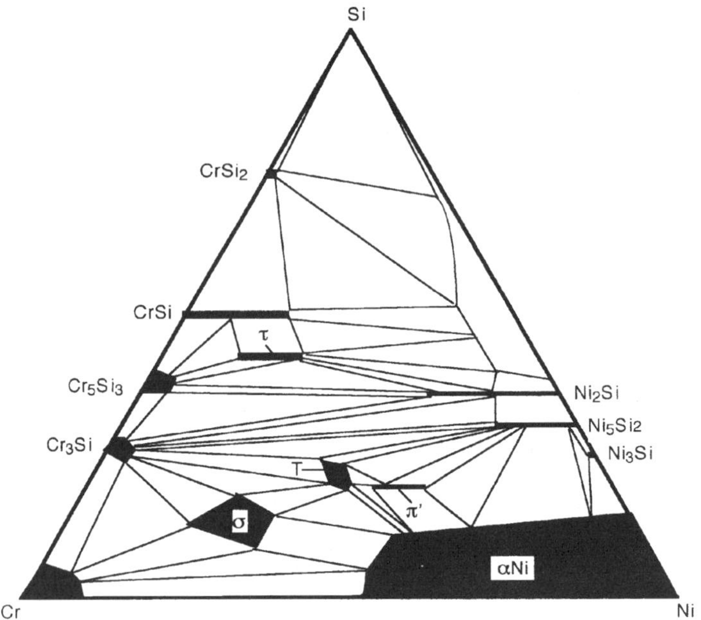

Figure 7.3 A 1050 °C isothermal section through the Ni-Cr-Si phase diagram. The solid black regions identify solid solutions and compounds. After Ceccone *et al.* (1995)

the Ni-Cr-Si system, shown in Figure 7.3, contains many compounds. However, the microstructures revealed by Morretto were relatively simple, as can be seen by examination of Figure 7.4. A reaction product layer of a $(Cr_{0.79}Ni_{0.12}Si_{0.09})N_{0.86}$ variant of CrN was formed on the ceramic surface and this impeded the ingress of Si sufficiently to ensure that an inner layer of Ni_5Si_2 adjacent to the bonding foil was formed only after a prolonged time. The formation of a layer of a nitride rather than a silicide immediately adjacent to the ceramic is noteworthy since reaction with the ceramic releases both N and Si in the ratio of 4 to 3, and the standard free energies of formation CrSi and $CrSi_2$ are more negative than those of CrN or Cr_2N. However, the microstructural observations demonstrate that the *chemical activity* of N dissolved in the Ni-20Cr must have risen more rapidly than that of Si, and this is in accord with the lower solubility of N in Ni, which at 1100 °C is less than 0.01% as compared to 8% for Si. Similarly, the free energy of formation of Cr_2N is more negative than that of CrN except at high N_2 pressures and therefore the observed formation of a variant of CrN shows that the high pressure used to achieve diffusion bonding both trapped and compressed the N_2 released by reaction of the Ni-20Cr with the Si_3N_4.

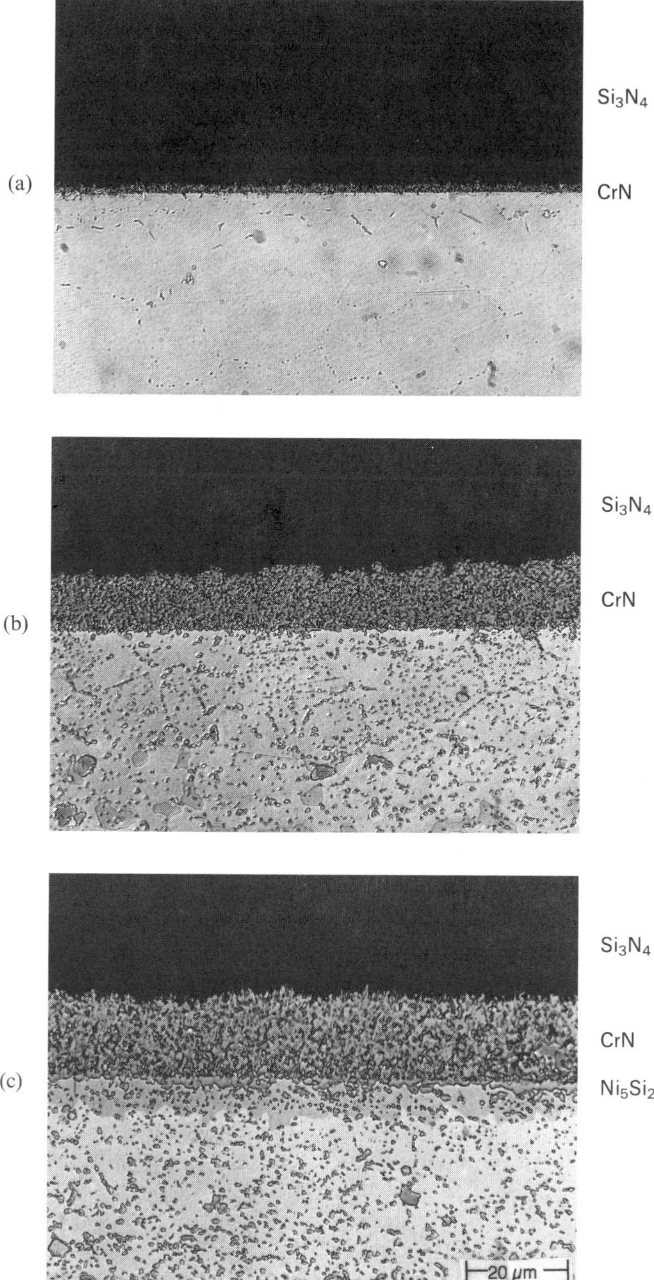

Figure 7.4 Optical micrographs of the interfaces formed by Si_3N_4 diffusion bonded in Ar with foils of Ni-20Cr using a pressure of 50 MPa applied at 1100 °C for (a) 15 minutes, (b) 25 hours and, (c) 100 hours. (Courtesy Dr S. D. Peteves)

Also noticeable in the microstructures shown in Figure 7.4 but not directly relevant to the bonding process is the presence of more precipitation within the Ni-20Cr as the bonding time was increased, so that the occasional Cr_2N precipitates present after 15 minutes at temperature were supplemented by more numerous and larger precipitates shown by EPMA studies to be N enriched variants of a Cr_3Ni_2Si phase that phase diagram studies show to be in equilibrium with the α(NiCrSi) solid solution, Ceccone *et al.* (1995).

Other work shows that strong joints can be produced by diffusion bonding Si_3N_4 with a Ni-20Cr interlayer for about an hour at 1100 °C, and Figure 7.4 suggests that this will be associated with a reaction product layer thickness of a few micrometres. Such thicknesses are commonly encountered during microstructural examination of well-bonded interfaces, but occasionally much thinner reaction product layers are formed. Thus the strong bonding of diamonds achieved by contacting them with liquid Cu-Cr alloys at 1150 °C for 20 minutes was associated with the formation of a 0.16 μm thick reaction product layer of Cr_3C_2, while the strong interfaces produced by diffusion bonding with Cu-0.27Cr at 800 °C for 4 hours was associated with a reaction product layer thickness of only about 0.01 μm, 10 nm. Why success was achieved with reaction product layers so much thinner than those usually encountered is not known for certain and the effects of interfacial chemistry and microstructure on the mechanical properties of joints is an important and current field of materials research.

The microstructures shown in Figure 7.4 were achieved by diffusion bonding at 1100 °C in Ar and similar results are produced by bonding at 1200 °C, although the growth of the reaction product layer is more rapid. However, a notably different microstructure is produced if bonding at the higher temperature is performed in an evacuated chamber, as illustrated in Figure 7.5, Peteves and Nicholas (1992). The reaction product layer adjacent to the ceramic is similar in composition to the $(Cr_{0.79}Ni_{0.12}Si_{0.09})N_{0.86}$ shown in Figure 7.4, but the inner interdiffusion layer is a N enriched variant of Ni_3Cr_2Si and extending further into the Ni-20Cr foil is a zone composed of islands of α(Ni,Cr,Si) in a matrix of Ni_5Si_2. Thus there had been a much more vigorous ingress of Si and this can be attributed to the evacuation of the bonding chamber since this will have pumped away most of the N_2 released by dissociation of Si_3N_4 while the Si will have remained at the interface. The evacuation of the chamber, therefore, changes the relative proportions of Si and N available to diffuse into and react with the Ni-20Cr interlayer. In this case the change was sufficient to permit formation of a series of brittle low melting temperature Ni-Cr-Si compositions.

The combined thickness of the reaction product layer and the interdiffusion layer shown in Figure 7.4 increases as does $(\text{time})^{1/2}$ because the

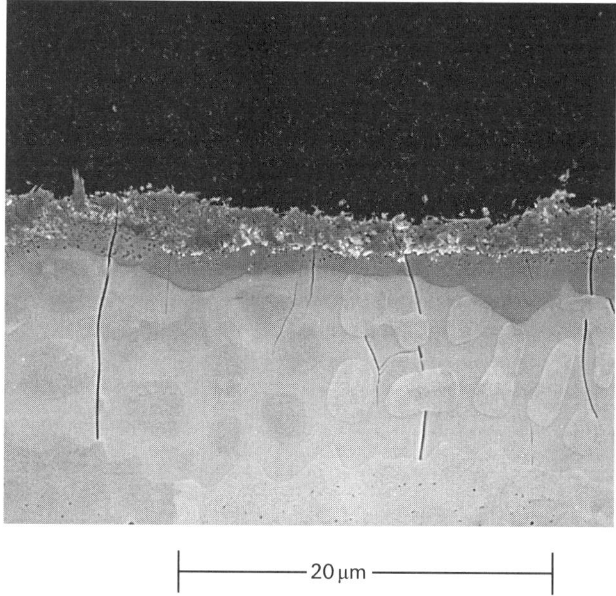

Figure 7.5 Cross-section of a Ni-20Cr/Si_3N_4 interface produced at 1200 °C in an evacuated chamber. (Courtesy Dr S. D. Peteves)

rate of thickening is controlled by the supply of Cr and hence by the diffusion kinetics of Cr within the Ni-20Cr interlayer. Such a time dependence of the reaction product layer thickness is observed for many, indeed for most, interacting systems but the actual rate of thickening depends on the temperature and on the detailed chemistry of the interaction. The activation energy for growth of the CrN variant reaction product layers shown in the figure is 266 kJ.mole^{-1}, in excellent agreement with the 264 kJ.mole^{-1} for the diffusion of Cr in Ni-20Cr, and this indicates that the growth rate will be decreased a thousand times when the temperature is lowered to 800 °C.

The variation of growth rates with the chemistry of the reaction products is difficult to predict, but sometimes it is possible to discern systematic trends in experimental observations. Thus Table 7.1 shows the relationship between the thicknesses of layers of intermetallic compounds formed at the surfaces of various metals and alloys diffusion bonded with Al foils by applying a pressure of 50 MPa for 30 minutes at 600 °C. Al reacts with many metals to form binary intermetallic compounds whose formulae, such as $FeAl_2$, $FeAl_3$, Ni_2Al_3, $NiAl_3$, CuAl and $CuAl_2$, are reflected in the EPMA derived compositions of the reaction products reported in the table. The layer thicknesses of the different compounds can vary by a factor of ten, and the diffusion

Table 7.1 Thicknesses of the intermetallic layers formed by some materials diffusion bonded using Al foils. Bonds were produced in an evacuated furnace at 600 °C by applying a pressure of 50 MPa for 30 minutes. Nicholas and Crispin (1982)

Component material	*Intermetallic layer Composition and thickness,* μm	
BS310, (Fe-24Cr-20Ni)	$(Fe_{0.54}Cr_{0.24}Ni_{0.19}Mn_{0.01}Si_{0.02})Al_{2.08}$	4
	$(Fe_{0.52}Cr_{0.25}Ni_{0.14}Mn_{0.01}Si_{0.05})Al_{3.15}$	4
BS431, Fe-16Cr-2Ni	$(Fe_{0.81}Cr_{0.14}Ni_{0.03}Si_{0.02})Al_{3.12}$	10.5
Inconel 600, Ni-16Cr-7Fe	$(Ni_{0.85}Fe_{0.11}Cr_{0.02}Si_{0.02})Al_{5.02}$	5
	$(Ni_{0.68}Cr_{0.19}Fe_{0.09}Si_{0.04})Al_{3.44}$	7
Ni	$NiAl_{1.71}$	10
	$NiAl_{3.31}$	8
BS321, Fe-17Cr-9Ni	$(Fe_{0.71}Cr_{0.18}Ni_{0.07}Mn_{0.02}Si_{0.02})Al_{2.63}$	11.8
	$(Fe_{0.68}Cr_{0.16}Ni_{0.09}Mn_{0.02}Si_{0.05})Al_{3.25}$	8.8
Invar, Fe-36Ni	$(Fe_{0.62}Ni_{0.37}Si_{0.01})Al_{3.01}$	20
	$(Fe_{0.50}Ni_{0.47}Si_{0.03})Al_{4.70}$	2
Cu	$CuAl_{1.08}$	17
	$CuAl_{2.25}$	7

coefficients range from 5×10^{-13} to $5 \times 10^{-15}\,m^2.s^{-1}$ for the two intermetallics formed by Invar. The reason for such a difference is unknown but the data for the steels show a trend for the reaction product layer thickness to increase as their Cr content decreases, and it is noteworthy that the behaviour of the Inconel 600/Al couple also fits this trend.

Eutectic structures

The joint structures produced by brazing are similar to those formed by diffusion bonding using interlayers, but generally the microstructures are made more complex in detail by the greater multiplicity of elements in braze compositions. Thus brazes are usually variants of eutectic compositions that often have microstructures containing brittle materials. Al brazes for example are usually Al rich variants of the Al-13Si eutectic composition and cooling such a liquid braze will at first cause rejection of αAl containing up to 1.6Si and then eutectic solidification of the remaining liquid. Thus the microstructure produced by cooling an Al-6Si alloy is shown in Figure 7.6 to consist of dendrites of relatively soft αAl, that will have been nucleated as soon as the temperature fell below the liquidus at 622 °C, embedded in a eutectic mixture formed by near instantaneous solidification of the residual liquid as the temperature fell below 577 °C. If the Si content of the alloy had been somewhat lower, more αAl would have formed and the dendrites would have linked up to become the matrix phase with the eutectic mixture being confined to isolated islands. Whether present as a matrix or as isolated islands, the structure of the eutectic mixture itself is a matrix of αAl matrix in which are embedded small platelets of Si, and the smallness of these platelets is important because Si is quite brittle and may crack when stressed.

Not all brazes are simple binary eutectic compositions but microstructural studies can still reveal useful information. For example, the introduction of Mg promotes the flow of Al-Si braze alloys that are frequently used as a thin cladding on sheets of a constructional Al alloy

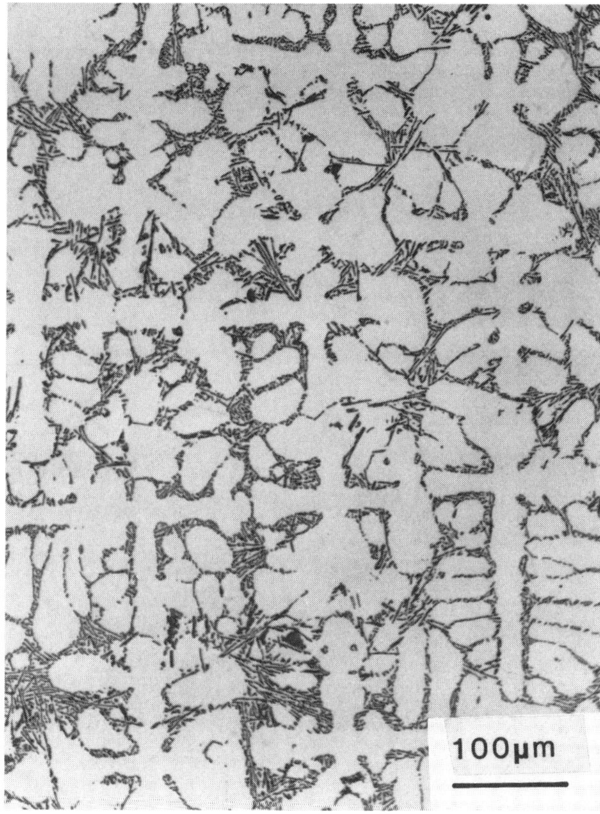

Figure 7.6 Microstructure produced by cooling an Al-6Si alloy at 4.5 °C.s^{-1} from the molten state. (Courtesy of Dr R. N. Grugel)

such as 3003. Phase diagram studies have identified an Al-12.9Si-4.9Mg eutectic composition that melts at 555 °C, lower than the 577 °C melting temperature of the Al-Si eutectic composition. Thus, complex microstructural changes occur when the braze is heated during a vacuum brazing cycle as shown in Figure 7.7 by the micrographs of quenched samples. The braze cladding in the as received condition has an Al solid solution matrix populated with inclusions of Si and Mg_2Si, but the sample quenched from 555 °C contains bubbles that are associated with Mg rich regions and which grew in size and number as the braze was heated to the melting temperature of the Al-Si eutectic. However, no bubbles or Mg_2Si inclusions are present in the sample that had been quenched from the recommended use temperature of 590 °C and EPMA studies revealed only a trace of Mg to be present. The interpretation of these structures and analyses is that Mg had vaporized to form bubbles once the ternary eutectic temperature had been exceeded and some liquid was produced, and that this vaporization was so intense that virtually all the Mg was lost

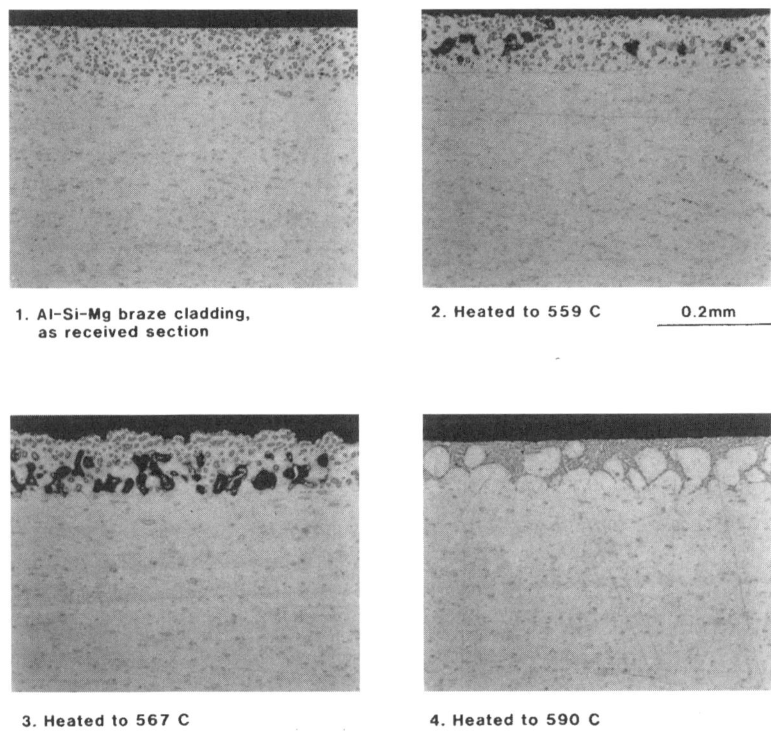

Figure 7.7 Micrographs of Al-Si-Mg braze alloy cladding in the as-received condition and after being quenched from various temperatures. (Courtesy Dr J. C. Ambrose)

by the time the use temperature was reached. Thus the unevenness of the free surface of the braze cladding shown in some of the micrographs can be related to the formation of bubbles that have not yet escaped from the cladding. This interpretation of the microstructural studies has been confirmed by real-time observations of the melting of Al-Si-Mg brazes using a hot stage mounted in a scanning electron microscope. As shown in Figure 7.8, heating caused the surface of the braze cladding to be disrupted by eruptions of a pasty liquid–solid mixture once a temperature of 555 °C was exceeded, and mass spectroscopy of the gaseous environment revealed the sudden presence of Mg, and of MgO formed by reaction with residual O_2 within the evacuated chamber, as the temperature was raised from 555 to about 600 °C.

Such evidence from microstructural and other studies enables the melting behaviour of Al-Si-Mg alloys to be understood reasonably well and provides a clear framework for interpreting the effects of compositional variations and changes in process parameters such as cleaning procedures, vacuum quality, and heating rate. The essential requirements for achieving good flow of Al-Si-Mg brazes is for the surface oxide to be thin, for the time spent between the ternary and binary eutectic temperatures to be the minimum needed to achieve vaporization of the Mg and for the vacuum to be good enough to prevent the surface becoming coated with a thick layer of MgO.

Many brazes are based on eutectic compositions and structures similar to that shown in Figure 7.6 are observed quite commonly with binary and more complex formulations. Thus the Cu-Ag-P system contains three binary eutectic compositions which form valleys in the liquidus surface leading to a ternary eutectic Cu-20Ag-6.9P composition that melts at

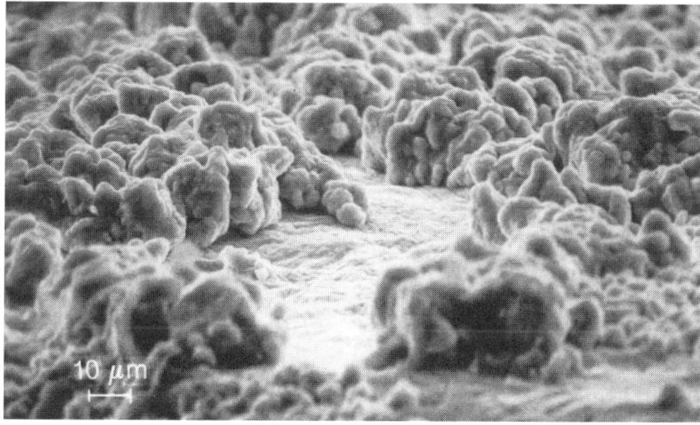

Figure 7.8 Scanning electron micrograph showing the eruptions on the surface of an Al-9.3Si-1.45Mg alloy held at 560 °C for 10 minutes in an evacuated chamber. (Courtesy Dr J. C. Ambrose)

646 °C. These phase relationships are exploited by the BCuP-5 braze which has a Cu rich hypo-eutectic composition of Cu-14.5Ag-4.5P that solidifies as dendrites of Cu-2Ag-1.5P embedded in a matrix of the eutectic mixture. Similarly, although used at significantly higher temperatures and intended for markedly different service conditions, the microstructures of Ni brazes based on the Ni-B, Ni-P and Ni-Si eutectics resemble those of the Al-Si eutectic braze family in that they too contain brittle phases embedded in ductile matrices. In practice the Ni braze compositions are usually markedly hypo-eutectic so that the matrix phase is a solid solution that encases the embrittling phases. Thus the as-received microstructure of a BNi-5, Ni-18Cr-10Si, braze is a matrix of an α(Ni-24Cr-8Si) solid solution populated with inclusions of Ni_5Si_2 and the Cr_3Ni_5Si or T phase, in accord with Figure 7.3.

Microstructural changes in joint cores

Because of the complex chemistries of many braze alloy compositions, interdiffusion at the interfaces formed with the components being joined often produces microstructural changes not only at the interfaces themselves but also within the interiors of the joints. As when diffusion bonding, the microstructural changes at the braze/component interfaces can blur definition of joint boundaries or cause intermetallic layers to form. Thus when brazing AISI 316 components with a BNi-5 braze, interdiffusion of Ni, Fe and Cr after 4 hours at 1050 °C caused the original interface with the steel to be replaced by a broad zone that EPMA showed to be a γ(Ni, Fe, Cr) solid solution, Reid *et al.* (1994). In contrast, Figure 7.9 shows that using the same braze and brazing conditions to join W components caused major dissolution of them in the braze that resulted in the formation of thick layers of an intermetallic compound similarly shown to be a Cr enriched variant of the hexagonal intermetallic $(Ni_{0.625}W_{0.375})_4Si$ phase and of intrusions of a Cr enriched $(Ni_{0.375}W_{0.625})_4Si$ phase. The consequence of these changes is that the core of the joint formed with the AISI 316/AISI 316 couple became enriched in Si and solidified on cooling to produce hard cracked inclusions containing Ni_5Si_2 and the $Cr_3Ni_5Si_2$, π', phases while consumption of Si caused the core of the joint formed with the W/W couple to convert to a seam of ductile αNi.

As indicated already, the presence of large brittle inclusions within joints is mechanically detrimental and, therefore, the propensity of Ni brazes for intermetallic formation is of technological importance. This problem is well recognized and considerable attention has been paid to the effects of composition modifications to prevent the formation of continuous layers of intermetallic compounds at the centre lines of joints and

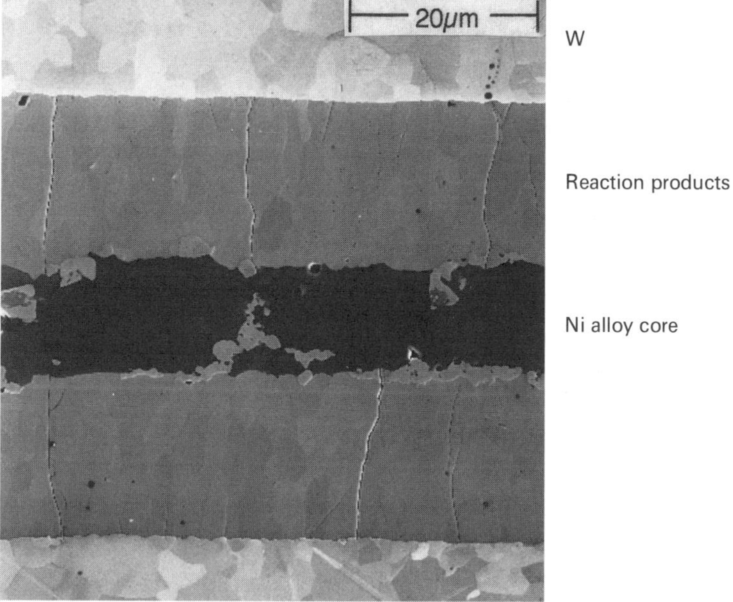

Figure 7.9 Scanning electron micrograph of the joint produced using an 80 μm thick foil of BNi-5 to vacuum braze W for 4 hours at 1050 °C. Reid *et al.* (1994)

more particularly to the effectiveness of prolonged heat treatment as a means of breaking up the layers into small fragments. This second approach has been successfully applied to disrupt the continuity of the intermetallic layers to produce an archipelago of islands that diminish in their size and number until none is left. The joint is then composed of relatively ductile αNi while the released Si, B or P is dispersed at low concentrations within the large reservoir provided by the component material.

An example of such behaviour is illustrated by the two scanning electron micrographs presented as Figure 7.10. These show the structure of joints between Ni-20Cr components produced using a BNi-3, Ni-4.5Si-3.1B braze. Holding at the fabrication temperature of 1110 °C for 20 minutes resulted in a joint containing a continuous central layer of Ni_3B and Ni_3Si inclusions, but prolongation of the holding time permitted the B and Si to diffuse into the component material and produced a joint that was free of inclusions. In this particular case, it should be noted that precipitation of Ni_3B and Cr_3B but not of Ni_3Si or Cr_3Si had occurred within the components close to the joint. This differentiation reflects the greater solubility of Si in Ni and Cr, 8 and 3% compared to less than 0.1% for B in both Ni and Cr.

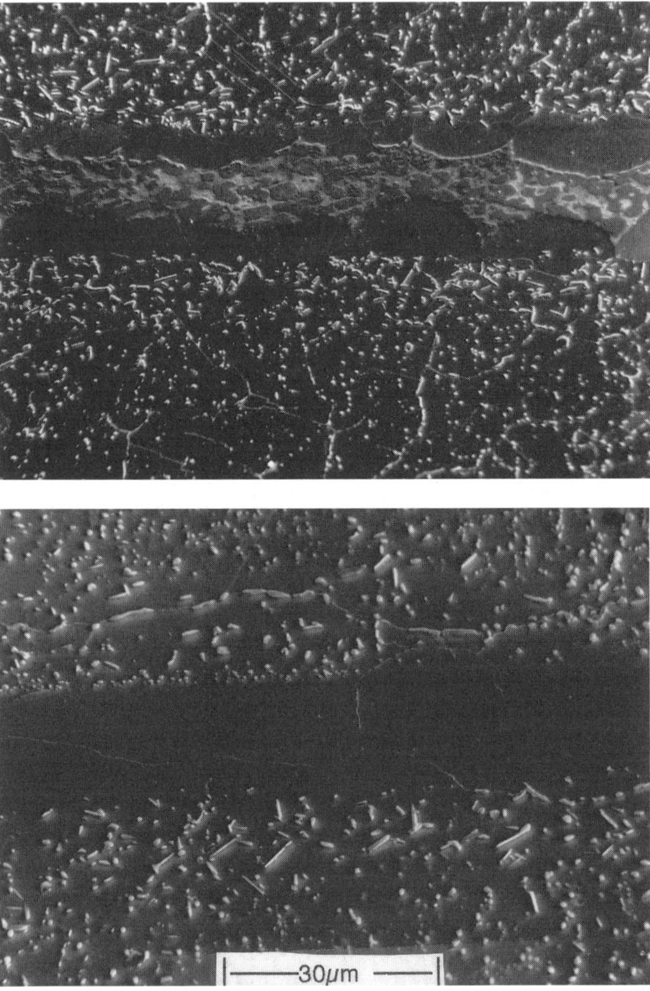

Figure 7.10 Scanning electron micrographs of Ni-20Cr joints produced by brazing with Ni-4.5Si-3.1B using holding times of 20 minutes, top, and 120 minutes, bottom. (Courtesy Dr W. F. Gale)

Microstructures produced by brazing ceramics

Not all eutectic alloys contain hard brittle phases and one of the prime attractions of the Ag-28Cu alloy is that its structure is composed of two ductile phases, a matrix of face centred cubic αAg containing up to 8Cu in which are embedded inclusions of face centred cubic αCu containing up to 8Ag. Additions of other constituents such as Cd, Pd, Ti or Zn degrade the ductility to some extent but enhance other characteristics. Thus the

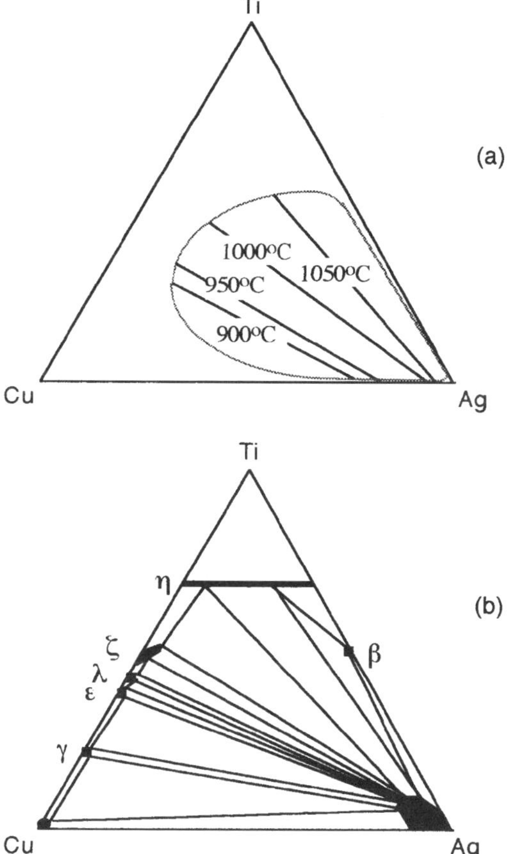

Figure 7.11 Phase diagram of the Ag-Cu-Ti system: the isotherms for 1300 °C (a), and 700 °C,(b). The black regions identify solid solutions and compounds. After Chang *et al.* (1977)

addition of small concentrations of Ti to Cu rich variants of the eutectic composition is the basis of the active metal brazes able to wet and bond ceramic components. The concentration of Ti in these alloys is usually restricted to about 2% since this is the maximum that can be contained within liquid Ag-28Cu without producing two immiscible liquid phases, Figure 7.11a, which will complicate the flow behaviour without achieving any enhancement of the ability of the alloy to form wettable reaction product layers.

The presence of Ti also affects the phases present in the solidified brazes, although as shown in Figure 7.11b it does not result in the formation of additional ternary phases. The basic structure of such solid brazes is of dendrites of αCu embedded in a matrix of the Ag-28Cu eutectic, of a mixture of αCu and αAg. However, EPMA studies show the Ti to be largely associated with the αCu dendrites and an uneven partition is to be expected since the chemical activity of Ti must be equal in all the phases and the solubility of Ti at the eutectic temperature of 780 °C is twice as large in Cu as in Ag.

Ti is even more soluble in Ni, to the extent of about 9% at the use temperatures of Ag-Cu-Ti active metal brazes, and this can be of importance when making joints between ceramics and Ni base alloys. Thus the use of a Ag-35Cu-1.75Ti braze to join Si_3N_4 to Inconel 909, an Fe-38Ni-13Co-4.7Nb-1.5Ti alloy, at 850 °C resulted in preferential reaction of Ti with the metal alloy to form variants of Ni_3Ti and this diminished the ability of the braze to wet and bond to the ceramic (Ljungberg, 1993).

The wetting of ceramics by brazes containing Ti is associated with the formation of reaction product layers and thermodynamic modelling described in Chapter 3 has suggested they should be hypostoichiometric variants of TiO, TiC and TiN. Microstructural and other studies of braze/ceramic interfaces have provided support for the thermodynamic predictions in that hypostoichiometric, metal rich, reaction products are formed, but not all detailed aspects of the microstructural and microchemical observations are susceptible to simplistic thermodynamic analysis. Thus the interaction of a Ag-35Cu-1.6Ti braze with Si_3N_4 results in excellent wetting, and X-ray diffraction analysis of fracture surfaces confirmed that TiN was formed adjacent to the ceramic and Ti_5Si_3 adjacent to the braze in accord with thermodynamic expectations. Again, scanning electron micrographs shown in Figure 7.12 demonstrate that the excellent wetting of ZrO_2 by a Ag-4Ti alloy is associated with the formation of a hypostoichiometric TiO, but the similar wetting by a Ag-26.5Cu-3Ti alloy is associated with the formation of other and more complex metal rich Ti-O reaction products.

While the Ag-Cu-Ti brazes have found successful applications in the manufacture of electrical, automobile and other mass produced components, their low melting temperatures mean that they cannot be used to fabricate components that exploit the good high temperature mechanical characteristics of the SiC and Si_3N_4 engineering ceramics. Ni brazes are obvious candidates for such applications and many studies have established that they wet these ceramics very well, but the joints produced are usually weak and brittle.

Ni forms eutectic compositions with Si that solidify to produce dense populations of Ni_3Si and other brittle intermetallic compounds and this is just as undesirable in brazed joints between ceramics as in joints between metal components. However, a chemical reaction requires the presence of an activity gradient as a driving force, and hence saturation of Ni with Si and N will prevent it from interacting to cause dissociation of Si_3N_4. Such

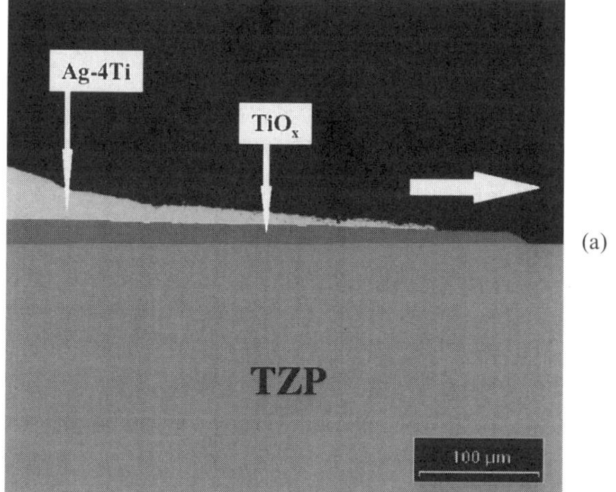

(a)

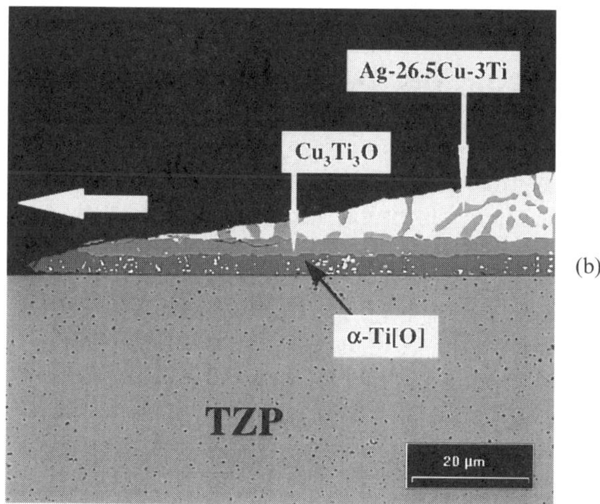

(b)

Figure 7.12 Scanning electron micrographs of cross-sections through sessile drops of active metal braze alloys on ZrO_2 substrates: Ag-4Ti in contact for 10 minutes at 1000 °C, (a), and Ag-26.5Cu-3Ti in contact for 10 minutes at 900 °C, (b). (Courtesy Dr S. D. Peteves)

considerations have resulted in attention being paid to the usefulness of Ni brazes containing Si for the joining of SiC and Si_3N_4, and in particular to the performance of the BNi-5, Ni-18Cr-10Si, braze whose Cr content confers the additional advantage of enhanced high-temperature corrosion resistance. However, the results of laboratory studies of joint microstructures have been disappointing.

Ceccone *et al.* (1995) have shown that producing joints between Si_3N_4 samples by vacuum brazing with 90 μm thick foils of BNi-5 for even 5 minutes at the recommended temperature of 1200 °C results in severe embrittlement of the joint due to a high population of inclusions of brittle intermetallic compounds. Producing such joints by holding for a more practical time of 15 minutes converts the joint into a matrix of hard and brittle Ni_5Si_2 that is enriched with 8.4Cr and in which are distributed inclusions of Ni_2Si and N enriched Cr_3Ni_5Si T phase. Such a conglomeration of brittle phases permits cracking to occur and propagate readily, so the ingress of Si to equilibrate the composition produced detrimental mechanical changes as well as affecting the phase structure of the joint. Similar degraded structures were revealed by examination of joints produced at 1150 °C using BNi-5 and also in joints produced using experimental brazes containing varying concentrations of Cr and Si.

Comparison of these microstructural observations with the phase diagram for the Ni-Cr-Si system shown earlier as Figure 7.3 suggests that brazes based on these alloys are inherently incapable of forming ductile joints with Si_3N_4 ceramics. Thus brazing temperatures of 1150 or 1200 °C, will cause the Si_3N_4 to dissociate in a vacuum and its surface to be covered with Si. However, the phase diagram shows that Si is in equilibrium only with $CrSi_2$ and a liquid phase whose composition incorporates that of NiSi. Thus the chemical interaction with Si_3N_4 of BNi-5 and similar braze alloys will continue until the joints are composed of brittle Si rich intermetallic compounds.

Similar limitations in usefulness have been revealed by microstructural examination of joints formed with SiC. McDermid *et al.* (1989) used BNi-5 to join reaction bonded SiC, ceramic grains held in a Si matrix, by heating in a vacuum to 1170 °C. Contact with the braze caused dissociation of the ceramic and release of C to react and form Cr_7C_3. The main outcome of contact with the braze, however, was penetration of the ceramic to a depth of several millimetres even after a holding time of a mere 5 minutes to decompose the SiC grains and form a melt containing about 30% of Si which subsequently solidified to form a porous structure of a Cr and C enriched variant of the brittle NiSi intermetallic.

Thus there is a clear need for new Ni braze formulations that wet the SiC and Si_3N_4 and react to form continuous layers of refractory products that not only bond the engineering ceramic to the braze but also protect it from excessive interaction and particularly from interactions that result in the formation of low melting temperature compositions. The difficulties of identifying suitable braze compositions is one of the main reasons why diffusion bonding using ductile interlayers is often regarded as the logical choice when the need arises to form joints that can withstand high service temperatures.

7.3 MECHANICAL PROPERTIES

7.3.1 Characteristics

An almost certain requirement in any joint specification is that it should be 'strong' or 'tough' and earlier in this book successful optimization of joining process parameters was usually judged in terms of the ability of the joints produced to satisfy such descriptive terms. In practice, what is usually meant by these terms is that the joint is able to survive the mechanical stressing that will be encountered in normal service conditions. These conditions are different for each application and hence so too are the minimum levels of mechanical characteristics required to yield a satisfactory service performance. Thus the stresses encountered by a brazed joint used to fabricate an item of jewellery are normally far less than those experienced by a joint in an aircraft engine component. Furthermore, these mechanical characteristics may have to be displayed at very different temperatures and environments.

It is good practice to choose locations and designs of joints that minimize the stresses they must endure, but nevertheless the ability to demonstrate that a joint is 'strong' or 'tough' generates confidence about its integrity and suitability for most applications. Partly by design and partly of necessity or accident, joints are frequently not subjected to simple tensile or shear stresses and in fact the only satisfactory test of the mechanical suitability of a joint is often a test that simulates service conditions. Unfortunately such tests usually yield results that cannot be interpreted in a fundamental manner or be used to build a generalized database that can include information derived from other sources or types of component. There are, however, a few exceptions that demonstrate that the design and selection of component materials influence the mechanical properties of joints as well as the choice of joining materials and process parameters.

7.3.2 Assessment

There are numerous mechanical properties that can be specified and hence many ways of making quantitative assessments but those of most concern and most commonly quoted for brazed or diffusion bonded joints are their ability to withstand

- brief but increasingly severe stresses during tensile, bend or shear tests;
- prolonged steady stresses at high temperatures during creep or stress rupture tests;
- varying or even oscillating stresses during fatigue tests;
- suddenly imposed stresses during impact tests;
- indentation caused by hardness tests.

Standardized testing procedures have been developed that are suitable for the assessment of monolithic materials, and those unfamiliar with them will find more information given in Appendix D.

Adaptation of such standardized tests to generate information about joint characteristics has resulted in many individual variants but some are used quite commonly. Thus ultimate tensile strength (UTS) values for metal–metal butt joints are commonly derived from the (failure load/ cross-sectional area) ratio of a test sample that has a narrow central portion in which the joint is located, as sketched in Figure 7.13. In practice the shapes of the joint tensile strength test pieces seldom adhere rigidly to those specified in the standards for monolithic samples because failure will

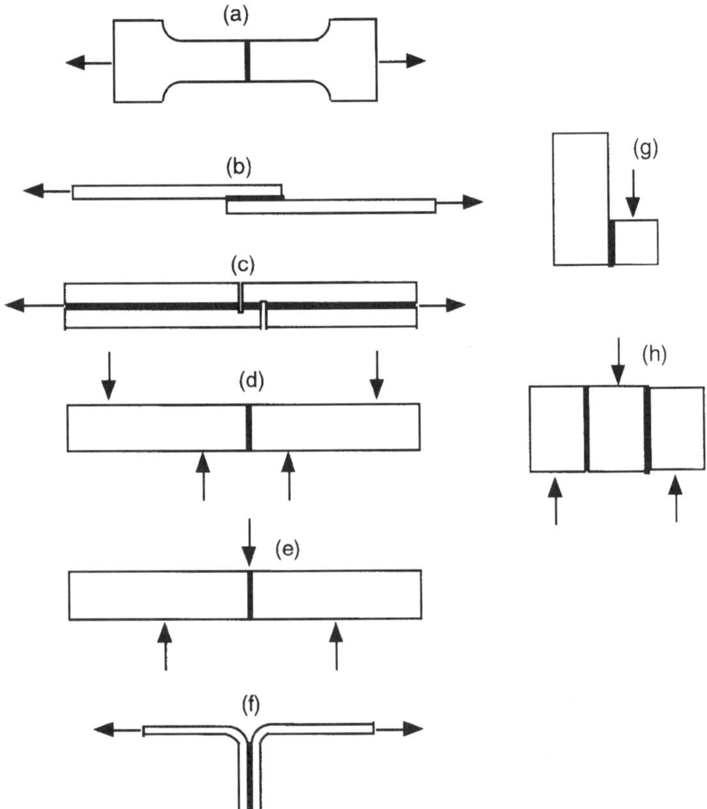

Figure 7.13 Sketches of some types of test pieces that have been used to characterize the mechanical properties of joints. From the top, they show cylindrical or rectangular section tensile test pieces, (a), lap test pieces, (b) and (c), and three four point bend test pieces, (d) and (e), a peel test piece, (f), chair shear test piece, (g), and a three ply shear test piece, (h)

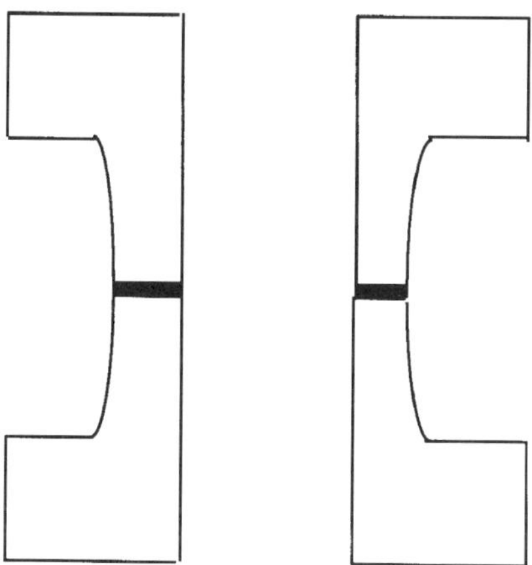

Figure 7.14 ASTM Tensile test piece for the strength of brazed and metallized Al_2O_3. After ASTM F19-64

occur within weak and even moderately strong joints using test pieces with modest non-standard narrowing of their central portions. In addition to the UTS, standard tensile test pieces can be used to derive ductility characteristics such as the elongation at which failure occurs and the reduction in the cross-sectional area of the narrow central portion before fracture occurs. While these additional parameters could be derived also from test pieces used to assess the tensile strength of brazed or diffusion bonded joints, this is seldom done because deformation of the test piece may not be confined to the joint region and it is difficult, therefore, even to rank the ductility characteristics of joints. However, such deformation of the test piece away from the joint and particularly failure away from the joint is a dramatic demonstration that the bonding process has been very successful.

Various configurations of wasp waisted test pieces have been used also to derive data on the stress rupture and fatigue characteristics of metal–metal butt joints, while peel tests have been used to assess the mechanical quality of metal–metal lap joints between thin sheets. In contrast to this profusion of non-standard tests that are used to characterize metal–metal joints, there was an early standardization of the configuration of tests to be used to assess the tensile strength of brazed joints between metallized Al_2O_3. The configuration selected for this particular application, sketched in Figure 7.14, has since been adopted for studies of joints produced by active metal brazing or diffusion bonding with an interlayer and even for ceramic–ceramic joints that include an insert. However, recently most mechanical evaluations of joints between ceramic components have employed bend tests. Tensile bend strength values for strongly bonded joints have been derived from notched test pieces similar to those used in the standard tests without the need to generate cracks at the notch roots before testing, and for weak or moderately strong joints it is common practice to use unnotched test pieces.

Shear testing has been used to assess the influence of processing parameters on the strength of brazed and particularly of diffusion bonded joints which can be produced readily in lap, chair and sandwich configurations such as those shown in Figure 7.13. Finally mention needs to be made of the use of micro-hardness testing to study the variation in mechanical properties within a joint. This technique is used in particular to assess the influence of heat treatment schedules on the homogenization of joint microstructures produced by Ni brazing or the transient or partial transient liquid phase bonding techniques.

7.3.3 Observations

Not surprisingly, the vast bulk of reported mechanical property data relate to measurements made at room temperature and to the stresses needed to break tensile, shear or bend test pieces. In general, joints between metal components are strong and tough, and perhaps because of this satisfactory situation the availability of mechanical property data for metal–metal joints is not as abundant as might be imagined given the vast number of joints that are produced every working day. In contrast, joints made to ceramics can be weak and lacking in toughness unless care is taken to optimize their design and the selection of materials and process parameters, and this frequent potential failing has resulted in relatively numerous reports of mechanical properties of ceramic joints.

One of the difficulties in using reported mechanical property data is the variety of tests from which they are derived. Fortunately data obtained from differing tests and procedures usually generate similar rankings of sample strengths, and hence can be used to assess the relative benefits of changing designs, materials or process parameters. While the quantitative results obtained from one type of test cannot be compared rigorously with those from another, a number of semi-empirical correlations exist that can provide some guidance. Thus monolithic or perfectly bonded metal test pieces break in shear at stresses that are generally 60–70% of those needed to cause failure in tension. Comparison is more difficult in the case of ceramic–ceramic and ceramic–metal joints because of their sensitivity to the presence of flaws, but Suganuma *et al.* (1986) made a direct comparison of the failure stresses in tensile and 3-point bend tests, for Si_3N_4 brazed to Invar using Al, in which fracture occurred in the intermetallic layers formed at the Al–Invar interface. The stresses needed to fracture the bend test samples were 2.5 times higher than those for the tensile test pieces and this was ascribed to the smaller volume of material that was stressed and hence to the lower likelihood of it containing a flaw that would initiate premature failure.

As might be expected, the majority of available mechanical property values are for joints made by brazing or diffusion bonding, but there is some information characterizing joints produced by hybrid processes. It would be impractical to comment on all the strength data that have been reported, but an attempt can be made to illustrate some of the principal effects caused by design, material and process parameters.

Design

Two of the main objectives of joint design are to minimize the stresses imposed on a joint by locating it in a low-stress area and to provide the maximum joint area. Design details can then be used to influence the capability of a joint to resist stresses. Thus causing the path of a brazed joint to be stepped or serrated will increase the failure load, while, more simply, increases of two to seven times in shear strength values have been achieved with super plastic formed diffusion bonded joints between the IMI 318 Ti alloy and a BS 304 stainless steel by machining grooves in the surface of the steel (Partridge and Baker, 1991), and a similar beneficial effect of undulations may contribute to the strengths achieved by explosively welded components.

More generally, the strongest joints have been found to be produced using small joint gaps between the components being bonded, as illustrated by the data for diffusion bonded Al_2O_3-Al joints shown in Figure 7.15. This effect of decreasing the joint thickness is caused by the relatively rigid components progressively exerting more transverse constraint on the deformation of the ductile material within the joint. According to slip line theory (Hill, 1950), non-elastic deformation of perfectly bonded joint material will not occur until a stress of σ_J has been exceeded where

$$\sigma_J = 0.5\sigma_Y(W^*/h^* + 3) \qquad \{7.1\}$$

and σ_Y is the tensile yield stress of the joint material, W^* and h^* are the width and thickness of the joint. In practice, failure of joints occurs soon after they start to deform and hence it is noteworthy that the curve drawn in Figure 7.15 was derived from equation {7.1} by assigning a value of 7.9 MPa to σ_Y.

Even though σ_Y values for joining materials are usually low, equation {7.1} predicts that the strength of thin joints can be very high and implies that the strength of the component materials will be the ultimate limiting factor. The maximum strength of the Al_2O_3-Al diffusion bonds produced by Klomp, therefore, is predicted to be the 90–100 MPa at which tensile failure of the ceramic occurs. Equation {7.1} predicts this would have

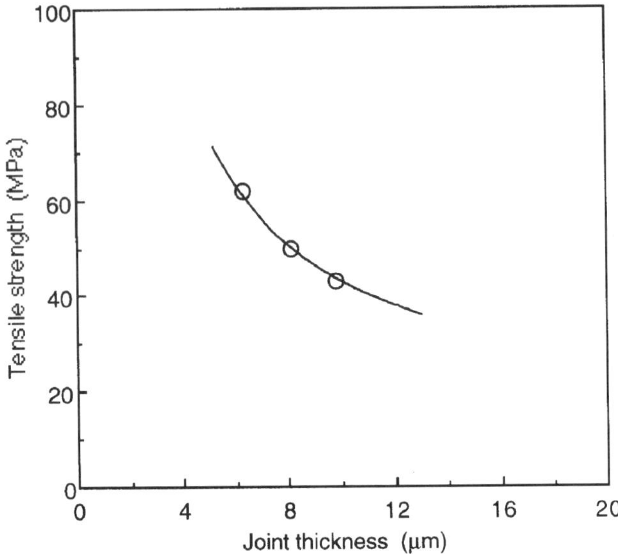

Figure 7.15 The tensile failure strengths of Al_2O_3-Al joints produced by diffusion bonding at 605 °C when using 1.0 mm diameter rings of Al as interlayers. After Klomp and van de Ven (1980)

been achieved when their thickness was decreased to 3.7 μm or less and similar critical thicknesses can be predicted for other materials systems. These thicknesses are generally very small and in practice may be too small to produce reliably. Thus it is difficult to machine the matching pairs of surfaces to the flatness needed to achieve the uniformity of gap separation that is required for the effective sweeping away of oxide debris and flux residues by the advancing liquid metal when brazing. Similarly, it can be impractical to achieve uniform mechanical and sometimes even chemical surface preparation of very thin interlayer foils when diffusion bonding.

Failure to achieve perfect bonding when using very thin gaps will result in lower strengths, and hence joint strengths will peak at the smallest joint thickness at which it is possible to achieve perfect bonding. Peaking of the strengths of brazed joints is well known and an early example of this is illustrated in Figure 7.16. The strengths of joints between perfectly bonded stainless-steel rods should increase as the joints are made thinner until they reach the steel UTS of 400 → 500 MPa for a joint that is 0.01 mm thick, but in fact the strengths peaked at a value of 105 MPa when the joint had a thickness of 0.05 mm. Once again, the curve showing the progressive decrease in strength at greater thicknesses was derived from equation {7.1} using a σ_Y of 26.9 MPa.

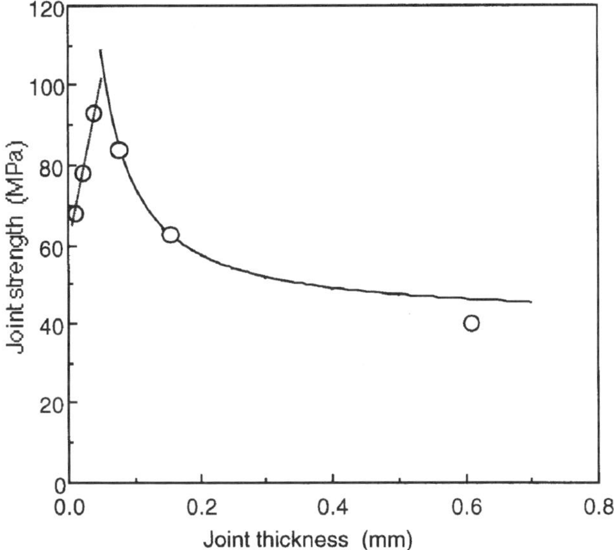

Figure 7.16 The tensile strengths of butt joints between austenitic stainless-steel rods brazed with a fluxed Ag-Cu-Zn-Cd alloy. After Udin *et al.* (1954)

The data presented in Figures 7.15 and 7.16 demonstrate that joints should be thin if they need to be strong. However, the tensile strength is not the only mechanical property to be affected by joint thickness. Thus toughness measurements with Cu brazed butt joints between low alloy steel rods show their toughness to *increase* progressively with the thickness of the joint (Baker, 1987). Hence the thinnest joints may be the strongest but they are also the most likely to fail prematurely because of crack propagation. Unless it can be guaranteed that no cracks or crack forming flaws are present, and it seldom can, it is more prudent to specify and accept a joint that is weaker than the maximum possible but which has some tolerance for flaws because of its toughness.

The strength data plotted in the previous two figures were derived from tensile tests performed at room temperature and the observations have been related to room temperature values of σ_Y. However, yield strengths almost invariably decrease if the testing temperature is raised and therefore, according to equation {7.1}, so should the stresses at which joints fail. However, in practice the effect of test temperatures on failure stresses is more complicated even for simple systems in which there is no significant effect of temperature on the chemistry or phase structure of joint interfaces. An example of the behaviour of such a simple system, Al_2O_3 brazed with Al, is presented in Figure 7.17. This shows data for both thick and thin joints with the failure stresses of the test samples with

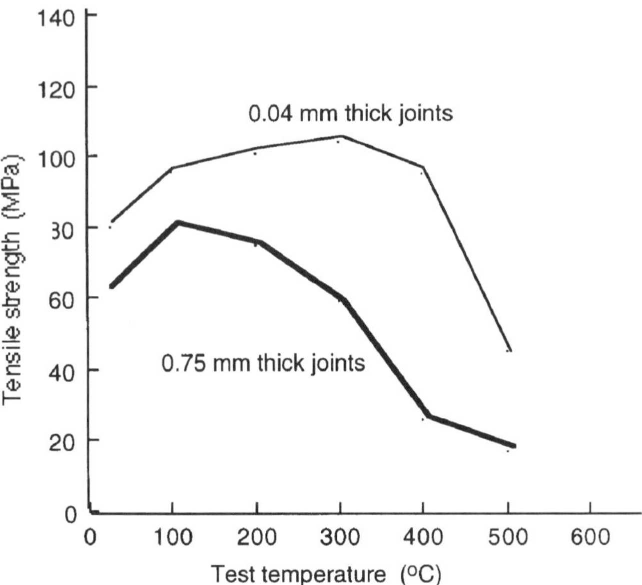

Figure 7.17 The tensile failure stresses of brazed Al_2O_3-Al samples plotted as a function of the testing temperature. The samples were truncated cones with bonding surface diameters of 5 mm brazed in vacuum for 10 minutes at 1000 °C. After Iseki and Nicholas (1979)

thinner joints being higher in accord with the expected greater constraint exercised by the rigid Al_2O_3 components. However, the experimental observations also show that the initial effect of raising the testing temperature is to increase the failure stress and this is not in accord with expectations generated by equation {7.1}. Further, the measured failure stresses do not decrease to values significantly less than those measured at room temperature until the testing temperature is raised to about 350 °C for the thicker joints and to 450 °C for the thinner joints. The reason for this behaviour is that the coefficient of thermal expansion of Al is much larger than that of Al_2O_3 so that the metal within the brazed joints will be stressed in tension as the samples cool and this will decrease the externally applied tensile stress required to cause failure of the joint. The initial effect of reheating towards the test temperature is to diminish this residual stress by causing the Al to expand more than the Al_2O_3, and hence to increase the external stress that has to be applied to cause failure of the joint. Ultimately the weakening of the Al at high temperatures, $0.60 \rightarrow 0.75\,T_M$, more than compensates for the benefit derived from the relaxation of residual stresses and the externally applied stress that is needed to cause joint failure decreases.

The data selected so far to illustrate effects of joint thickness and testing temperature have been for relatively simple systems and testing procedures. However, the effects are not limited to such systems, as is shown by these other examples.

- The impact resistance of stainless-steel rods vacuum brazed with a Au-18Ni alloy was increased from 72 to 100 J by diminishing the joint gap from 200 to 20 μm, Christensen and Sheward (1982).
- The shear strengths achieved when using Zn interlayers to promote diffusion bonding of an Al-Zn-Mg-Cu alloy increased from 86 to 228 MPa when the interlayer was thinned from 6 to 3 μm but then decreased to 132 MPa when the interlayer was further thinned to 1.5 μm (Livesey and Ripley, 1993).
- An increase from 0.1 mm to 1.0 mm in the thickness of Fe-13Cr BS 405 steel inserts diffusion bonded to Si_3N_4 by applying a pressure of 100 MPa at 1200 °C for 30 minutes decreased the tensile strengths of the samples from 50 to 9 MPa (Suganuma and Okamoto, 1987).
- When brazing low C steel with Cu, decreasing the gaps between 4 mm thick sleeves and 7 mm diameter rods from 0.08 mm to a nominal −0.05 mm increased the shear strength from 152 to 220 MPa (ASM, 1973).
- Nimonic 90 samples with 25 μm thick joints vacuum brazed with a Ni-Cr-Si-B alloy were fractured after 1000 hours by a stress of 60 MPa at 800 °C, but only 30 MPa were needed at 900 °C (Christensen and Sheward, 1982).
- The bend strengths of Si_3N_4 samples diffusion bonded with Fe-15Cr interlayers decrease about 15% as the testing temperature is raised to 850 °C, 0.62 T_M, but fall dramatically thereafter (Peteves, 1995).
- The bend strengths of Si_3N_4 joined by the PTLPB technique using Au coated Ni-20Cr foils are only 68% of the room temperature value at a test temperature of 400 °C, 0.4 T_M, and decrease somewhat more rapidly thereafter Ceccone *et al.* (1995).

Materials

The strengths of joints can be dominated by design factors, but the properties of the joining and component materials and the effects of joining material/component material interactions must also be considered when selecting a system that must form strong joints. Usually there is more opportunity to vary the choice of the joining material and equation {7.1} indicates that they have a direct effect on the mechanical properties of joints, so it is not surprising that most attention has been paid to the effects of these materials.

The most obvious effect of joining materials on the mechanical performance of bonded components is that they limit the service temperature. Clearly the melting temperature must not be exceeded, so that for example joints brazed with alloys based on the Al-Si eutectic or Ag-Cu eutectic compositions have maximum potential service temperatures of 577 °C or 780 °C, and many members of these families of brazing alloys will have lower melting temperatures because of the influence of ternary and quarternary alloying additions. In practice, service temperatures are usually much lower than those required to melt the braze or diffusion bonding interlayer and reflect the very low yield strengths of metals at temperatures above about $0.5\,T_M$ which degrade the strength of even very thin joints, as shown in Figure 7.17.

At modest temperatures below $0.5\,T_M$, the mechanical properties of joining materials can exert a direct influence on joint strengths, in accord with equation {7.1} which predicts that the strongest joints will be formed by the strongest metals if the joint interfaces are perfect or of comparable perfection. In practice, perfection or even comparable perfection of joint interfaces can be difficult to achieve and this is what often determines the mechanical properties. Sometimes the apparent perfection suggested by mechanical measurements can be an illusion because the presence of a few percent of small flaws often has a negligible effect on the tensile or shear strengths of joints. Early work on the diffusion bonding of Al suggested that the presence of insoluble fragments of surface oxides present at the bonded interface affected room temperature shear strengths in accord with the relationship

$$(B_J/B_0) = (1 - A_U)(1 + A_U) \qquad \{7.2\}$$

where B_J is the strength of the joint, B_0 is the strength of the parent metal and A_U is the fraction of the interface that is unbonded. More recent work has shown that diffusion bonded joints with Ti alloys are able to tolerate up to 20% of the joint interface being unbonded without significant degradation of their mechanical properties, in fair accord with the 4% decrease suggested by equation {7.2}. However, the fatigue strengths of joints can be degraded significantly by even a small percentage of voids and other flaws if they are clustered, and the impact resistance of joints is often sensitive to the presence of isolated flaws with diameters of a micrometre or even less. This sensitivity of fatigue and particularly of impact characteristics is due to the possibility of easy failure by the nucleation of cracks at flaws and their propagation unless blunted by plastic deformation. Hence values of the impact resistances of joints can be markedly influenced by the deformation characteristics of the material through which cracks have to propagate.

An example of the effects of material parameters on the impact resistance of joints is provided by a comparison of the properties of diffusion bonded

Table 7.2 Effects of material properties on the mechanical behaviour of diffusion bonded Ti alloy joints. After Baker and Partridge (1991)

Alloy	*IMI 318*	*IMI 550*
Composition	Ti-4Al-6V	Ti-4Al-4Mo-2Sn-0.5Si
UTS MPa	1050	1200
0.2 PS, MPa	990	1070
Ra, μm	0.20–0.27	0.20–0.27
Voids, %	2.1	2.4
Average size, μm	2.6	3.4
Fatigue strength, MPa	460	580
Impact resistance, J	7.0	4.2

rods of IMI 318 and IMI 550, for which data are summarized in Table 7.2. Applying a pressure of 0.7 MPa at 950 °C for 2 hours to carefully ground surfaces produced well-bonded joints containing only isolated small voids and with tensile strengths equal to those of the parent metals. The fatigue strengths were also unaffected by the isolated porosity and were similar for the two parent metal alloys, the stresses to cause failure of the joints after 10^7 cycles being 95–96% of the values for the parent metals. However, the impact resistances were low and different for the two alloys, the joints formed by the IMI 318 alloy being 35% as resistant as the parent metal while those formed by the less readily yielding IMI 550 alloy were only 22% as resistant. Similar changes in impact resistance that can be associated with material properties have been observed for brazed joints. Thus, high-quality joints between stainless-steel samples produced using a Au-18Ni braze alloy and 25 μm braze gaps had an average room temperature impact resistance of 100 J while those similarly produced with the stronger and stiffer BNi-2 or BNi-5 brazes had resistances of only 50 or 20 J.

The impact resistance values quoted above were derived from tests conducted at room temperature, but the most critical conditions for impact testing are at even lower temperatures where the component and joining materials can have diminished ductilities. Fortunately, service requirements for bonded components to endure such conditions are relatively rare, but in contrast requirements for joints to survive service at high temperatures are becoming commoner and are frequently encountered in advanced technology applications such as aircraft engines and power plants. Comment has been made already on the effect of temperature on σ_Y values but attention must be paid also to creep, the slow but continuous deformation of a material subjected to a low stress at a high temperature.

The creep behaviour of joining materials is not well characterized for two main reasons. First, it is good practice to use designs that place joints in cool regions. Second, the chemistries of joining materials can often be complex and can be changed by interaction with the component materials so that it is difficult to define test conditions that are relevant to more than a few specific applications. Nevertheless, there are some data available for joining materials such as the Al or Ni-20Cr foils used to promote diffusion bonding, and for the Ag-28Cu braze alloy. In the case of the braze alloy, finite element modelling has shown that creep can have a significant effect in decreasing the residual stress caused by mismatched coefficients of thermal expansion. For example, Stevens *et al.* (1991) calculated that creep of a Ag-28Cu braze alloy halved the residual stressing of a 0.05 mm thick joint between a Kovar end cap and an Al_2O_3 insulator tube. While creep in that case will have the beneficial effect of reducing the likelihood of the ceramic component cracking as it cools from the brazing temperature, the price that has to be paid is loss of rigidity by the bonded assembly as a whole.

In addition to the properties of the joining materials at low and high temperatures, those of the component materials can exert an influence on the mechanical characteristics of joints, despite the fact that they are not assigned any obvious role in equation {7.1}. This influence is due to two separate factors. The first of these is the effect of changes in the mechanical properties of the component materials on the constraint they exert on the deformation of the joint material. Equation {7.1} assumes that the component material is perfectly rigid at the stress levels used in the test, but this can be markedly at variance with reality when the components are made of metals or alloys. If the component deforms, the constraint it exerts on the deformation of the material in the joint will be less severe than anticipated. This can be an important effect, but usually of even more importance is the differing chemistry of component materials which promotes interdiffusion with the joining material. Since the joints are small in volume compared to the components, the effect of interdiffusion can produce marked changes in the chemistry and mechanical properties of the joining material as well as producing reaction product layers of new phases at the joint interfaces.

An illustration of the effects of changes in joint chemistry produced by interdiffusion is provided by the behaviour of Ni brazes used to join steels or Ni alloys. As described already, holding bonded components at the brazing temperature, or even at somewhat lower temperatures, can cause the microstructure of the joint to become single phase because the B, P, Si constituents diffuse into the components. This change in chemistry also causes changes in the mechanical properties of the joining materials with the αNi matrix becoming softer as more of the B, P or Si diffuses into the component material. Thus Figure 7.18 shows that the hardness of the

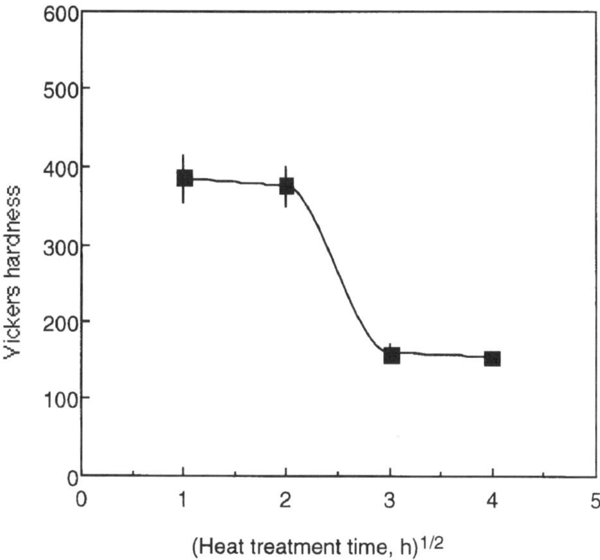

Figure 7.18 Microhardness values for the centre line of joints between AISI 316 samples vacuum brazed with BNi-5 for 15 minutes at 1150 °C and then held for varying times at 1050 °C. After Reid *et al.* (1994)

matrix phase in a BNi-5 joint between austenitic stainless steel fell sharply once saturation levels of Si were no longer maintained because the Ni_3Si and Cr_3Si inclusions had been dissolved. However, it should be noted that the softening of these joints is often accompanied by an enhancement in some mechanical properties rather than a degradation because it is associated with the removal of cracks that can cause premature failure at low externally applied loads.

The removal of B, P or Si from joints formed by Ni brazes means that the component materials become alloyed with embrittling species, albeit only transiently at mechanically significant levels. The effects of such alloying depend on the identity of the mobile elements, and the measurements of the UTS and fatigue strengths of AISI joints formed with various Ni braze alloys are shown by the data assembled in Table 7.3 The brazing conditions used varied from alloy to alloy but were selected to ensure that there were no centre line seams of intermetallic compounds and the quoted strength values were all for samples that failed within the steel components close to the interfaces formed with the braze alloys. Joints formed using the BNi-1a braze were as strong as the AISI 316 in both simple tension and fatigue, but the presence of B or Si on their own in BNi-5 and HTN9 produced degradation of fatigue properties and the P present in BNi-7 and its high Cr variant caused marked decreases in both the tensile and fatigue strengths.

Table 7.3 Some mechanical properties of Ni brazed AISI 316 joints. After Christensen and Sheward (1982)

Braze	*Composition* Cr	P	Si	B	Fe	*Brazing Conditions**	*UTS* MPa†	*Fatigue strength* MPa†§
BNi-7	14	10	–	–	–	1070/10	224	–
–	25	10	–	–	–	1070/10	245	202
BNi-5	19	–	10.2	–	–	1200/60	404	225
HTN9	15	–	–	3.5	–	1175/10	537	220
BNi-1a#	14	–	4.5	3	3	1175/10	597	315
AISI 316							603	302

A low C variant of BNi-1.
* Temperature °C/Time minutes.
† For 25 μm wide joints.
§ Fails after 10^7 cycles.

Such reports of the good mechanical properties of joints with desirable single phase microstructures are found relatively frequently in technical literature, but reports of the degraded properties of joints containing centre-line layers of intermetallic compounds are seldom published, if only for reasons of self-esteem. Nevertheless, the presence of intermetallic compounds can cause joints to have very poor impact resistance and fracture toughness characteristics, with values sometimes less than 10% of those of the relevant single-phase structures. The presence of centre-line layers of intermetallic compounds also affects the fatigue strengths of joints, albeit less dramatically, but the UTS values may not be significantly different from those achievable with joints having single-phase structures, just as UTS values are relatively insensitive to the presence of flaws and modest unbonded areas at the component/joining material interface.

Ni brazes are unusual in that they are brittle in their as-received form but can be rendered ductile by appropriate braze and heat-treatment cycling. Most brazes and diffusion bonding interlayers are ductile materials that can then be degraded by interaction with the components they are used to join. This can be caused by a change in the overall chemistry of the joining material or more commonly by the formation of brittle reaction product layers at their interfaces formed with the component surface. Hence the tendency of Al braze alloys to form intermetallic compounds with a wide range of component metals helps restrict their use largely to joining Al.

The mechanical behaviour of joints between ceramic–ceramic or ceramic–metal components can be influenced by many physical and chemical factors, only some of which reflect the fact that bonding may

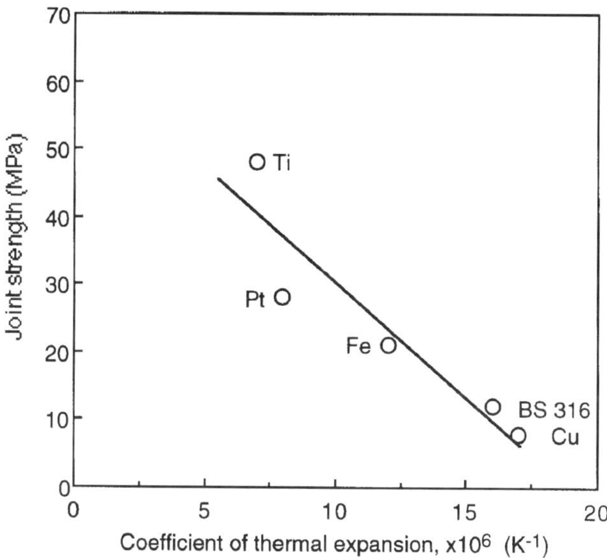

Figure 7.19 The tensile strengths of metal–Al_2O_3 samples diffusion bonded using Al interlayers. After Nicholas and Crispin (1982)

have been promoted by or associated with the formation of a reaction product layer at the joining material/component interface. Thus, the effects of mismatched coefficients of thermal expansion ceramics and metals are seemingly unaffected by the presence of a reaction product layer. Reference has been made already to the effect of such mismatches on the residual stresses present in joints, illustrated in Figure 5.6, but of more direct concern to the user of a joint is the stress at which it will fail. This too can be related to the difference between coefficients of thermal expansion as shown in Figure 7.19 for samples of various metals and alloys diffusion bonded to Al_2O_3 using Al foils as interlayers. Again, effects of joint thickness and testing temperature are seen for samples produced by the reactive joining of ceramics that are similar to those described earlier for apparently inert systems. Thus the bend strengths of Si_3N_4/Si_3N_4 joints diffusion bonded using Fe interlayers are highest for the thinnest interlayers; increasing the thickness from 50 to 100 and then to 250 μm decreases the room temperature bend strength from 640 to 560 and then to 470 MPa while the fracture toughness increases from 5.5 to 6.8 MPa.m$^{1/2}$, as compared to a bend strength of 789 MPa and a fracture toughness of 6.49 MPa.m$^{1/2}$ for the ceramic, (Peteves *et al.*, 1998). Similarly, the effect of testing temperature on the bend strengths of diffusion bonded Si_3N_4/Si_3N_4 joints produced using interlayers of Fe-15Cr, which has a solidus temperature of 1530 °C, were slight until it exceeded a temperature of $0.6\,T_M$, decreasing from 490 MPa at room temperature to 410 at 800 °C and then to 215 MPa at 900 °C (Peteves, 1995).

Other aspects of joint behaviour, however, reflect the chemical reactivity of ceramic–metal systems. It is well established that increasing the concentrations of reactive constituents in both brazes and diffusion bonding interlayers can enhance the strengths of joints formed with ceramics. For example, the peel strengths of Kovar vacuum brazed to Al_2O_3 with sintered or lapped surfaces using Ag-Cu-Ti alloys increase from 100 to 140 N when the braze Ti content is raised from 1 to 3% (Mizuhara and Mally, 1985). This compositional change also increases the hardness of the alloy, from 120 to 186 VHN, and hence it is noteworthy that raising the Ti concentration to 3% *decreases* the peel strengths of samples produced using Al_2O_3 with ground surfaces, probably due to easier crack nucleation at the rough and possibly damaged ceramic surface compounded by easier propagation within a harder and stiffer joint material. Similar effects have been seen with diffusion bonded samples with Peteves (1995) showing that increasing the Cr contents of Ni foils from 5 to 20% increased the bend strengths of joints formed with Si_3N_4 from 10 to 40% of the ceramic strength of 905 MPa.

Continuing to increase the concentrations of reactive elements ultimately results in the formation of a reaction product layer that is not only thick but also fragile because of its porosity or microscopic cracking, but prediction of how or under what circumstances such detrimental features first appear is not yet possible. Whether or not a thick layer is formed will depend not only on the chemistry of the joining and component materials but also on the temperature and duration of the interaction, and these and other process parameters can have marked effects on the mechanical properties of joints formed by both reactive and relatively inert systems as will be elaborated in the next sub-section.

Process parameters

Many of the effects of process parameters on the mechanical properties of joints have been implied or described in passing in previous sub-sections or chapters. Together with other data to be presented in this sub-section, they permit a number of general conclusions to be drawn about common effects of variations in parameters such as surface preparation techniques, the environment that is used and particularly the temperature and duration of the process.

The influence of preparation techniques arises because of their effect on both the surface chemistry and topography. It is good practice to clean the surfaces of components immediately before they are brazed or diffusion bonded and very poor cleaning procedures can be identified readily because bonding is not achieved, the braze does not run into a capillary gap or the components separate without the stressing at the end of a diffusion bonding cycle. However, sometimes the effects are not so clear-cut but nevertheless important as illustrated by the data summarized in Table 7.4 for diffusion bonded steel components (Kazakov, 1985). Understanding why CCl_4 was a better cleaning solvent than C_2H_5OH or CH_2COCH_2 in that work would require detailed chemical characterization of the bonding surfaces and of any fracture paths that followed the original interface. That information is not available, but nevertheless the data illustrate the fact that not all cleaning solvents are equally effective. In that particular case, CCl_4 was best, but this need not be the case for other materials or indeed for the same steel after a different history of contamination and experimentation is almost always needed and always prudent.

The effect of surface topography on joint strengths can also be significant and can be rationalized to some extent. Thus rough surfaces are usually better wetted by brazes and hence joints should be better filled, unless the roughness is so coarse that it grossly alters the width of the joint gaps. High mechanical strengths should be achievable with well-filled joints between roughened components and slight undulations of the joint/component interfaces could even provide additional keying when shear stresses are applied. However, mechanical degradation of both brazed and diffusion bonded joints will be caused if the roughening of the component surfaces generated microscopic cracking that can propagate when the bonded assembly is stressed, as in the case referred to previously of Al_2O_3 brazed to Kovar using Ag-Cu-Ti alloys (Mizuhara and Mally, 1985).

Similarly, the nature of the gaseous environment can affect the mechanical quality of joints. In general, it is better to use environments that have low O_2 contents such as reducing flames when brazing with a gas torch. When furnace brazing, satisfactory joint strengths can be achieved using reducing or inert gases or vacuum so that the flow of liquid braze into joints will not be impeded by surface films of oxides. Similarly, it is good practice to use high-quality vacua or inert gases when diffusion bonding to prevent the formation of thick films of oxide that are difficult

Table 7.4 The influence of cleaning procedures on joint strengths of Fe-0.45 C steel diffusion bonded at 1200 °C using a pressure of 20 MPa applied for 10 minutes. After Kazakov (1985)

Treatment	*Tensile strength,* MPa
Not degreased	320
Degreased with C_2H_5OH	460
Degreased with CH_2COCH_2	455
Pickled in acid	460
Degreased with CCl_4	570

to disrupt and hence result in lower joint strengths, as was illustrated in Figure 6.7 for the joining of Fe-0.45C (Kazakov, 1985). However, there are exceptions to these generalizations. Thus the diffusion bonding of Si_3N_4 with foils of Ni-20Cr at high temperatures is best accomplished using an environment that has a significant partial pressure of N_2, which can react with many metals, otherwise excessive dissociation of the ceramic occurs and the metal interlayer is embrittled. Similarly, the presence of O_2 causes a protective flux to form when brazing with Cu-P and Cu-Ag-P alloys, although the need for a flux is precisely because of the presence of O_2 which would otherwise form deleterious oxide films. Evidence of more positive effects of the presence of O_2 is provided by laboratory studies of the wetting and diffusion bonding of oxide ceramics by metals such as Cu or Ag. Low concentrations of O_2 in the gaseous environment or of O in the metal promote bonding by forming, for example, Cu–O clusters that segregate to decrease the interfacial energy and hence promote wetting. Although not understood as well, the presence of O as a solute can also affect the strength of diffusion bonded Cu-Al_2O_3 joints, as illustrated in Figure 7.20 (Ambrose *et al.*, 1993).

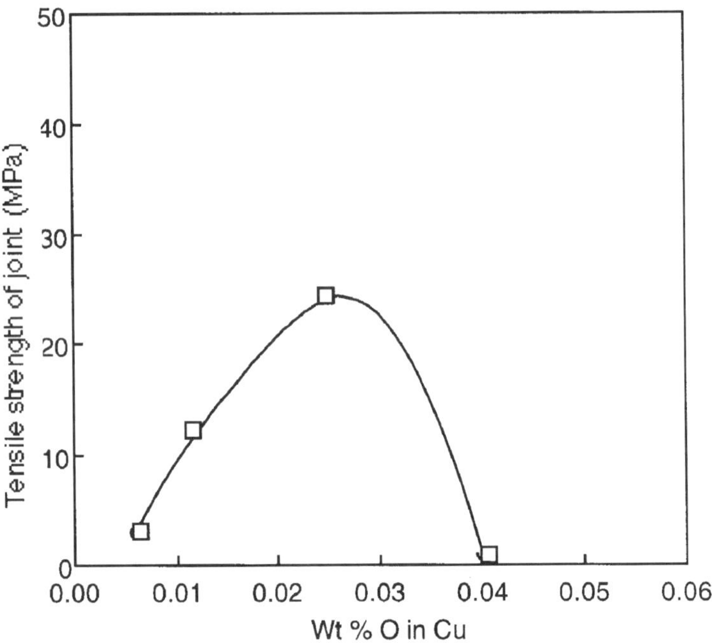

Figure 7.20 The effect of the O content of Cu on the strengths of diffusion bonds formed with Al_2O_3. ASTM test pieces were bonded at 800 °C in vacuum using a pressure of 50 MPa maintained for 30 minutes. After Ambrose *et al.* (1993)

The other process parameters of prime importance when considering effects on the mechanical properties of joints are temperature and time. These are the most commonly quoted and easily adjustable parameters and many of the earlier references to strength values have included direct or implied specification of fabrication temperatures and times. These previous citations have demonstrated that using a low temperature or short time can result in failure to produce a bonded joint let alone a strongly bonded joint. However, while the use of high temperatures or prolonged times will increase the extent of joint formation, the practice also has drawbacks. Fabrication costs are increased. The mechanical properties of the joint material can be degraded by interdiffusion as can those of the component by coarsening of its grain and phase structure. There are, therefore, upper as well as lower limits for effective bonding temperatures and times and hence windows of opportunity within which strong joints can be fabricated reproducibly and economically.

The existence of such windows is very clear when brazing because both standards organizations and braze alloy suppliers recommend minimum and maximum brazing temperatures. As was illustrated in Table 6.5, these ranges for joining metal components are quite narrow and seldom allow superheats above liquidus temperatures of more than 100 °C. Further, the times needed for brazes to flow into metal–metal gaps are significantly shorter than the typical 15 to 60 minutes needed to ensure that assemblies of components have reached a uniform temperature. Thus the brazing of identical metal components by different fabricators is generally accomplished using very similar temperatures and times, and the selection of good joint designs ensures that this uniformity of processing can result in the consistent achievement of joints that are as strong in tension or shear as the component materials.

Potential exceptions to this idyllic situation can occur when components are brazed with Ni alloys containing embrittling additions of B, P or Si. Reference has been made earlier to the beneficial structural and mechanical effects of the migration of such elements from Ni alloy brazed joints into the component materials that can be achieved by using prolonged fabrication or heat treatment times, and to the detrimental effects of the ingress of additional Si from SiC or Si_3N_4 components. However, beneficial effects due to dispersal of embrittling elements from Ni joining materials are not restricted to brazed joints. Thus when using the transient liquid phase bonding technique, increasing the holding time at a temperature of 1200 °C from 1 to 2 and then to 10 hours increased the room temperature tensile strengths from 815 MPa to 890 and then to 1180 MPa for joints formed by a complex single crystal Ni alloy, Ni-10.5W-9Cr-5.7Al-3.2Ta-1.2Ti-1Mo, that had been bonded using 16 μm foils of Ni-4.2B-2.9Si (Nakao *et al.*, 1990).

Even more common than such homogenization effects promoted by interdiffusion is the formation of reaction product layers at component

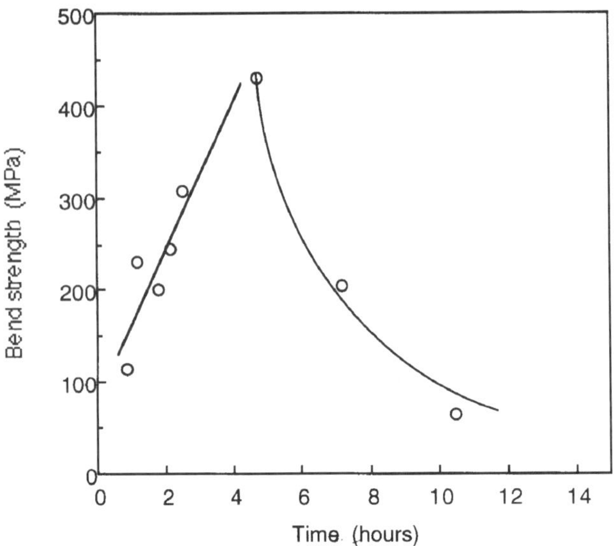

Figure 7.21 The effect of the time at temperature on the joint strengths of Si_3N_4/Ni-20Cr/Si_3N_4 samples. Diffusion bonding was achieved at 1100 °C in Ar using a pressure of 50 MPa. After Peteves (1995)

surfaces. This is an essential stage in the joining of ceramic components by active metal brazes but needs to be controlled. Layers of such compounds more than 1 or 2 μm thick are often fragile because of the presence of growth flaws and hence progressively raising the brazing temperature or prolonging the time ultimately causes a weakening. Thus the optimum combination of these variables is that which produces a continuous layer of an unflawed reaction product and this can sometimes be expressed as an Arrhenius relationship

$$t_{opt} = A' \exp.(Q_{RPL}/RT_{opt}) \qquad \{7.3\}$$

where t_{opt} and T_{opt} are the optimum time and temperature, A is a constant and Q_{RPL} is the activation energy of the process that controls the rate of formation of the reaction product. An example of such an effect was provided by the peaking of strength values of heat treated Al-4.5Mg/Al_2O_3 samples shown in Figure 6.15. In this plot, the peak strengths varied little and assuming them to be associated with the same degree of interaction permits interpretation of the optimum times to derive an activation energy of 121 kJ.mole^{-1} which is close to the 116 to 119 kJ.mole^{-1} for the diffusion of Mg in an Al-5Mg alloy. Similar increases in strength followed by decreases have been observed for chemically reactive diffusion bonded systems, thus Figure 7.21 shows the effect of holding time on the strengths of Si_3N_4-(Ni-20Cr) joints.

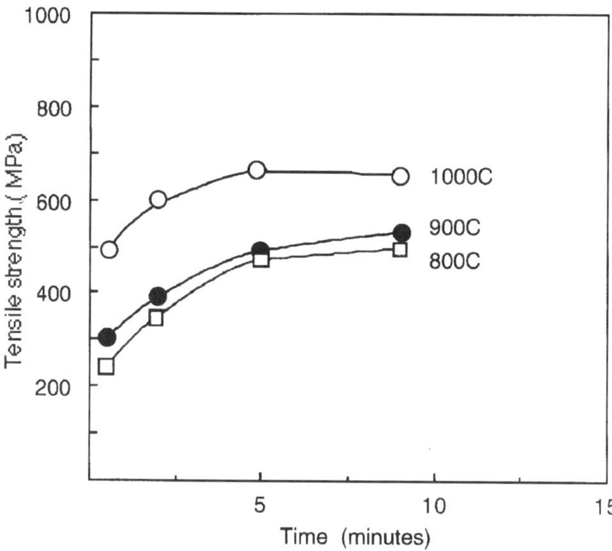

Figure 7.22 The effects of joining temperature and time on the strengths of diffusion bonded joints between Fe-0.45C. After Kazakov (1985)

If a system is not chemically reactive, the effect of prolonging the time or raising the temperature used to diffusion bond is to increase the area that is bonded and hence the strength. Thus the steady increase with time in the bonded area of Fe diffusion bonded at 700 °C shown in Figure 3.21 will have been accompanied by a strength increase in accord with equation {7.2}, and some such increases are shown in Figure 7.22 as a function of both time and temperature for the diffusion bonding of Fe-0.45C (Kazakov, 1985).

Summary

Mechanical properties are affected by many factors since they reflect both the structures and chemistries of joints. Of prime importance for the production of the strongest joints is the achievement of bonded interfaces that are free of voids and unbonded inclusions such as flux residues. It is also of crucial importance if a joint is to be tough and have a good impact resistance that it is free of brittle compounds located at seams within the core of the joint or as thick reaction product layers at the joint/component interfaces. The best joints are those achieved by direct diffusion bonding of components with identical compositions or more generally by using a thin layer of a single-phase ductile metal or alloy. Such joints can have mechanical properties that approach or even slightly exceed those of the component materials being joined.

7.4 OTHER PROPERTIES

The acceptability and successful application of joints can depend on characteristics other than the structure and mechanical strength. The most important other characteristic is a low cost, and indeed this may be the most important characteristic of all. In addition, there are a number of technical considerations that can be of importance in particular applications. In particular, it may be necessary to pay attention to the colour of joining materials, to their conductivity characteristics and more commonly to their chemical stability in service environments.

7.4.1 Colour

The colour of a brazed joint can be of major importance to the end user. Inappropriate colouration of joining materials and visible joints can cause rejection of items of domestic decoration and utility, such as dishes, cutlery and kitchen ware. Thus the use of a ruddy Cu-based braze to join assemblies of stainless-steel components is often unacceptable because the joint looks dirty or may even be confused with rust. Of even more importance is the matching of component and joining material colours when making items of jewellery whose design calls for the use of a fillet or for the production of a visible witness. While the colour characteristics of brazes and of component materials can be defined by spectral analysis, in practice it is often the aesthetic sensibilities of the purchaser that define the acceptability limits for colour variations.

In the case of jewellery, it is not only necessary that the braze alloy is aesthetically acceptable because of its colour and because it can form well shaped fillets of a pleasing size when the design calls for that type of joint but also it must be technically effective. In fact, little pure Au is used to make jewellery not only because of cost but also because of its softness, and alloying with Cu is used to improve its mechanical characteristics. The fineness of gold items of jewellery is usually characterized in terms of carats, each carat representing a Au concentration of a 24th of that of the pure material, so that 24 carat gold is 100% Au, 18 carat gold is Au-25Cu and 9 carat gold is Au-62.5Cu. This introduction of Cu has an effect on the colour, progressively causing it to assume a ruddy hue and the need to match these varying hues has led to the development of a series of Au-Ag-Cu-Zn-Cd brazes These brazes have liquidus temperatures of 705 to 730 °C and match the colour of the various purities of gold because of their varying Au content; 75% Au braze is a good match for 18 carat gold, a 58.3% Au alloy matches 14 carat gold and a 37.5% Au braze matches 9 carat gold.

7.4.2 Conductivity

It may be necessary for bonded assemblies of components to possess a certain degree of electrical or thermal conductivity and hence the values of these characteristics for joining materials and for joints can be important. In practice the characteristics of joints and joining are seldom responsible for the failure of bonded assemblies to attain the desired levels of conductivity. Thus Table 7.5 compares the values of the electrical and thermal conductivity for a number of component and joining materials and it can be seen that the widespread use of braze alloys based on Ag, Al or Cu ensures that the conductivity of the joining material is usually better than that of the component. In the case of diffusion bonding, a similar conclusion can be reached because of the use of Ag, Al ,Ni and other pure metals as interlayers. Further, while the conductivity of the Ni-20Cr alloy used as a diffusion bonding interlayer is relatively poor, it is still markedly better than that of the ceramics and equivalent to that of the high-strength Ni component materials it might be used to join.

The formation of reaction product layers at joint/component interfaces will usually degrade both the electrical and the thermal conductivity of the joint, but the use of thin joints that is desirable for high strengths also ensures that the thermal impedance caused by the joint will be modest. Thus in one of the case histories described in Chapter 8, the thermal conductivity of a sample produced by vacuum brazing 1.34 mm thick

Table 7.5 Some electrical and thermal conductivity values for component and joining materials

Material	*Electrical resistance* $\mu\Omega m$	*Thermal conductivity* $W.m^{-1}K^{-1}$
Ag	1.63	425
Ag-28Cu	2.1	
Al	3	218
Al-12Si	4.6	142
Au	2.2	315.5
Au-37Cu	14.6	
Cu	1.7	399
Cu-1Cr	2.2	167
Ni	9.5	74.9
Ni-20Cr	110	
Pb	20.6	34.9
Sn	12.6	73.2
Sn-40Pb	15.0	
IMI 318	168	5.8
Inconel 600	103	14.8
Stainless steel	69.4	15.9

Cu-1Cr and 1.64 mm thick Inconel 600 plates with Ag-28Cu was measured and found to be 30.7 $W.m^{-1}K^{-1}$, so the presence of the 0.08 mm thick seam of braze alloy was equivalent to no more than an additional 0.02 mm of Inconel 600. Similarly, the relatively poor electrical conductivity of the Sn-40Pb solder as compared to that of Cu does not prevent it and its derivatives from being used to produce millions of electrical connections every day.

7.4.3 Chemical stability

Requirements

Chemical interactions of the metallic or ceramic component materials with their joining materials or with the environment can be of crucial importance in determining whether or not a joint can be fabricated, but their post-fabrication chemical reactivity of the bonded assembly is also important and can have dramatic effects on the service conditions that can be endured. The ideal situation is that the joint is not subject to any form of chemical instability and hence to changes and possible degradation from the optimum characteristics that the selection of materials, design and process parameters was intended to achieve. In practice, almost every environment is reactive to some extent, even air at ambient temperatures, and some can promote galvanic interactions between the joining and component materials. The realistic characteristic of an acceptable system, therefore, is that the rate of chemical attack is slow and does not result in significant degradation within the normal service lifetime.

The range of environmental conditions that may need to be endured by joints is vast but those that may be experienced most commonly by joints usually fall into one of four broad categories:

- oxidation by gases and mixtures such as O_2, air, CO_2/CO and H_2O/H_2;
- non-oxidizing reactions to convert the joint material to a carbide, nitride, sulphide by contact with gases such as incompletely burnt hydrocarbons, N_2 or NH_3, H_2S or SO_2;
- corrosion by molten or solid salts such as Na_2SO_4, V_2O_3 or $CaSO_4$ deposited by burnt fuel oil or gas;
- corrosion by salt water and other aqueous solutions.

Other relatively rare but potentially very destructive environment-induced conditions that can in principle degrade the integrity of a joint include corrosion by liquid metals.

The primary requirement for a joint is that it should not be directly or indirectly weakened by overall loss of material or by surface pitting that can act as crack nucleation sites. In principle, and sometimes also in

Table 7.6 The electrode potentials of a number of metals

Metal and electrode potential, volts			
Mg/Mg^{++}	2.37	Cu/Cu^{++}	−0.34
Al/Al^{+++}	1.76	Ag/Ag^{+}	−0.80
Zn/Zn^{++}	0.76	Au/Au^{+++}	−1.42
Fe/Fe^{++}	0.44		
Cr/Cr^{+++}	0.41		
Co/Co^{++}	0.28		
Ni/Ni^{++}	0.23		

practice, it is possible to avoid such damage by overcoating the exposed joint surfaces with a chemically inert material such as Au. This can provide a complete technical solution to the chemical stability problem – provided there is no interaction between the overcoat and its substrate, that the overcoat is free of pin holes and other imperfections and that it is not eroded by abrasion or disrupted by stresses and strains encountered in service.

The last category of degrading environment listed above, aqueous solutions, can result in particularly severe chemical degradation that is caused by the environment but by acting as an electrolyte to permit interaction between the component materials. The driving force for this type of degradation is the difference in the electrode potential of the joining and component materials, the voltage that must be applied to ionize the metal in the presence of an electrolyte, and values of this characteristic are given for some metals of interest in Table 7.6. If an electrolyte bridge can be established between two metals or alloys, that with the larger or more positive potential will act as an anode and hence will be dissolved.

Assessment

Whatever the evaluation technique employed, it is essential that the test environments and conditions closely simulate those that will be encountered in service since it is seldom possible to do more than interpolate between experimentally measured data. For example, there is evidence that the rate of attack by oxidizing environments such CO_2/CO or H_2O/H_2 depends on their precise composition and not just on their O_2 activity, and the rate of attack by salts deposited by burnt fuel depends on their rate of deposition as well as their chemistry so that the gas velocity or at least the rate of supply of the reactant salt must be simulated.

The most frequently studied form of chemical instability of those listed above is oxidation, whose progress is usually measured by the weight gained as metal is converted to oxide. Measurements can be made at regular intervals by interrupting an experiment or can be made continuously by using a balance fitted with a, usually, high-temperature environmental cell filled with O_2, air or other oxidizing mixtures. Other methods of measurement include monitoring the gas composition or pressure changes that occur within a closed system, deriving the thickness of the surface layer of reaction product by ellipsometry and measurement of the natural frequency of oscillation of very thin samples. These and other techniques can also be used to follow the progress of reactions of chemically aggressive gaseous environments with constituents such as SO_2.

The simplest to use of the techniques mentioned so far is measurement during an interrupted exposure, but this can give a distorted impression of the reaction kinetics because the temperature changes on removal from or return to the furnace may cause spallation and hence an underestimate of the extent of reaction. However, even if the weight of the sample being tested is monitored continuously and isothermally, it is prudent not to base interpretations only on weight gains. Chemical analyses are needed to identify the reaction product being formed and hence to relate the weight gain due to the incorporation of O_2 or other gaseous reactant with the loss of metal by the sample. Similarly, microstructural studies can reveal invaluable information such as whether or not several products are formed, such as CuO and Cu_2O, and whether there is preferential attack on grain boundaries or particular phases that may be present in alloy systems. Such microstructural and micro-chemical assessments also can be particularly important in the evaluation of changes wrought by interactions with aggressive salts deposited from burnt fuel.

Assessing the extent of attack or degradation caused by non-gaseous reactants requires other and more specialized techniques and equipments. Thus tests simulating the effects of exposure to an electrolyte that facilitates galvanic corrosion require that the sample shall not be in electrical contact with the containment structure. The test may be a simple exposure to a simulation of the liquid that will be encountered in service conditions, although defining its composition can be a difficult task if it arises adventitiously. This is followed by post mortem examination of the joint surface and microstructure to assess the extent of the attack, as would be done if there was cause to assess damage to a joint that had been in service. Alternatively, the progress of the attack process can be monitored continuously during a laboratory test by making the sample containing a joint part of an electrical circuit and measuring current flows.

Laboratory tests to assess the effects of attack by liquid metals require very specialized equipment constructed of materials that are immune to attack. The liquid metal whose containment and attack characteristics have been of particular concern in recent years is Na because of its use as a heat exchange fluid in the development of fast breeder reactors. Fortunately, Na is not a particularly aggressive metal unless in contact with sources of O_2 or H_2O and hence there is an extensive literature describing techniques for its safe handling. Assessing the extent of attack caused by exposure usually involves destructive examination of microstructures and micro-chemistries.

Special care must be taken to ensure the safety of operating personnel when conducting laboratory simulations of exposure to chemically aggressive service conditions since some of the possible environments can be toxic, explosive or lethal.

Observations

The chemical stabilities of joining materials exposed to service conditions exhibit the same general characteristics as those of the wider class of component materials (Saunders and Nicholls, 1996). Thus the progress of oxidation and similar reactions can be slowed by the formation of layers of reaction product that impede the flow of reactants. The degree of this beneficial effect depends on the thickness of the reaction product layer and also on its chemistry and lattice and microstructure. In the case of oxidation, formation of layers of Al_2O_3, Cr_2O_3 or SiO_2 can be a particularly effective means of diminishing corrosive attack and the addition of Cr in concentrations sufficient to ensure formation of a protective oxide is used to confer oxidation resistance to Ni brazes. In fact, Al_2O_3 and particularly SiO_2 are better barrier materials than Cr_2O_3 and Al brazing alloys can be effectively oxidation resistant at 400 or even 500 °C, well in excess of any temperature they might be expected to endure in service. However, the use of Al or Si as alloying additions to confer improved oxidation resistance to joining materials is rarely adopted because they can degrade wetting and bonding characteristics, but Cr can promote the bonding process as when Ni-20Cr foils are used to diffusion bond Si_3N_4.

Most joining materials are alloys and hence the nature of their reactions with aggressive environments depends on the chemical activities of their constituents. If the reaction products are mutually soluble, the presence of other metal constituents will modify the lattice structure and hence the effectiveness of the product as a diffusion barrier. Where the products are not soluble in each other, a composite layer structure will result. This may be a sequence of products reflecting the chemical affinity of the metal for the external reactant or the formation of a complex product such as the spinel oxides $MgAl_2O_4$ or $NiCr_2O_4$. Similarly, the chemical interactions can be more complex if the environment contains two or more reactive components. Thus a gaseous environment may contain not only O_2 but also SO_2 or a chloride and hence the resultant reaction product and rate of attack can be affected by the generally faster diffusion rates in sulphides and carbides than in the corresponding oxides, and the volatility of chlorides. However, the affinity of Cr, as of Al and Si, for O_2 is such that exposure to mixed gaseous environments generally results in the formation of an oxide layer, although sulphides and carbides may still be formed within the metal or alloy where the O activity is very low. For thorough descriptions of these and many other important aspects of the interaction of metals and alloys with chemically aggressive environments the reader is recommended to consult authors such as Kofstad (1988) or Kubaschewski and Hopkins (1967).

As in the case of mechanical properties, information about the chemical stability of joints in service conditions is generally commercial property and what data have been published are widely scattered. Further, it may not be sufficient to use a joining material that resists attack by aggressive environment because the production of a joint can change the composition of the joining material. Corrosion resistance and other measurements of the chemical stability of joints, therefore, need to be done with actual joints or at least with the joining material in contact with the component metal or ceramic. An interpretive review of such data as were available for brazed joints was produced some time ago when Cole (1979), showed that they are not in general sites for excessively vigorous corrosion. No such exercise is known to have been conducted for diffusion bonded joints but it is probable that they exhibit similarly satisfactory characteristics. This in part reflects common use of joining materials based on Ag, Al, Au, and Ni-Cr that have inherently good resistance, at least in the as-received condition, to attack by many aggressive environments, and the design of joints to minimize exposure of joint surfaces

The oxidation resistance of joints made with Ag alloys is generally good as might be expected, and laboratory studies have shown this to be true for Ag base active metal braze alloys. Thus Moorhead and Kim (1991) coated the surface of coupons of Al_2O_3 or ZrO_2 partially stabilized by additions of MgO with Ag-Cu-Sn-Ti or Ag-Cu-In-Ti alloys and found the growth of oxides on the free surfaces of the alloys to be a slow parabolic function of time when they were exposed to O_2 or Ar-O_2 mixtures at temperatures of 400 to 700 °C for times of up to 100 hours. Some of the results obtained are listed in Table 7.7. The Ti played only a minor role in the oxidation process which generally resulted in the formation of surface films of oxide that were several microns thick. In the case of a Cu-41.1Ag-3.8Sn-7.2Ti coated ZrO_2, the growth of a thin layer of TiO reaction product at the braze/ceramic interface occurred due, it was thought, to the easier transport of O through the ceramic.

The upper limits to the service temperatures of Ag-Cu alloys are usually no more than about 400 °C because of their yielding but a number of Au, Ni and Pd braze alloys have been developed specifically for high

Table 7.7 The corrosion resistance of some active metal braze alloys heated in Ar-20% O_2 for 50 hours at 400 °C. After Moorhead and Kim (1991)

Braze	*Reaction product*	*Weight gain* mg.cm^{-2}
Ag-37Cu-11In-1.4Ti†	In_2O_3	0.18
Ag-48Cu-9.7Sn-5Ti	$SnO_2 + TiO_2$	0.22
Ag-44Cu-8.4Sn-0.9Ti	CuO	0.37
Ag-41Cu-3.8Sn-7.2Ti	CuO	0.49

† Incusil ABA, trade name of Wesgo Metals.

temperature applications. Some of these are particularly suitable for use in oxidizing environments. Thus Christensen and Sheward (1982) reported that Ni alloy or stainless steel with butt joints produced by vacuum brazing with Au, Ni-Cr or Pd alloys can be oxidation resistant in air at temperatures of up to at least 800 °C. In some but not all cases the resistance of the joint can equal or be better than that of the component material. The resistance of Ni brazes containing 8 or 10% of Si or 19% of Cr was particularly good and that of the Ni-19Cr-10Si braze, BNi-5, was the best of those evaluated. Joints made with this braze also performed well when exposed to more complex environments, as indicated by the data assembled in Table 7.8.

An important and growing use of Al alloys is in the production of heat exchangers that have to operate in marine environments and hence the susceptibility of their joints to galvanic corrosion is of importance. In

Table 7.8 The effects of exposure for 5000 hours to two corrosive gaseous environments on the penetration of brazed joints between stainless steel components. After Christensen and Sheward (1982)

Braze	*Joint penetration*
CO_2 + 5% CO, 900 °C	
Ni-11P	Complete
Ni-13Cr-10P	Variable
Ni-19Si-10P	< 0.003
Steam, 700 °C	
Ni-11P	< 0.013 cm
Ni-13Cr-10P	0.013 cm
Ni-19Cr-10Si	0.0013 cm

practice this is not often a major problem because Al alloys are also used to make the joints and hence the potential difference between the component and joining materials is moderate and gross attack is unlikely. However, the minor compositional and phase structure difference between the component and brazing alloys can result in localized pitting. However, the electrode potential of Al is greater than that of most of the minor constituents of the braze alloys such as Si, Zn or Cu, while the Mg present in the braze alloy that has an even higher electrode potential can be removed by vaporization during the brazing process, so that component acts as an anode and it is dissolved rather than material from the joint. Thus while it is necessary to be aware of possible galvanic effects, they may not be of great importance in normal service conditions for brazed joints.

The brazing of Al components is a somewhat special case because components of other metals are often brazed with alloys that have markedly dissimilar electrode potentials as can be seen by examination of Table 7.6. For a reasonable galvanic compatibility it is desirable for the electrode potential values to differ by no more than about 0.5 volt and this limit is exceeded when Ag or Cu, or their alloys, are used to braze Fe or Ni, or austenitic stainless steel or Ni-Cr alloy, components. However, the joint materials will be cathodic in such systems and hence it is the component that will be attacked, but at a slow rate if, as is usual, its exposed surface area is much larger than that of the joint.

Also of interest is the chemical stability of diffusion bonded joints, but very few data are available. Thus, the potential usefulness of Al and Al alloys as interlayers to promote the diffusion bonding of ceramics has been examined and shown to have considerable promise by many research groups, but only one report of the influence of aqueous environments on joint strength is known. In that work by Suganuma *et al.* (1988) it was found that Si_3N_4 brazed with Al failed in room temperature bend tests conducted in an atmosphere of Ar failed at an average stress of 465 MPa, virtually identical to the 470 MPa required for the monolithic ceramic. The average failure stress fell to 410 MPa when the ceramic was tested in water, but that of the brazed samples fell to 350 MPa.

Reports of the resistance of joints to corrosion by liquid metals are also very rare but it has been observed that the dissolution of noble metals is relatively rapid. The attack on Ni alloy brazes by alkali metals depends on the O content of the liquid as well as factors such as temperature, time and flow rate but joints made with the BNi-2 or BNi-5 brazes have been found to withstand exposure to Na or K for 7000 hours at 700 °C while BNi-4 was significantly degraded at temperatures above 650 °C (Christensen and Sheward, 1982).

REFERENCES

Ambrose, J. C., Perkins, R., Airey, R. and Nicholas, M. G. (1993) The influence of oxygen on copper–alumina bonding. In S. D. Peteves (ed.), *Designing Ceramic Interfaces II*, Commission of the European Communities, Luxembourg.

American Society for Metals (1973) Metallography, structures and phase diagrams, *Metals Handbook*, **8**.

American Society for Testing Materials (1964) *Standard Method for Tension and Vacuum Testing Metallized Ceramic Seals*, ASTM F19-64.

Baker, T. J. and Kavishe, F. P. L. (1987) The fracture toughness of brazed joints, *BABS 5th International Conference of High Technology Joining*, British Association for Brazing and Soldering, Abington, Cambridge, paper 21.

Baker, T. S. and Partridge, P. G. (1991) Fatigue and impact strengths of diffusion bonded titanium alloy joints. In R. Pearce (ed.), *Diffusion Bonding*, Cranfield Institute of Technology, Bedford.

Ceccone, G., Nicholas, M. G., Peteves, S. D., Kodentsov, A. A., Kivilahti, J. K. and van Loo, F. J. J. (1995) The brazing of Si_3N_4 with Ni-Cr-Si alloys, *Journal of the European Ceramic Society*, **15**, 563–72.

Chang, A. A., Goldberg, D. and Neuman, J. P. (1977) Phase diagram and thermodynamic properties of the of ternary copper–silver systems, *Journal of Chemical and Physical Reference Data*, **6**, 621–67.

Christensen, J. and Sheward, G. E. (1982). Characteristics of brazed joints in high temperature materials. In T. G. Gooch, R. Hurst, H. Kroeckel and M. Merz (eds), *Behaviour of Joints in High Temperature Materials*, Applied Science Publishers, London, 117–67.

Cole, N. C. (1979) *Corrosion Behaviour of Brazed Joints*, Bulletin 247, Welding Research Council.

Derby, B. (1981) Theoretical model of diffusion bonding, Ph.D. Thesis, Cambridge University, Cambridge.

Greaves, R. H. and Wrighton, H. (1967) *Practical Microscopical Metallography*, 4th edn, Chapman & Hall, London.

Hill, R. (1950) *The Mathematical Theory of Plasticity*, Oxford University Press, Oxford, 230.

Iseki, T. and Nicholas, M. G. (1979) The elevated temperature strengths of alumina–aluminium and magnesia–aluminium samples, *Journal of Materials Science*, **14**, 687–92.

Kazakov, N. F. (1985) *Diffusion Bonding of Materials*, Pergamon Press, Oxford.

Klomp, J. T. and van de Ven, A.R.C. (1980) Parameters in solid-state bonding of metals to oxides materials and the adherence of bonds, *Journal of Materials Science*, **15**, 2483–9.

Kofstad, P. (1988) *High Temperature Corrosion*, Elsevier Applied Science, London.

Kubaschewski, O. and Hopkins, B. E. (1967) *Oxidation of Metals and Alloys*, Butterworths, London.

Livesey, D. O. and Ripley, N. (1993) Diffusion bonding of superplastic alloys using a transient liquid phase interlayer (zinc). In D. J. Stephenson (ed.), *Diffusion Bonding 2*, Elsevier Applied Science, London.

Ljungberg, L. (1992) Joining of ceramics to metals by brazing, PhD thesis, Chalmers Technical University, Gothenberg, Sweden.

McDermid, J. R., Pugh, M. D. and Drew, R. A. L. (1989) The interaction of reaction-bonded silicon carbide and Inconel 600 with a nickel-based brazing alloy, *Metallurgical Transactions A*, **20A**, 1803–10.

Mizuhara, H. and Mally, K. (1985) Ceramic-to-metal joining with active brazing filler metal, *Welding Journal*, **64**, 27–33.

Moorhead, A. J. and Kim, H.-E. (1991) Oxidation behaviour of titanium-containing brazing filler metals, *Journal of Materials Science*, 4067–75.

Moretto, P., Moulaert, M., Glaude, P., Frampton, P., Ceccone, C. and Peteves, S. D. (1993) Interfacial reactions and kinetics between Si_3N_4 and Ni-Cr alloys. In S. D. Peteves (ed.), *Designing Ceramic Interfaces II*, Commission of the European Communities Directorate General XIII, Luxembourg, 519–39.

Nakao, Y., Nishimoto, K., Shinozaki, K. and Kang, C. Y. (1990) Transient liquid insert metal diffusion bonding of nickel base super alloys. In T. H. North (ed.), *Advanced Joining Technologies*, Chapman & Hall, London, 129.

Nicholas, M. G. and Crispin, R. M. (1982) Diffusion bonding stainless steel to alumina using aluminium interlayers, *Journal of Materials Science*, **17**, 3347–60.

Partridge, P. A. and Baker, T. O. (1991) The application of SPF/DB to dissimilar metal joints. In D. J. Stephenson (ed.), *Diffusion Bonding 2*, Elsevier Applied Science, London, 242–50.

Peteves, S. D. (1995) Joining nitride ceramics. In P. Vincenzini (ed.), *Advances in Science and Technology*, **3C**, 2179–90.

Peteves, S. D. and Nicholas, M. G. (1992) Interface microchemistry of silicon nitride/nickel–chromium joints, *Metallurgical Transactions A*, **23A**, 1773–82.

Peteves, S. D., Paulasto, M., Ceccone, G. and Stamos, V. (1998) In presses of *Acta Materialia*.

Reid, C. G., Peteves, S. D. and Nicholas, M. G. (1994) Si dilution mechanisms in BNi-5/metal joints, *Journal of Material Science Letters*, **13**, 1497–1500.

Saunders, S. R. J. and Nicholls, J. R. (1996) Oxidation, hot corrosion and protection of metallic materials. In R. W. Cahn and P. Haasen (eds), *Physical Metallurgy*, 4th edn, Elsevier Science, London, 1291–1361.

Stevens, J. J., Burchett, S. N. and Hosking, F. M. (1991) High-temperature creep properties of eutectic and near-eutectic silver–copper alloys: application to metal/ceramic joining. In M. J. Cieslak, J. H. Perepezko, S. Kang and M. E. Glicksman (eds), *The Metal Science of Joining*, Metallurgical Society.

Suganuma, K., Okamoto, T., Koizumi, M. and Shimada, M. (1986) Relationship between the strength of ceramic/metal joints in tensile and three point bend testing, *Communications of the American Ceramic Society*, C-235–236.

Suganuma, K. and Okamoto, T. (1987) Interlayer bonding methods for ceramic/metal systems with thermal expansion mismatches, *Studies in Physical and Theoretical Chemistry*, **48**, Elsevier, Amsterdam, 71–89.

Suganuma, K., Niihara, K., Fujita, T. and Okamoto, T. (1988) Stress corrosion in ceramic/metal joints. In M. Doyama, S. Somiya and R. P. H. Chang (eds), *Proceedings of the MRS International Meeting on Advanced Materials*, 113–19.

Udin, H., Funk, E. R. and Wolff, J. (1954) *Welding for Engineers*, John Wiley Inc., New York.

Application case histories

8

Several earlier chapters have been concerned with optimization of individual aspects of joining processes but in reality it may not be possible to reach a firm decision about any one of them without reference to the decisions made about others. Thus the specification of an optimized joint fabrication procedure for a particular application may require a repetition or perhaps several repetitions of the procedures for selection of materials, design of joints and definition of process parameters. Experience is of great assistance when making even tentative decisions but it is sometimes also possible to facilitate the decision-making process by referring to the published account of successful developments of joining processes for particular applications.

The setbacks encountered during the development of a joining process, and also often the ultimate achievements are usually undisclosed proprietary information but in some cases sufficient information has been released to allow an appreciation to be gained. Thus ten examples are described in this chapter largely using information drawn from published accounts. Particular attention is paid to developments concerned with the joining of ceramic components because they usually present an even bigger challenge than that posed by metal components. All the case histories relate to brazing or diffusion bonding although one doubtful or hybrid case is described in section 8.2.3. This particular application used a braze-clad sheet of Al alloy to promote bonding to Al_2O_3 at a temperature sufficient to melt the cladding but required pressure to achieve contact between the ceramic and the molten alloy; in this chapter it is classified as brazing while those who did the work described it as diffusion bonding.

8.1 THE ACTIVE METAL BRAZING OF A CERAMIC AUTOMOBILE TURBOCHARGER

The oil shocks caused by the arbitrary price increases imposed by OPEC in the 1970s, and a continuing concern about the stability of the environment, have prompted many efforts to develop more efficient use of hydrocarbon fuels. In the field of automobile engineering, the development of lightweight automobiles with better power/weight ratios has accelerated and this has resulted in a significant replacement of steel by Al alloys and plastics. Another important result has been the development of internal combustion engines that are more fuel efficient and produce cleaner exhaust gases. This has been achieved by the optimization of design, including the introduction of ceramics that have decreased the engine weights, increased their maximum permissible service temperatures and improved wear and corrosion resistance. Potential benefits such as these led to a great interest in the application of engineering ceramics in the 1980s when, in particular, the research laboratories of Japanese industries were gripped by a 'ceramic fever' and secured over 200 patents on ceramic joining techniques in 1985 alone. One of the first, and still arguably most notable, products of this push to use ceramics was the development of a ceramic bladed turbocharger unit by the Nissan Motor Company and the NGK Spark Plug Company Limited in 1985.

The function of a turbocharger is to push air into the piston chamber by compressing it, using a turbine driven by the exhaust gases. Thus the turbine blades must be able to withstand high exhaust gas temperatures and sometimes aggressive chemistries. A material found to have both of these desirable characteristics is Si_3N_4, which also has a low density and low coefficient of thermal expansion. The design of the Nissan turbocharger rotor involves a monolithic ceramic hub and blade structure attached to each end of a metal shaft, one driven by the exhaust gases and the other driving the inlet air. The optimum design of the attachment involves the use of brazing to make the joint between the ceramic and metallic structures as sketched in Figure 8.1.

The Nissan/NGK development has been described by Suga (1989), Ito and Taniguchi (1993) and Itoh and Kato (1993) among others who identified the performance targets for the joint as including a failure strength of 400 MPa even at temperatures of 400 °C. Laboratory studies selected a number of ferritic steels as being potentially suitable constructional materials for the shaft with the best results being achieved with a Fe-3Cr-3Ni-0.5Mo-0.4Mn-0.3C alloy. Similar evaluations led to the selection of a non-porous grade of sintered Si_3N_4 with a density of 3.23 $Mg.m^{-3}$ for the bladed hubs. This ceramic had a coefficient of thermal expansion of $2.8 \times 10^{-6} K^{-1}$, its average bend strength was 900 MPa at room temperature and 850 MPa at 800 °C, and its room temperature fracture toughness, K_{IC} was $6 \rightarrow 7$ $MPa.m^{1/2}$.

It was necessary to introduce an insert between the ceramic and the steel shaft because of their markedly differing coefficients of thermal expansion. Materials considered during detailed modelling and experimentation included Fe, Nb, Ni and Kovar, Fe-29Ni-17Co, but the best results were achieved using a multiple buffer layer comprising a 1.5 mm thick disc of a

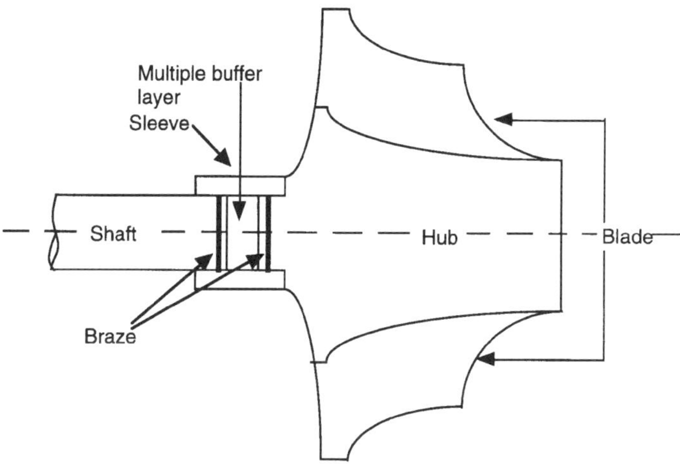

Figure 8.1 Schematic arrangement for the attachment of a monolithic bladed ceramic hub to one end of the shaft of a turbocharger rotor

W-5(Fe, Ni) alloy whose end faces were coated with 0.25 mm thick layer of Ni. For the braze alloys, a 50 μm thick foil with a composition of Ag-30Cu-5Ti was chosen for the ceramic–buffer layer joint and a similar foil with a composition of Ag-28Cu was used to effect the shaft–buffer layer joint. In addition to the use of Ni layers to make the multiple buffer layer, a 5 μm layer of Ni was plated on to the end face of the steel shaft to enhance its wettability. Finally, to provide extra assurance about the integrity of the joint, a sleeve of Incoloy 909, a relatively low expansion alloy with a composition of Fe-38Ni-13Co-4.7Nb-1.5Ti-0.4Si was used to encase the joint area.

Joining was effected by brazing at 900 °C for about 20 minutes in a chamber evacuated to a residual gas pressure of about 2×10^{-4} mbar, typical conditions for the industrial use of a Ag-Cu-Ti active metal braze. When used to make a joint between Si_3N_4 ceramics, these brazing conditions would have resulted in the formation of a relatively thick, 3 μm, layer of brittle reaction product. In practice, the observed thickness was only 1 μm and this smaller value is presumed to be due to the affinity of Ti for the Ni present on the surfaces of the shaft and the inserts. The principal tool used to assess the quality of the joints was mechanical testing conducted at room and elevated temperatures.

Room temperature bend tests on Si_3N_4/steel rods joined using a multiple buffer layer showed that increasing the total thickness of the insert from 2 to 4 mm increased the failure stress from 400 to 425 MPa, while decreasing the insert thickness to 0.25 mm caused the strength to drop to 180 MPa. Room temperature tests on sleeved turbocharger rotors

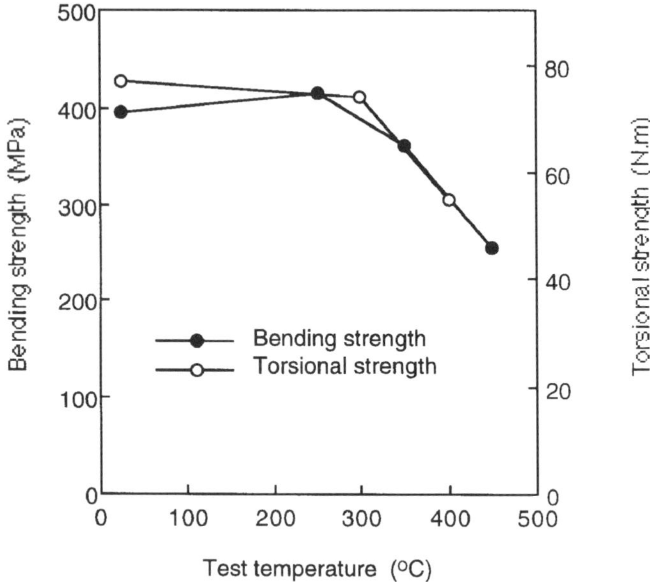

Figure 8.2 The effect of test temperature on the strengths of turbocharger joints

gave an average joint bend test strength of 390 MPa and a torsional strength of 78 Nm, and similar values were maintained up to temperatures of about 400 °C, as illustrated in Figure 8.2. Other more engineering oriented evaluations at low temperatures or in conditions simulating rapid acceleration also gave satisfactory results.

The real proof of the success of the joining process is provided by the fact that about 100 000 turbocharger units a year were being produced by the late 1980s and the NGK Spark Plug Co. Ltd were also supplying other ceramic turbocharger components to diesel engine manufacturers, while similar design features have been incorporated into the production of turbomolecular pumps by the Japanese Atomic Energy Research Institute and Mitsubishi Heavy Industries Ltd.

8.2 ACTIVE METAL BRAZING OF OTHER CERAMIC COMPONENTS

The selection of a Ag-Cu-Ti active metal braze for the joining of the ceramic turbocharger rotor reflects not only the technical suitability of the material for that particular application but also, probably, the fact that it was and still is the only fully commercialized family of brazes that can be used with ceramics without prior metallization. It is, therefore, not surprising that these alloys have found other applications during the push to widen the use of ceramics as engineering materials.

8.2.1 Automobile engine tappet

A research programme sponsored by the Commission for the European Communities resulted in a novel design of a 23 mm diameter hollow steel tappet that was faced with a ceramic. The advantages of this development were perceived as being reductions in the weight of the tappet, in its wear by contact with the rotating cam, and in the friction losses between it and the rotating cam.

In describing the development of the tappet design and joining technology, Bucklow *et al.* (1992) identified the principal problem as accommodation of the difference in the coefficients of thermal expansion of the steel tappet and possible ceramic face materials. The preferred ceramic was Sialon, a Si-Al-O-N material that is known to have excellent wear resistance and has a coefficient of thermal expansion of $3.0 \times 10^{-6}\,K^{-1}$, about a quarter of that of the steel. Since the service temperature of the tappet can range from $-20\,°C$ to $+150\,°C$, mechanical attachment was impractical and joining by vacuum brazing with an Ag-Cu-Ti alloy was selected as the most suitable joining technique.

The use of brazing to produce a ceramic–metal joint requires a relatively high process temperature and this will subject the joint to severely mismatched straining as it cools to room temperature. Design and cost constraints made it unattractive to accommodate this mismatch by using a substantial ductile or multiple buffer layer and an alternative approach of using an insert with a compliant configuration was pursued, the principle of which is illustrated by the sketch shown in Figure 8.3. This particular type of joint configuration was adopted after experiments with a honeycomb structure failed to provide sufficient braze area at the T junctions between the honeycomb struts and the ceramic face to give the needed shear strength. The final design of the compliant insert was developed by applying finite element analysis techniques used the flexing of one circular corrugation to accommodate lateral mismatches of contractile strains. The realization of this used a 1 mm thick sheet of Fe that was bonded to a 5 mm thick ceramic disc.

The braze used to make the joints was a Ag-24.5Cu-10.51n-1.3T1 alloy which has a liquidus temperature of about 715 °C. Strips of this braze 25 mm in width and 40 μm thick were manufactured as part of the development programme by tape casting, a rapid solidification process in which liquid metal is squirted on to the surface of a chilled drum to produce a non-adherent thin solid layer. Brazing process parameters were not reported by Bucklow but other work suggests that they probably

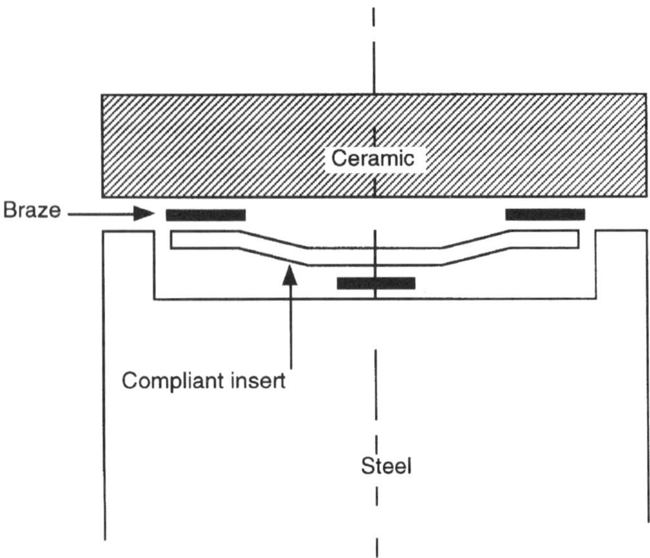

Figure 8.3 Sketch illustrating the principles involved in the design of a valve tappet using a compliant interlayer

involved a temperature of 800 to 900 °C maintained for 10 to 30 minutes in a vacuum of 2×10^{-4} mbar or better. Whatever conditions were used, failure of the tappets when subjected to shear testing occurred either by tearing of the compliant interlayer or by tearing followed by cracking of the ceramic. In no case did failure occur at the braze/ceramic interface.

8.2.2 Vacuum interrupter

Vacuum interrupters are electrical engineering devices used to switch off short-circuit currents of up to 40 000 amperes at potentials of up to 24 000 volts in 'medium voltage' systems. Their internal construction is complex and includes bellows and massive movable electrical contacts but their outer construction consists essentially of a thick-walled insulating alumina tube whose ends are closed by brazing on thin walled metal caps, such as was shown earlier in Figure 5.4. Ideally, the end caps should be of stainless steel with the substantial mismatch in the coefficients of thermal expansion of the metal and the ceramic being accommodated by flexing of the thin metal. One of the principal requirements for the joint between the end cap and the insulating tube is that it should be strong enough and compliant enough to stay attached to the ceramic when the end cap flexes in service due to temperature and stress variations.

The problems of joint integrity that can be caused by flexing of end caps was recognized for many years and its solution by using strongly bonded braze fillets to attenuate the resultant stresses was the target of a number of development programmes. Thus, van Esdonk and van der Sluis (1983) described an assessment conducted on behalf of the Dutch Institute of Welding to compare the vacuum tightness and strengths of joints produced by three techniques using the Ag-Cu braze alloys. The greatest success was achieved using the most ductile braze, Ag-28Cu, and reference will be made only to results obtained with this composition.

The three joining techniques employed by van Esdonk were:

1. the established process of moly-manganese metallization of siliceous Al_2O_3 followed by overcoating with Ni and brazing;
2. an activated joining process effected by painting the ceramic surface with a suspension of $<$ 60 mesh TiH_2 powder and $<$ 40 mesh Ti powder held in an aqueous binder of ethylene glycol, $HOCH_2CH_2OH$, followed by brazing;
3. what was at that time the new process of active metal brazing by using a Ti cored Ag-28Cu alloy without prior treatment of the ceramic surface.

The specific design used in this assessment employed 31 mm high tubes of a 96% pure siliceous Al_2O_3 with internal and external diameters of 70 and 83 mm, while the 75 mm diameter 0.5 mm thick end caps were composed of alloys such as the low expansivity Fernico, Fe-28Ni-23Co, and a ferritic AISI 430 steel with the composition Fe-17Cr-2.5Ni-1Mn. Before being used in the brazing trials, the Fernico caps were Cu plated to diminish the likelihood of mechanical degradation being caused by liquid metal embrittlement induced by contact with molten braze while the AISI 430 caps were Ni plated to promote their wetting.

Trials using the moly-manganese coated ceramics in a vacuum of 10^{-5} to 10^{-6} mbar and a brazing temperature of 830 °C produced strong bonding of Al_2O_3 with Fernico using a five minute holding time and with AISI 430, when holding times of up to 60 minutes were used. The joints produced were sound with leak rates of no more than 10^{-8} torr.l.s^{-1} and failed within the braze fillets when stressed to destruction by imposing external loads of 15 to 19 kN. In contrast, weaker samples were produced with the Ti/TiH_2 painted ceramic surfaces with the average loads required to cause failure being 11.5 kN when Fernico was used and only 3 kN when the metal was AISI 430. Using the cored Ag-Cu braze required an increase in the brazing temperature to 970 °C to effect dissolution of the Ti and the resultant joints were the weakest of any of those produced in these trials, those with Fernico failing at the braze/ceramic interfaces at an average load of only 0.7 kN.

The results of these trials show that strong joints could be produced using ceramics whose surfaces had been metallized by the moly-manganese process provided attention was paid to details such as coating of the metal end caps. The poorer results achieved with Ti/TiH_2 activation and particularly with the Ti cored active metal braze are noteworthy and probably attributable to excessive chemical interaction at the braze/ceramic interfaces that resulted in the formation of thick and fragile Ti based reaction product layers. Of even more significance is the fact that, even after this programme, up to 30% of the vacuum interrupters had to be rejected because they leaked or were in other ways unsatisfactory. This high rate was due in part to the fact that the braze alloys required a small joint gap of about 0.05 mm and ensuring a flatness of better than 0.025 mm for the large ceramic insulator as well as the metal end cap was a very demanding task.

Further work on the development of brazing technology for vacuum interrupters was reported eight years later by a new team, one of whom had been involved with the 1983 programme, van der Sluis and Schellekens (1991). This time attention was focused on braze alloys that could be used to achieve joining using wide joint gaps, the best of which was found to be a Ag-32Cu-10Pd alloy that had a melting range of 824–855 °C. The metal components were either Fernico or AISI 316L stainless steel, Fe-17.5Cr-12.5Ni-2.75Mo1Mn-0.015C, but the ceramic chosen was again a siliceous Al_2O_3 whose surfaces had either been metallized by the moly-manganese process and then coated with Ni or had been painted with a TiH_2 suspension. Test samples were produced using furnaces evacuated to 10^{-5} to 10^{-6} mbar and the best brazing temperature was identified as 875 °C.

The wetting of the component materials by the Ag-Cu-Pd braze alloy was good and tensile tests using Al_2O_3 ASTM test pieces bonded to 3 mm thick inserts of Fernico using 0.5 mm thick braze gaps showed that vacuum tight joints with strengths of up to 90 MPa could be achieved with metallized ceramic components. The joints produced using TiH_2 painted ceramic surfaces were again weaker by about 30 to 50%. Joints produced using AISI 316L as the metal had poorer strength values, averaging 35 MPa, but were vacuum tight.

The achievement of vacuum tight joints with good or adequate strength values in the test programme using metallized ceramic samples led to the adoption of the technology for the production of vacuum interrupters. This required the design of the stainless steel end cap to be slightly modified by bending to provide a small lip or skirt where it contacted the Al_2O_3, much as shown for modified T joints in Figure 5.2, so that a wide gap configuration could be used. The immediate advantage of this change in technology was that the machining standards for the ceramic and metal components could be relaxed and finishing costs were reduced by 70%. Additionally the use of laboratory proven brazes enhanced the quality of the joints, so that they remained leak free even when cycled between −200 °C and +400 °C.

Thus success was finally achieved by changing from conventional braze alloys with narrow melting ranges to sluggish materials that permitted joints to be produced that were substantial enough and sufficiently compliant to allow flexing of a stainless steel end cap without transmission of stresses large enough to cause failure of the braze/ceramic interface or within the ceramic itself.

8.2.3 Battery seal

Development of the Na-S battery received a boost from the 1970s oil shocks referred to earlier but many technical problems had to be overcome, not least of which was how to make seals between the battery parts that would withstand contact with molten Na and S. In essence, the battery is composed of a number of cells each of which consists of a closed steel tube that is joined to but electrically isolated from a beta-alumina, $Na_2O.xAl_2O_3$, electrolyte tube filled with Na while the space between the tubes is filled with S. The layout of a Na-S cell that is sketched in Figure 8.4 illustrates the role of the robust Al_2O_3 collar in protecting the mechanically fragile electrolyte tube from the stresses imposed by the greater coefficient of thermal expansion of the steel casing.

In deciding how to make the joints between the ceramic collar and the casing, note has to be taken of the fact that Ag, Cu and Ti all react with S so the commercially available active metal braze alloys are not suitable. Al also forms a stable sulphide but this reaction can be impeded by the

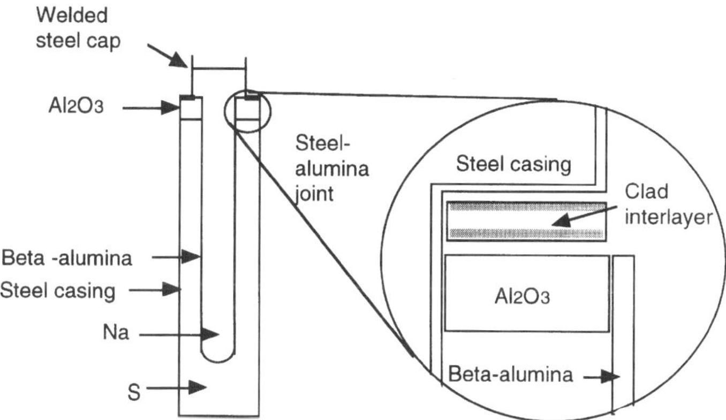

Figure 8.4 A typical arrangement of a Na-S battery cell with the Al_2O_3-steel casing joint shown in more detail as an insert

tenacious and chemically stable films of Al_2O_3 present on Al surfaces. Thus Al is a contender as a joining material and Yamada *et al.* (1989) described a successful application using Al sheet clad with an Al-Si as the interlayer between the steel casing and the Al_2O_3 collar. Sheets such as this are used in the fabrication of Al alloy heat exchangers and typically comprise a core of the Al alloy 3003, Al-1.2Mn, clad on each side with a layer of Al-12Si or an Al-Si-Mg braze alloy that has a thickness of 5 to 10% that of the core.

Early trials identified the optimum conditions for joining as the use of a 2 mm thick interlayer and application of a pressure of 5 MPa for 30 minutes at 600 °C in a chamber evacuated to a residual gas pressure of 1×10^{-4} mbar. The joints produced had a room temperature, He, leak rate of less than 10^{-10} torr.l.s^{-1} and an average tensile strength of 25 MPa. Decreasing the interlayer thickness to 1 mm caused the average joint strength to fall to 4 MPa, but increasing it to 5 mm had no beneficial effect. Similarly, increasing the bonding temperature by 25 °C, or decreasing it to just below the 577 °C melting temperature of the Al-13Si eutectic, caused a 30 to 50% decrease in joint strength. At 600 °C, the Al-Si cladding will have been fully molten and hence the application of a 5 MPa pressure would disrupt the oxide film on the cladding surface and then disperse it by squeezing excess liquid out of the joint. Finally, it is notable that while additions of small concentrations of Mg were found to promote disruption of the oxide films present on the surfaces of Al-Si clad sheets of Al alloy used for their original application of fabricating heat exchangers by vacuum brazing, additions of Mg were not found to be significantly beneficial in this particular application.

8.3 THE REPAIR BRAZING OF A FERRITIC STEEL HEAT EXCHANGER

The construction and development of the prototype fast reactor at the Dounreay Laboratory of the United Kingdom Atomic Energy Authority, UKAEA, required many joining problems to be overcome, not all of which had been anticipated. The reactor was used to produce 250 MW of electricity by transferring heat from the Na pool in which the fuel elements were submerged to raise steam and hence drive the power generating turbines. One of the unexpected problems was the occurrence of leaks in all three of the secondary exchangers where heat was transferred from Na to steam. Each exchanger consisted of three sections relating to different stages of the steam cycle – the evaporator, the superheater and the reheater – and leaks occurred first in the superheater after a few years of operation and then in the evaporator section. The application of concern at present is the repair of the evaporator section which consisted of a

Fe-2.25Cr-1Mo steel pot with a diameter of 1.8 m and a height of 10.6 m that was filled with Na. Within the evaporator were partially submerged tubes shaped in the form of a U through which passed steam. The stout lid of the exchanger, the 400 mm thick 'tube plate', was penetrated by a close packed array of 1000 holes each with a diameter of 20 mm. The 500 U tubes of Nb stabilized Fe-2.25Cr-1Mo had outside diameters of 24.9 mm and hence did not pass through the tube plate but were welded to its bottom face to cover the holes as sketched in Figure 8.5.

Initially, U tubes with leaking weld joints were sealed off by plugging with Incoloy 800 to prevent direct entry of steam into the Ar cover gas above the Na pool and hence ultimately resulted in the production of chemically aggressive NaOH. While providing safety, this procedure diminished the electricity generating capacity of the reactor, as did the down-time necessary to seal a tube, and after a while it was decided to develop a technique for returning the leaking tubes to service. It was both technically difficult to repair the leaks by rewelding to the bottom of the tube plate and economically unacceptable to shut down the exchangers while this was being done. Instead, the approach adopted was to repair by fitting internal sleeves that would shield the cracked welds. These sleeves were explosively welded to the tube plates near their top surfaces and were brazed to the U tube just below the level of the tube plates as indicated in Figure 8.5.

It was decided to use brazing as a repair sleeve joining technique because it had been successfully employed during the manufacture of critical

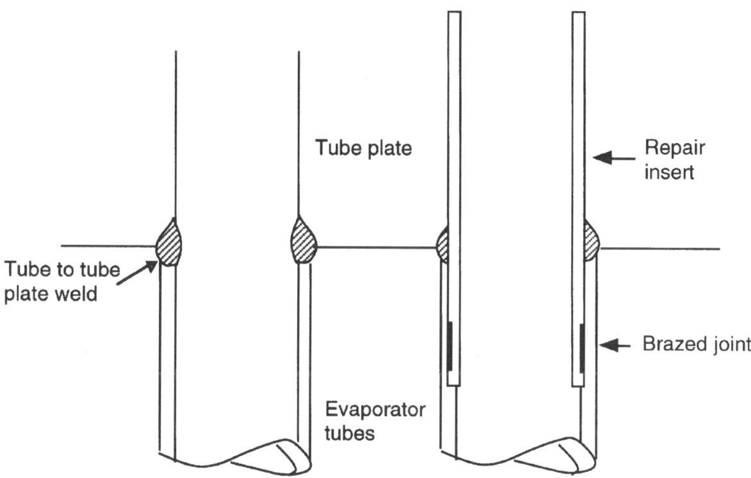

Figure 8.5 Schematic arrangement of the bottom of the evaporator tube plate showing the original attachment of the steam tubes by welding and the configuration of the repair insert

reactor components such as Na flow gauges and liquid level detectors, shut-off rods and filters, and even more pertinently during the earlier repair sleeving of leaking welds on the tops of the tube plates of the superheater sections of the exchangers. While this earlier repair work had been successful, the problems faced in designing a brazing programme for the repair of the leaking evaporators were more numerous and formidable because the welds were on the bottom of the tube plate and hence it was impossible to fit external sleeves or use external heat sources. Aspects of these repairs of first the superheater sections and then the evaporator sections have been described by Taylor and Hayden (1979) and by Sheward and Baron (1983) and those of particular relevance to joining processes can be summarized as follows.

One of the first problems requiring solution when contemplating the use of brazing is usually the selection of a suitable braze alloy but in the case of the evaporator repairs this was perhaps the least of the problems. The component material was an Fe-2.25Cr-1Mo steel and a number of braze alloys and procedures have been proven to be effective when joining this, and other ferritic alloys using furnace or closely controlled manual techniques. However, the repair brazing had to be done about 500 mm down a 20 mm diameter hole on site. Thus surface preparation was peculiarly difficult to perform or qualify, and both the heating and provision of a protective gaseous environment was rendered especially difficult because it had to be done using devices that would fit within the tube being repaired. The close packing of the tubes presented difficulties of access and the ambient temperature of about 46 °C, 115 °F, made the conditions for the operators very fatiguing.

It was essential that neither the insert sleeve material nor the braze alloy should be corroded by Na, steam or Na–steam reaction products. This was satisfied for the sleeve by choosing Fe-2.25Cr-1Mo, while BNi-4, a Ni-3.5Si-1.7B1Fe alloy was chosen as the braze because trials conducted in connection with superheater repair had shown it not only to wet and bond adequately to Fe-2.25Cr-1Mo but also to produce joints with mechanical and corrosion properties that should be adequate for a 20 year service. In the evaporator repair work the braze was used as 0.5 mm thick sheets and some flow problems were encountered in early material assessment trials. The specified liquidus temperature of BNi-4 is 1066 °C but while the compositions of the braze sheets always fell within the specification for BNi-4, their liquidus temperatures ranged from 1060 to 1150 °C, apparently because the varying configurations of boride inclusions in the as-received sheets resulted in different dissolution rates to release B and hence depress the melting temperature. Rather than initiate a braze composition development programme, the pragmatic approach was adopted of using braze material only from sheets shown by melting tests to have a liquidus temperature close to the specified value. Circlets were formed from satisfactory sheets of the brazing alloy and then were attached to the outside of the repair sleeve by tack welding. After having been inserted into the evaporator U tubes, the repair sleeves were expanded to produce close fits and the desired joint gaps.

The insides of the evaporator tubes had been exposed to steam for several years and were covered by a layer of Fe_3O_4 which would have prevented flow of the braze alloy had it not been removed by the machining to accommodate the repair sleeve. However, this surface oxide also contained some adsorbed H_2O and lubricant residues arising from both the tube enlargement and the original machining prior to welding. Evolution of O_2, H_2 and H_2O from tube regions adjacent to the sleeve that were heated during the brazing operation initially caused problems but these were overcome by cleaning the inside wall of the entire U tube. This was achieved by using tight-fitting cylindrical plugs of absorbent material that were either dry or moistened with acetone, CH_2COCH_2, and were driven from the entrances to the exits of the U tubes by compressed air.

In principle, a brazing temperature of about 1100 °C should be sufficient to ensure good flow of BNi-4 braze alloy with a liquidus temperature of 1066 °C, and this was achieved in laboratory tests using coupons of Fe-2.25Cr-1Mo. However, the area to be brazed was deep within a narrow tube and close to the massive tube plate so that considerable conductive heat losses could occur. The requirement for the heating source, therefore, was specified as being able to produce a temperature of up to 1200 °C in the braze zone within 90 seconds. Achieving this using an induction heating device that could be accommodated within the repair sleeve was a challenging task that required a separate development programme. In fact it was also necessary to develop a second device with differing coil configuration to temper, soften, both the insert sleeve and the original evaporator tube because they were of a ferritic steel that transformed on being cooled from such a high brazing temperature to produce a hard and somewhat brittle structure. The power needed for brazing was generated by a 30 kW unit operating at a frequency ranging about 300 kHz which minimized the wasteful resistive heating of the sleeve and evaporator tube and allowed power to be supplied to the coil along water-cooled leads using a low current at a high voltage. The coil itself was encased in a castable ceramic to support and protect it, and channels within the ceramic body directed an Ar-5%H_2 mixture to purge and shield the materials in the brazing zone.

The final schedule and procedures for brazing repair sleeves were used in trials with tube–tube plate test pieces that replicated a section of the evaporator. After purging the replica tube with Ar-5%H_2 for 10 minutes, power was supplied to the induction coil at a constant level to produce a brazing cycle that lasted 5 minutes and caused the repair sleeve to attain a temperature of about 1160 °C. Satisfactory joints were produced providing that the brazing temperature never rose above the 1180 °C at which liquation was found to occur due to the formation of an Fe-3.5B eutectic. Laboratory assessment of the replica test samples showed their joints to be leak tight and able to withstand simulations of the stress levels that would or could be encountered in service conditions. Thus high-temperature fatigue tests using a minimum load of 20 kN produced no discernible degradation after 20 000 cycles. Similarly, the resistance of BNi-4 brazed joints to corrosion by both static and flowing Na at 650 °C was confirmed, and accelerated tests demonstrated their resistance to attack by Na-steam reaction products at 350 and 500 °C.

Having achieved such successes, a campaign was mounted to repair the evaporators. The integrity of every brazed repair sleeve joint was assessed by leak testing and ultrasonic inspection. Of the 175 joints repairs effected in one campaign, one was of marginal quality and the other 174 had integrities well within the specifications. The actual in-service mechanical and corrosion characteristics of the repair sleeve brazed joints are not known, but leaking of evaporator heat exchangers ceased to be a problem.

8.4 THE SELECTION OF A BRAZE ALLOY TO JOIN INCONEL 600 TO Cu-1Cr

Many brazing problems are solved every day by applying the recommended practices enshrined in national or international standards, and the advice available from suppliers of braze alloys. Often the solution requires merely a single decision, such as changing to another identified braze alloy but sometimes the process is more complex as in the selection of a braze alloy to join Inconel 600 to Cu-1Cr for an application with demanding service requirements.

The present frontier of the development of fusion reactors for power generation is represented by the Joint European Torus plasma physics facility constructed at the Culham Laboratory of the UKAEA. Its development posed and its operational programme poses many material problems, one of which was the design of a limiter, a device to provide a solid surface that defines the edge of the plasma and protects the walls of the containment vessel from bombardment by the plasma. The realization of limiter devices called for the joining of 1 m × 1 m plates of Inconel 600 alloy, to a water-cooled backing chamber of Cu-1Cr. Many joining techniques were considered and several were evaluated but vacuum brazing was the final choice for technical, design and economic reasons. A programme of work was needed, therefore, to select a suitable brazing alloy and process conditions and aspects of this have been described by Crispin *et al.* (1983).

The effectiveness of the limiter depends in part on good thermal contact between the plate and the water-cooled backing. The ingress of heat from the plasma is substantial and the design assumed that the Inconel 600/Cu-1Cr joint would have to withstand a temperature of 500 °C. At this high temperature, the joint should be stress-free but stresses of up to 200 MPa could be encountered at a temperature of about 300 °C during start up or shut down stages. Factors such as melting temperature, volatility, viscosity and potential interactions with the component materials led to the preliminary selection of two precious metal brazes for evaluation, Ag-28Cu and Au-37.5Cu. The programme to make a final selection assessed the interaction of these brazes with the component materials either separately or as dissimilar sandwiches were evaluated in terms of wetting behaviour, strengths at room and elevated temperatures and joint thermal conductivity values.

Excellent wetting of degreased samples of both component materials, as defined by contact angles of less than 20°, was achieved with both braze alloys by heating a vacuum of 10^{-5} mbar. The temperatures required when using a 15-minute hold ranged from 880 °C when using Ag-28Cu in contact with Cu-1Cr to 975 °C when using Au-37.5Cu in contact with either Cu-1Cr or Inconel 600. However, achievement of good joint formation when brazing sandwiches of the two component materials generally required temperatures approaching or about 1000 °C.

Simple tensile test samples were produced at a temperature of 990 °C by brazing dissimilar pairs of truncated cones of the component materials, and the resultant joint strengths that were measured at room and elevated temperatures are shown in Figure 8.6. There was little to chose between

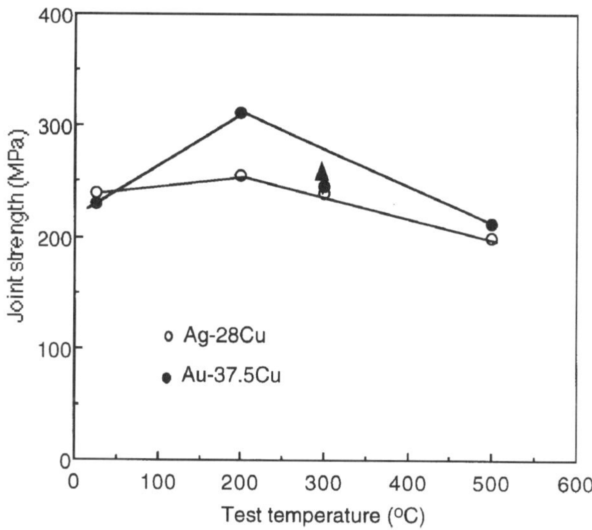

Figure 8.6 The effect of test temperature on the tensile strengths of Cu-1Cr/Inconel 600 samples vacuum brazed with Ag-28Cu or Au-37.5Cu

the strength values at room temperature or 500 °C of joints that had been produced using either alloy, but joints produced using the Au-37.5Cu alloy were stronger at intermediate temperatures. Thus joints produced using either braze satisfied the 200 MPa strength criterion specified for the application.

No quantitative criteria were specified for the thermal conductivity requirement, but tests suggested that the joints would be adequate. Thus as commented earlier in Chapter 7, the thermal conductivity of a sample produced by brazing a 1.34 mm sheet of Cu-1Cr to a 1.64 mm sheet of Inconel 600 using the Ag-28Cu alloy was only 30.7 $W.m^{-1}.K^{-1}$ while that of a 1.34 mm sheet of Cu-1Cr brazed to a 4.24 mm sheet of Inconel 600 was an even lower 20.2 $W.m^{-1}.K^{-1}$. The calculated thermal resistances of the braze joints were equivalent to an additional 0.02 mm of Inconel 600 for the Ag-28Cu alloy, little more than the precision of machining the 1 m × 1 m sheets, and 0.7 mm of Inconel 600 for the Au-37.5Cu alloy.

Both the precious metal brazes satisfied the specified and implied performance criteria in terms of wetting behaviour, strength and thermal conductivity targets. The final selection, however, was based on a previously unstated concern about volatility. The vapour pressure of Ag at 1000 °C is 7.6×10^{-3} mbar while that of Au is 1.6×10^{-6} mbar and concern was felt about both possible contamination of the furnace by Ag vaporization when brazing and, more remotely, of the torus containment vessel during operation. The final preference, therefore, was for the Au-37.5Cu braze alloy.

8.5 DIFFUSION BONDING OF AN Al_2O_3/Ti ACCELERATOR TUBE

The construction of a 20 million volt van der Graaf accelerator at the Daresbury Laboratory of the UK Science and Engineering Research Council in the 1980s provided the nuclear physics community with a facility for studies of nuclear structures in advance of anything else in the world at that time. Its design incorporated many new features and its realization required the application of an exceptionally reliable technology that would enable the 10 000 dissimilar material interfaces within the structure to satisfy very demanding performance specifications, as described by Joy *et al.* (1987).

The basic design of an accelerator tube comprises a column made by bonding insulating rings to annular metal electrodes, as sketched in Figure 8.7. The function of the tube is to accelerate a beam of heavy ions along its length by causing it to pass through a series of relatively modest voltage gradients until it ultimately impinges on a target. Previous

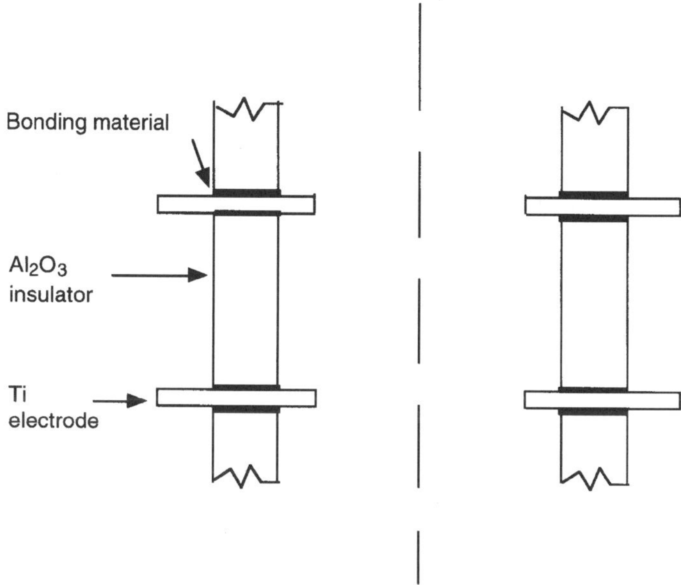

Figure 8.7 An illustrative sketch of the configuration of the electrode/insulator joint used in construction of the Daresbury Laboratory accelerator tube

designs of successful but smaller accelerator tubes had used glass insulators bonded to stainless-steel electrodes by an adhesive or ceramic insulators diffusion bonded directly to Ti electrodes. This second design provided a starting point for the facility planned for the Daresbury Laboratory, the chosen conceptual design of which specified an evacuated accelerator tube that was a 29 m long tube and comprised more than 2000 Al_2O_3 insulating rings bonded to annular Ti electrodes. Because it was impractical to construct such a tube as a single piece, it was actually built from more than 100 modules that were then joined to make a complete tube.

Effective acceleration of a collimated beam of heavy ions within the accelerator tube required that it should be evacuated to and maintained at a residual pressure of less than 10^{-7} mbar with a negligible concentration of hydrocarbon gases or other heavy molecules. Achieving this quality of vacuum was made a more challenging task because the tube was to be held vertically in a 45 m long and 8 m diameter steel pressure vessel filled with 8 bar of gaseous SF_6 to provide electrical insulation able to withstand a voltage of up to 30 MV. When in routine operation, the tube had to endure repeated mechanical cycling as the SF_6 was pumped out of the container vessel for brief periods to permit maintenance work on the accelerator. Additionally, the tube was subjected to DC voltage gradients of up to

$2\,MV.m^{-1}$ during normal operation and occasional high voltage discharges dissipating 10^5 J at 20 MV.

The programme to develop and realize a joining technology had three separate stages: initial scoping experiments to establish what type of joining process might be suitable using Al_2O_3 ASTM test pieces, laboratory construction of small tubular elements for mechanical, leak and electrical testing, and the construction and use of facilities for production of accelerator modules. The specification for the first scoping stage was qualitative, merely requiring that any joints formed should be strong and vacuum tight. The use of adhesive bonding was ruled out from the beginning because of the vacuum required within the tube and it was wished to avoid direct bonding of the electrodes to insulators because of the high fabrication temperatures required and the likely resultant distortion of the electrode shape, which would require post-fabrication manipulation or machining to remedy. Brazing of the electrodes to the insulators was not ruled out but some concern was felt about the effects of possibly variable braze flow and fillet shape on the electrical characteristics of the joints. The preferred technique, therefore, was diffusion bonding using a ductile metal interlayer as the joining material and confidence about this preference was gained by the end of the first week of laboratory scoping experiments during which vacuum tight joints were produced between Al_2O_3 and Ti annuli by using interlayers cut from Al cooking foil and diffusion bonding in an Ar environment. When stressed to destruction using high tensile loads, failure of these samples occurred partially within the ceramic. This satisfactory outcome of the first experiments was attributed in part to the fact that the coefficients of thermal expansion of Al_2O_3 and Ti are very similar so only modest contractile stressing should have been encountered as the samples cooled from their fabrication temperature.

Having made an encouraging start, systematic work was conducted to identify the optimum bonding parameters. Thus materials comparisons were made between different types and sources of Al_2O_3 and between different sources of commercial purity Ti. The process parameters were similarly varied with assessments of the effects of using different cleaning and surface preparation procedures, of using different thicknesses of Al foils, of varying the bonding pressure, temperature, time and environment. This led to the selection of a pore free 97% pure Al_2O_3 as the preferred ASTM test piece material not only because of its bonding behaviour but also its good high voltage characteristics and, not least, its availability in a wide variety of shapes and sizes. This particular ceramic had an average grain size of about 50 μm and a surface finish that had a usual average amplitude, R_a, of 0.50 μm and a wavelength of 47 μm. Before being used in bonding tests, the Al_2O_3 test pieces were fired in air at 1000 °C for 30 minutes to remove any traces of carbonaceous contamination. In practice there was no significant difference between the behaviour of the 1 mm thick Ti inserts obtained from different sources if the metal surfaces were carefully cleaned by abrading with fine Al_2O_3 cloth, to yield a surface finish with an amplitude of 1.21 μm and a wavelength of 64 μm, and were then degreased by ultrasonic agitation in Genklene for 5 minutes. Similarly, the surfaces of the Al foils were abraded to produce a surface finish with an average amplitude of 1.75 μm and wavelength of 99 μm before they were degreased.

Bonding trials were conducted by applying pressures of 10 to 100 MPa for between 10 minutes and 2 hours at temperatures of 450 to 650 °C. These trials were conducted in air in the open laboratory or in a chamber through which Ar flowed or which was evacuated to a residual gas pressure of 5×10^{-5} mbar. These trials demonstrated that bonding could be achieved reliably in an evacuated chamber by using a pressure of 50 MPa applied for 30 minutes at temperatures of 450 °C or above. A similar threshold was observed when bonding in flowing Ar, but that for bonding in air was higher. Bonding at temperatures above the threshold value produced progressively stronger joints, the average tensile strength of those produced in Ar at 600 °C using 1 mm thick Al interlayers was 51 MPa while similar samples produced in vacuum was 60 MPa. Growth of 1 μm thick reaction product layers of an intermetallic compound had occurred at the Al-Ti interfaces but this did not result in significant mechanical degradation of the joints because failure occurred at the Al-Al_2O_3 interfaces. These strong joints formed in an evacuated chamber were vacuum tight but those bonded in flowing Ar were often leaky and those produced in air were always leaky. The standard joint fabrication conditions used in future work, therefore, were defined as the application of a pressure of 50 MPa for 30 minutes at 600 °C in an evacuated chamber.

Of negligible importance in these trials was the fact that the 1 mm Al foils suffered considerable deformation, being compressed by more than 80% to produce a substantial flash. However, such extrusion of material would result in rejection if it occurred during the fabrication of accelerator modules so further trials were conducted with foil thicknesses as small as 25 μm. These thinnest foils suffered negligible compression but the resultant joint strengths were low and the flatness requirements for the Al_2O_3 and Ti surfaces were much more rigorous. This further set of trials resulted in the selection of a foil thickness of 0.1 mm as the most convenient to use, even though such foils were compressed about 25% and their use required quite high flatness standards.

These trials were repeated with 80 mm diameter samples, comprising two Al_2O_3 tubes and one Ti annular electrode, that were subjected not only to strength and leak testing but also to extensive evaluation of electrical characteristics. The strength and vacuum tightness of the joints produced using 0.1 mm thick Al foils and the standard joint fabrication conditions were satisfactory but the high-voltage performance of the electrodes depended critically on the completeness of bonding at the Al/Al_2O_3 interface and the configuration of the Al interlayer. The presence of even minute and isolated unbonded areas at the ceramic/interlayer interface was found to cause a dramatic decrease in the breakdown voltage of the test piece. Additionally, it was observed that breakdown could be initiated by applying a potential of about 20 kV when the Al interlayer projected or was recessed by more than its own thickness but electrodes bonded by interlayers whose edges provided a smooth continuation of the ceramic surface contours could withstand up to 70 kV before there was any breakdown.

These trials led to the definition of a series of requirements for joint characteristics that would have to be met during the production of accelerator modules, the basic parameters of which are summarized in Table 8.1. To be satisfactory, it was decided that the modules would need:

1. interfacial strengths greater than 15 MPa;
2. leak rates less than 10^{-8} torr.l.s^{-1};
3. interfaces that were free from microscopic voids;
4. strictly controlled interlayer configurations;
5. electrodes undistorted by the bonding process;
6. electrical characteristics not degraded by the bonding process.

These requirements had been met by the various test pieces in the earlier trials, but some additional problems that had to be overcome were identified during the module fabrication programme. Two of these are particularly noteworthy.

Table 8.1 Process parameters for the bonding of Daresbury Laboratory accelerator modules

Module length	196 mm
Insulators	96.7% pure Al_2O_3 16 per module, (2 for thermal balance) Length 10 mm Diameter, 190 mm o.d., 150 mm i.d. Surface finish, 1µm R_a
Electrodes	Commercially pure Ti Thickness, 1 mm Diameter, 203 mm o.d., 112 mm i.d.
Interlayers	0.1 mm thick Al

A large fabrication chamber was constructed and the first problem encountered was the selection of a bonding environment. Use of an evacuated chamber was found to be best in the laboratory trials but it was decided to use rough evacuation of the chamber followed by backfilling with purified Ar in the production runs for economic reasons. However, even using the best commercially available Ar containing no more than 1 ppm of O_2 or H_2O resulted in the formation of joints that were of variable strength and also some oxidation and nitridation of the Ti electrodes. The pumping facilities for the chamber, therefore, had to be upgraded so that it was first evacuated by an oil-free diaphragm roughing pump to less than 10^{-1} mbar and then by a 450 $1.s^{-1}$ turbomolecular pump to achieve a hydrocarbon free residual gas pressure of 7×10^{-6} mbar at the fabrication temperature.

Another unwelcome problem was the fact that the coefficients of thermal expansion for Al_2O_3 and Ti are similar but not identical. With the large module components, the small difference in contractions during cooling from the fabrication temperature caused failure of the end joints when the module was terminated with a metal plate. This problem had to be overcome by using two electrically redundant Al_2O_3 insulator rings at either end of the module to provide a thermal balance for the end joints.

Despite the difficulties and problems, the production programme was a clear success with a module rejection rate of less than 5% when producing the 144 modules required for the accelerator tube plus spares and variations. Each of these modules had a leak rate of less than 10^{-10} torr.l.s^{-1} when subjected to a pressure differential of 10 bar. None of the module joints failed during the more than 5 years of continuous operation of the accelerator, and examination of modules replaced after several years by others of an improved electrical design failed to reveal any sign of degradation.

8.6 DIFFUSION BONDING OF METAL COMPONENTS

The use of diffusion bonding as a means of fabricating critical metal–metal components for advanced applications is well established, but such developments are very commercially sensitive. Fortunately some organizations have allowed brief descriptions of particular applications and this subsection refers to three such examples that relate to the use of diffusion bonding to produce devices from assemblies of components of both identical and dissimilar metal alloys. Each of the examples chosen exhibits somewhat unusual features, in that one used resistance heating of the components to achieve the desired bonding temperature, another employed hot isostatic pressing to effect contact between the surfaces to be bonded and the third example employed superplastic forming to finish production of the bonded component.

8.6.1 Injector blocks

The process of extruding metals, and other materials, involves the use of dies or injector blocks that contain internal channels of sometimes complex configuration through which the metal or alloy is squeezed out, as is icing when decorating a birthday cake. To produce metal extrusions with an unchanging size and shape, it is essential that the internal channels within the injector block are wear-resistant and do not suffer distortion by the high pressures at elevated temperatures used in the process.

Many injector blocks are made of high-strength steels that can be machined to produce the fine surface finishes needed for the walls of the extrusion channels but the complexity of the internal channels required is usually such that it is impossible, or at least uneconomic, to machine the block from a solid piece of steel. While a variety of fabrication processes could be used in principle to produce blocks, such as casting or pinning together sections that are then fitted within a stout collar, diffusion bonding without the use of an interlayer was the method chosen by the Welding Institute in an application described by Spanswick and Nicholas (1987). Two of the main attractions of this technique were that no weaker or less wear resistant material was introduced into the structure and that it could be used to join large flat surfaces without affecting the contours of recessed half channels.

The components to be diffusion bonded were of a 50D steel with a composition of Fe-1.2Mn-0.15C-0.08(Nb, V) and were in the form of half blocks measuring 250 × 290 × 380 mm. The half blocks to be bonded were machined to produce the channel recesses and then the bonding surfaces were finished to an average amplitude, R_a, of less than 0.4 μm and thoroughly degreased.

A somewhat unusual feature of the joining process was that the half blocks were resistance heated to the diffusion bonding temperature using a current of up to 20 000 amperes drawn from two transformers. The components to be bonded were not in direct contact with the electrode platens but were separated from them by a stack of mild steel and graphite sheets. This practice produced a uniform current flow into the components, decreased the current density needed to achieve the bonding temperature and minimized the heat losses from the components to the platens.

The components were placed within the bonding chamber, a pressure of 4 MPa was applied, the chamber was evacuated and the blocks were heated at 4 °C.min^{-1} to 1050 °C and held at that temperature for 2 hours. The quality of the joints in the bonded components produced in development trials were assessed by ultrasonic inspection, pressure testing and in some cases by tensile testing and metallographic examination. These tests identified the ultrasonic characteristics of components with joint strengths comparable to that of the parent metal and hence enabled each of the several dozen injector blocks produced by this technique to be qualified before entering service.

8.6.2 Spallation Neutron Source

The ISIS Spallation Neutron Source at the Rutherford Appleton Laboratory of the UK Science and Engineering Research Council consists essentially of an array of 23 discs of depleted U that range in thickness from 7.7 mm to 26.2 mm encased in and bonded to Zircaloy-2, Zr-1.5Sn-0.1Cr. As described by Yiasemides (1987), each encased disc was then shrink-fitted into a stainless-steel surround into which cooling channels were machined before they were welded together to form a long stack

Diffusion bonding was chosen as the technique for joining the U to Zircaloy-2 because it had proven to be successful when used for the somewhat similar UKAEA requirement for UO_2 encased in Zircaloy-2. Similarly, hot isostatic pressing was again chosen as the joining technique because it applies a uniform pressure to the component surfaces and hence should result in uniform contact and diffusion bonding. Such uniformity is required to achieve unimpeded transfer of heat from the U discs to the cooling channels. Hot isostatic pressing also had the advantage of not changing the shape of the component being bonded so that post-fabrication machining was not required.

Before being assembled, the surfaces of the U discs were machined and degreased to remove old and often thick films of UO_2 and other contaminants, the average amplitude, R_a, of the finish being 0.8 μm. Each disc was then placed in a close-fitting Zircaloy-2 cup to which a lid was fused in an evacuated electron beam welder. These canned discs were then hot isostatic pressed at 850 °C for 3 hours using a pressure of 200 MPa followed by slow cooling at 1 °C.min^{-1}. These conditions were chosen both to ensure that bonding of the U to the Zircaloy-2 and also that an interdiffusion zone with a depth of about 60 μm was produced. Any markedly lesser depth was found to allow cracks to form at or near the interface when the discs were subjected to rapid temperature changes. Similarly, the exceptionally slow cooling rate was used to minimize residual stresses caused by the mismatched coefficients of thermal expansion of U and Zircaloy-2. After being bonded, the quality of the joints within the encased discs was evaluated by ultrasonic inspection which could detect not only unbonded regions but also those in which there had been insufficient interdiffusion. Using this procedure, three spallation neutron sources were fabricated that satisfied the customers' requirements.

8.6.3 Ti heat exchanger

Engines for fighter aircraft operate at very high temperatures and the fabrication of components such as their heat exchangers present formidable problems. As explained in earlier chapters, diffusion-bonded joints can have excellent high temperature characteristics, and diffusion bonding, followed by superplastic forming, was adopted by British Aerospace (MA) Ltd for the manufacture of ducting for the primary heat exchanger of the Tornado fighter aircraft, some aspects of which have been described by Bottomley and Ginty (1993).

The ducts are essentially ribbed structures of the type sketched in Figure 8.8 with the joints between the two S profile half ribs and the outer shell being made by diffusion bonding. The constructional material was IMI 318, Ti-6Al-4V, and the sheets of this alloy to be diffusion bonded were cleaned in acid and then screen printed with a pattern using Y_2O_3 stop-off powder to prevent bonding in regions subsequently to be inflated by superplastic forming. The cleaning procedure, the process conditions and the design of the manufacturing jigs used had been optimized by an intensive development programme. Thus trials showed that 100% bonding could be achieved at a temperature just above 900 °C by applying a pressure of 2.8 MPa for 20 minutes, or 2.1 MPa for 35 minutes or 1.4 MPa for 60 minutes. However, the production runs used a bonding time of 90 minutes to achieve an 'overkill' and this resulted in a reject rate of less than 1%.

Subsequent to bonding, the integrity of the joints was evaluated using ultrasonic techniques and then superplastic forming was used at 925 °C to inflate the duct structure at the optimum strain rate of $1.5 \times 10^{-4}\,s^{-1}$. Destructive testing of samples of the production ducts confirmed their expected quality, with joints being resistant to peeling under both static and fatigue loading and having lap shear strength values equal to those of the parent material.

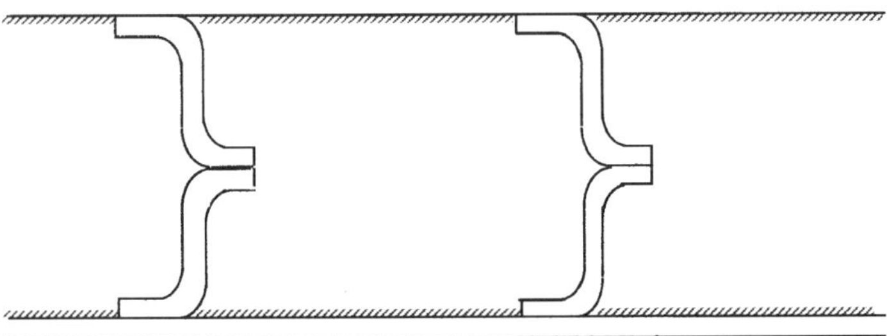

Figure 8.8 Part of the configuration of a diffusion bonded/superplastic formed Tornado heat exchanger duct. The areas that are covered by stop-off are identified by dashed lines

8.7 CONCLUDING COMMENTS

The first requirements for the fabrication of a device from an assembly of components are that intimate contact should be achieved either directly or indirectly via some interlayer of foreign joining material and that such contact should result in the formation of a permanent bond. Establishing a satisfactory process that results in high-integrity permanent joints can be a complicated and very drawn out exercise, as in the case of the vacuum interrupter development.

Even when strong bonding can be achieved readily, as was the case for $Al_2O_3/Al/Ti/Al/Al_2O_3$ joints during the first week of the development programme for the accelerator tube, the programme can still be arduous due to other problems. Indeed, every example cited in this chapter except that of the vacuum interrupter contains some obstacle that was more difficult to surmount than that of achieving strong bonding of the interfaces between the component and the braze or diffusion bonding interlayer. Thus the design of inserts to accommodate mismatched coefficients of thermal expansion was the crucial stage in the development of ceramic turbochargers and of ceramic-faced tappets, while the design or application of an unusual heat source is the most notable feature of the work to install heat exchanger repair sleeves or to manufacture injector blocks. Again considerable effort was spent during the development of successful procedures for the manufacture of accelerator tubes and of Tornado heat exchangers in optimizing the conditions for the production of a component containing multiple joints or of the simultaneous production of many components containing but one joint. Finally, it should be noted that the characteristic eventually determining the selection of a braze to be used to produce limiter devices was not wetting or bonding behaviour but volatility while the puncturing of surface films of oxide on a molten braze layer was the critical step in the development of a sealing process for Na-S batteries.

These brief and necessarily partial accounts of ten successful applications requiring joints to be made by brazing or diffusion bonding illustrate some of the complexities that can be encountered during a development programme. Such programmes are like a steeplechase which is not won until the last hurdle is surmounted. It is hoped that this book will assist its readers to identify and surmount at least some of the hurdles that may be encountered when their work requires the introduction of joining processes.

REFERENCES

Bottomley, I. E. and Ginty, B. (1993) The development of DB/SPF titanium technology for military aircraft applications. In D. J. Stephenson (ed.), *Diffusion Bonding 2*, Elsevier Applied Science, London, 232–41.

Bucklow, I. A., Dunkerton, S. B., Hall, W. G. and Chardon, B. (1992) A brazed ceramic-to-metal joint in a car engine tappet. In R. Carlsson, T. Johansson and L. Kahlman (eds), *4th International Symposium on Ceramic Materials and Components for Engines*, Elsevier Applied Science, London, 324–32.

Crispin, R. M., Hall, A. R., Nicholas, M. G. and Trevena, P. (1983) A laboratory study to select a vacuum brazing alloy, *Brazing and Soldering*, **4**, 4–10.

van Esdonk, J. and van der Sluis, H. H. (1983) Ceramic to metal seals by metallizing and or active brazing, *Brazing and Soldering*, **5**, 22–5.

Ito, M. and Taniguchi, M. (1993) Metal/ceramic joining for automobile applications. In S. D. Peteves (ed.), *Designing Ceramic Interfaces II*, Commission of the European Communities, Luxembourg, 205–21.

Itoh, M. and Kato, N. (1993) Brazing ceramics to metals and related application. In H. Krappitz and H. A. Schaeffer (eds), *Joining Ceramics, Glass and Metal*, Deutschen Glastechnischen Gesellschaft, Frankfurt, 321–31.

Joy, T., Crispin, R. M. and Nicholas, M. G. (1987) The accelerator tube for the Daresbury Laboratory nuclear structure facility: a successful application of diffusion bonding. In 5th International conference on *High Technology Joining*, British Association for Brazing and Soldering, Abington, paper 23.

Sheward, G. E. and Baron, M. (1983) An in-place repair method for a sodium-steam heat exchanger, *Welding Journal*, **62**, 39–48.

van der Sluis, H. H. and Schellekens, H. (1991) Wide gap brazing of ceramic to metal joints. In 6th International Conference on *High Technology Joining*, British Association for Brazing and Soldering, Abington, paper 12.

Spanswick, O. S. and Nicholas, E. D. (1987) Diffusion bonding of injector blocks. In R. Pearce (ed.), *Diffusion Bonding*, Cranfield Institute of Technology, Bedfordshire, UK, 245–53.

Suga, T. (1989) Current research and future outlook in Japan. In S. D. Peteves (ed.), *Designing Interfaces for Technological Applications*, Elsevier Applied Science, London, 247–65.

Taylor, A. K. and Hayden, O. (1979) High integrity steam generator tube/tube plate joints for the LMFBR, *Welding Journal*, **58**, 31–8.

Yamada, T., Yokoi, K., Kohno, A. and Kawamoto, H. (1989) Manufacturing of sodium-sulfur cell by diffusion bonding. In W. Kraft (ed.), *Joining Ceramics, Glass and Metal*, Deutsche Gesselschaft für Metalikunde, Oberursel, 147–55.

Yiasemides, G. P. (1987) Diffusion bonding using hot isostatic pressing. In R. Pearce (ed.), *Diffusion Bonding*, Cranfield Institute of Technology, Bedfordshire, UK, 259–69.

Appendix A Safety

An assessment of safety implications must be the first stage in the evaluating the potential suitability and usefulness of any joining process, with the bases for this evaluation being consideration for others and common sense. Advice should be sought from competent authorities such as the (UK) Health and Safety Executive or the (US) Department of Health, Education and Welfare or non-governmental experts before introducing any new process or changing an existing one. Some government bodies promulgate legally binding regulations, but merely satisfying statutory minimum requirements is seldom sufficient for a responsible attitude towards safety.

When considering safety, the question that needs to be asked about each and every proposed application of a joining is whether **I** would be happy to be the operative on the workshop floor. Clearly it is feckless to be satisfied about implementing a procedure unless the potential hazards involved in its use are defined and clear and proven actions have been taken to avoid the conditions in which hazards can occur. Having complied with statutory regulations, prudence dictates that the realities of the workshop should be considered to identify by inspection and consultation what additional effective steps can be taken.

Joining processes such as brazing or diffusion bonding always involve the use of high temperatures. Thus it should be standard practice to use tongs when removing component assemblies after bonding because they could be hot and for operatives to wear protective clothing when they are knowingly required to handle or be near hot material. The wearing of thermally protective aprons, gloves and goggles is particularly important when using a gas torch to braze, and the flame must always be directed away from the operator and anyone else who approaches. Protective clothing and face visors should also be worn when brazing involves the use of flux or metal baths.

Other potential hazards that are frequently an inherent part of joining processes include the use of high-voltage equipment as when heating by electron beams or induction coils and of dangerous gases such as H_2, CO or C_2H_2 when using controlled atmosphere furnaces or gas torches. Avoiding or at least minimizing the dangers created by these and other conditions is a matter of common humanity and of economic sense since any safety incident can lessen credibility in the eyes of customers and will at the very least cause disruption of the production process. Thus, high-voltage equipment requiring connection, disconnection or modifications should be handled only by properly trained and qualified electricians. Further particular care should be taken to ensure that no one touches cold but live induction heating coils, nor should watches, bangles or other metallic objects be worn by their operators. Similarly, reducing gases containing CO or H_2 or CH_4 used in controlled atmosphere furnaces should be vented and should usually be burnt safely as a jet where they emerge from the furnace to avoid the possibility of explosion or poisoning. Many of the dangers involved with the use of C_2H_2 and other combustible gases in brazing torches are manifest but care needs to be taken also to ensure the safe handling and storage of bottled gas regardless of whether they are combustible or inert since the internal pressures can be very high.

Particular care needs to be taken when brazing in the open workshop to avoid contact with fumes generated by the braze alloys or fluxes. Some of these can cause serious medical hazards so that, for example, the maximum level of ZnO fumes that can be tolerated during an 8 hour shift is only $5\,mg.m^{-3}$, while that for fumes generated by Cu is $0.2\,mg.m^{-3}$ and for Cd is only $0.05\,mg.m^{-3}$. Indeed so much concern is felt about the possibility of permanent respiratory damage being caused by fumes that can be emitted by Ag-Cu braze alloys containing Cd as a melting temperature depressant that the use of any braze alloy containing Cd is forbidden by some countries and avoided by some industrial companies. Similarly, the halide constituents of braze fluxes can be the source of harmful fumes containing, for example, HF or BF_3 for which tolerance levels are low. Minimizing the dangers associated with the inhalation of fumes requires that manual brazing should be done only in well-ventilated areas, usually areas better ventilated than a normal workshop and preferably in designated booths fitted with exhaust hoods. If these conditions cannot be achieved when brazing needs to be done on site, the operators should wear personal respirators.

Similar care needs to be taken not to inhale vapours of organic solvents or to come into contact with alkaline or acid pickling solutions. Again, contact with brazing flux powders can cause skin irritation, although this can be mitigated by prompt and thorough washing. A few people are sensitive to contact with Ni alloys and everybody can be endangered by unthinking use of Cu alloys containing Be if this results in ingestion. Similarly, extreme care must be exercised joining Be itself or some other special metals such as U or Th.

Quite apart from care taken over the handling of particular materials or types of equipment, safety can be degraded if attention is not paid to housekeeping. Sufficient space should be allowed for the safe operation of equipment and storage of its associated materials and tools. Inflammable liquids and gases should not be stored in the immediate vicinity of hot equipment. Tools and other small objects should not clutter the workspace but those that are necessary for the job should be accessibly located on racks or in bins or lockers. Lighting as well as ventilation should be good. While great scientific advances have been made in laboratories that presented vistas of stygian chaos, the reliable and safe execution of production runs requires order and system.

Having identified safe working practices, training in their implementation needs to be given to staff and there should be prominent display of notices giving safety information as well as exhortations. Finally, first aid facilities and trained personnel should be available in case all else fails.

Appendix B Phase diagrams

Phase diagrams contain information needed for the understanding of microstructures and hence of many of the properties of metal alloys and of metal–ceramic combinations are of interest to the joining technologist. A multitude of phase diagrams have been published for binary metal–metal systems such as the Ag-Cu or Al-Si brazes and the numbers of those available for more complex systems is increasing. The phase structures of component materials that are metal alloys or mixtures of ceramic compounds can also have profound effects on their characteristics. Thus the ability to use and interpret phase diagrams is an important tool for those concerned with joining processes.

The term 'phase' describes a homogeneous and physically distinct part of a system which is separated from other parts by a definite bonding surface. Thus H_2O can exist in three distinct phases: as solid ice, as liquid water or as gaseous water vapour. More than one phase can be stable in some conditions, thus both water and water vapour co-exist at temperatures of 0 to 100 °C and at 0 °C ice, water and water vapour can all co-exist. Further, the introduction of salt or an anti-freezing agent will decrease the temperature at which water ceases to be stable, while the pressure exerted by a skate causes localized melting of ice. Thus the stabilities of water and ice depend on temperature, purity and pressure and a phase diagram is merely a convenient means of illustrating the effects of such variables on phase stability. In practice, the vast majority of phase diagrams present the effects of only composition and temperature on phase stability. The effects of pressure are seldom presented except for gas–solid systems and presentation of the effects of other variables such as magnetic field are extremely rare.

It is thermodynamic stability rather than, for example, mechanical stability that is represented in a phase diagram and hence the form of the diagram reflects the values of thermodynamic data and operation of thermodynamic laws. Thus the conditions for the coexistence of ice, water and water vapour are more restricted than those for water and water

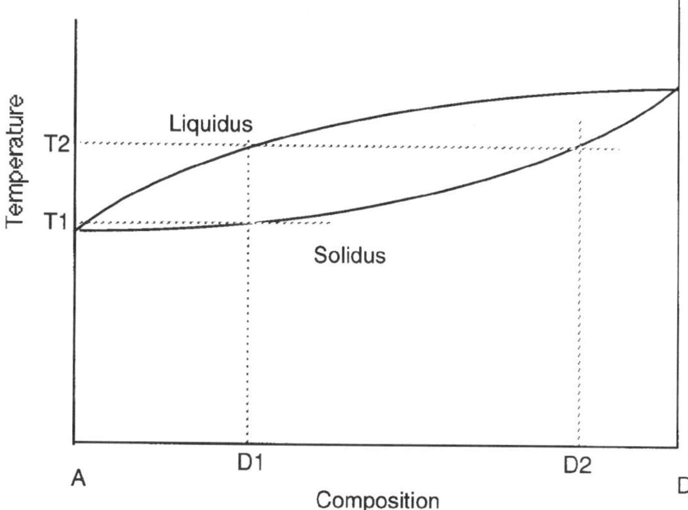

Figure B1 Phase diagram of a hypothetical system A-D showing the separation of liquidus and solidus temperatures

vapour, in accord with the phase rule enunciated by Gibbs (1961). Further, the chemical activities of the constituents of coexisting phases must be equal in equilibrium conditions, otherwise there would be a driving force for change. However the chemical activity of a certain concentration of solute is usually affected by whether the solvent is liquid or solid so that different concentrations are needed to achieve equilibrium at a liquid–solid interface. A consequence of this difference in compositions is that an alloy has different temperatures when it is completely liquid and completely solid, the liquidus and solidus, as illustrated in Figure B1. Thus the material with a composition of D1 has a liquidus temperature of T2 and a solidus of T1, and liquid D1 at temperature T2 is in equilibrium with a solid of composition D2.

One measure of the chemical activity of an alloy constituent is the ratio of its vapour pressure above the alloy as compared to that above the pure solute. If solute and solvent are completely soluble in the liquid state and completely insoluble when solid, an effect of introducing the solute will be to dilute the solvent concentration and hence diminish its vapour pressure. Vapour pressures also decrease as temperatures are lowered and hence the diluted solvent will be in equilibrium with that of pure solvent at a lower temperature. Thus the liquidus temperatures of both the constituents of a simple system such as that for G-J illustrated in Figure B2 will decrease as the alloying concentrations increase. At the eutectic composition and temperature, marked E in the figure, the two liquidus curves

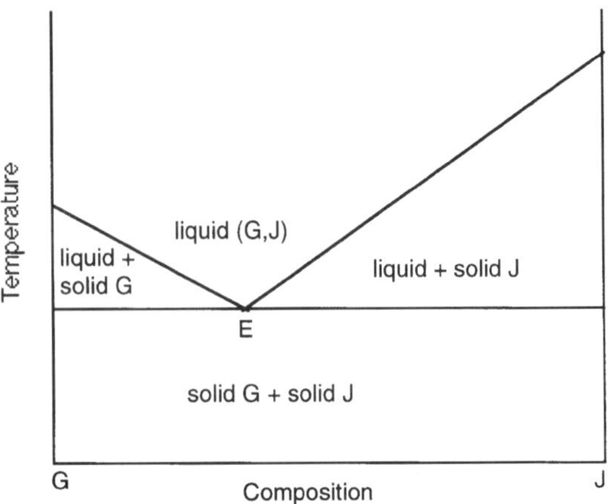

Figure B2 Phase diagram of the hypothetical system G-J showing phase constitutions and the location of the eutectic as E

cross. Thus cooling a liquid with a eutectic composition causes it to transform directly into two solids without passing through a two phase liquid–solid region. Cooling other liquid compositions will cause rejection of the solvent once the temperature falls below that of the liquidus and this rejection will continue until the eutectic temperature is reached, at which time the remaining liquid has a eutectic composition and transforms directly into the two solids. By analogy, a direct change on cooling from one solid composition to two different solids is called a eutectoid while a change on cooling from a liquid and solid to a single different solid is called a peritectic and from two solids to another single solid is called a peritectoid.

In practice, few systems are completely insoluble in the solid state, but nevertheless similar behaviour is exhibited by many systems of technical importance, and most braze alloys are based on eutectic compositions. The same is true for some component materials. Thus the phase diagram for the Fe-C system shown in the main text as Figure 4.1 shows that solid Fe can accommodate a significant concentration of C as a solid solution, but also exhibits a eutectic that produces a mixture of γFe and graphite or Fe_3C and a eutectoid with γFe transforming on cooling to a mixture of αFe and graphite or Fe_3C. The formation of compounds such as Fe_3C is of great technical importance because they can influence the mechanical and other properties of the system. Fe_3C is a stoichiometric, or line, compound in which the ratio of the atoms of its constituents does not vary significantly from (f/q) where f and q are low value integers. Other compounds can have compositions that vary significantly without causing the phase to become unstable, and of particular concern in this book is the

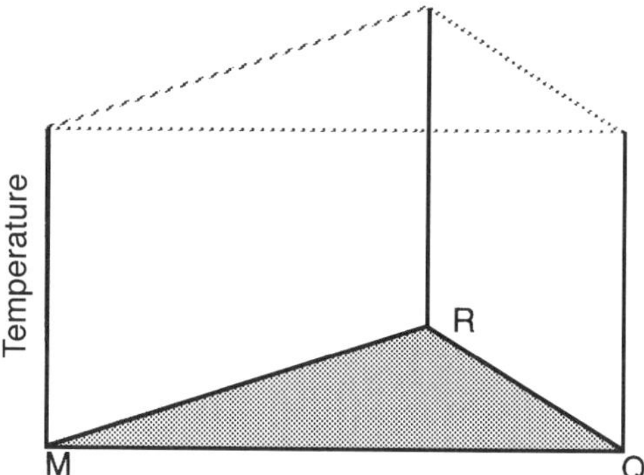

Figure B3 Framework for the presentation phase stability data as a function of temperature for the ternary system M-R-Q

existence of ceramic hypostoichiometric compounds in which the proportion of the non-metal present is significantly less than that in a simple stoichiometric structure. Thus the wetting of oxide ceramics by active metal brazes containing Ti is promoted by the formation of reaction products such as TiC which has an actual formula that varies between $TiC_{0.96}$ and $TiC_{0.61}$ at a brazing temperature of 900 °C. TiC has a face centred lattice structure and it is significant that hypostoichiometric variants contain an excess of vacancies on C sites, making the material more metallic on aggregate.

Some multiple component systems contain more than one compound and hence produce more than one reaction product layer when their constituents interdiffuse. Thus the microstructure produced by interdiffusion of a hypothetical M-R system containing three intermetallic compounds could be

$$\text{Solid solution } R_{\underline{M}} \rightarrow R_2M \rightarrow RM \rightarrow RM_2 \rightarrow \text{Solid solution } M_{\underline{R}}$$

Systems of technical importance are seldom composed of two pure elements and hence recourse has to be made to phase diagrams for ternary and more complex systems to understand the sequence of reaction products produced by interdiffusion.

Presenting phase stability data as a function of temperature for a system containing three constituents can be done by using the three-dimensional framework shown in Figure B3. Placing the phase boundaries within such a framework can result in a structure of considerable complexity that is difficult to represent as a two-dimensional illustration. Hence data for ternary systems are often presented as isothermal sections such as that for the hypothetical M-R-Q system shown in Figure B4. This very simple system contains only four phases: an α solid solution of R and Q in M, a γ

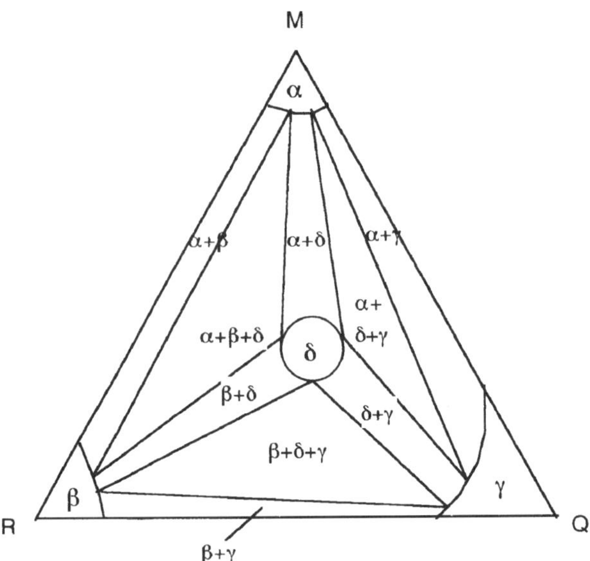

Figure B4 An isothermal section through the phase diagram of the hypothetical M-R-Q system showing the locations of single, two and three phase fields

solid solution of M and R in Q, a β solid solution of M and Q in R, and a δ phase that is a ternary MRQ compound. There are two phase regions linking the compound and each solid solution, and these act as the boundaries of three phase regions between the compound and pairs of the solid solutions. An alloy with a composition lying within one of the triangular three phase regions will have an equilibrium microstructure containing δ and two other phases and the microstructures developed by interdiffusion of, say, M and an R-Q composition can contain multiphase zones as well as reaction product layers of single-phase material. The essential characteristic of any microstructure developed by interdiffusion, however, is that there is no reversal in activity gradient, so that, for example, the activity of M must decrease progressively in the sequence of reaction products developed by its diffusion into a R-Q alloy.

The M-R-Q system is very simple in form compared to the Ni-Cr-Si system discussed in the main text and presented as Figure 7.3 but nevertheless even such complex diagrams can provide invaluable guidance. On occasion, ternary system data are presented in the form of a pseudo-binary composition-temperature diagram by fixing the ratio of, say, Q to R and then varying the M content of the system. If the liquidus surface of a ternary system is of particular interest, this can be presented as a series of isotherms on a M-R-Q plot to reveal thermal valleys and the locations of features such as eutectics.

Finally, it should be noted that the difficulties of formulating and interpreting graphical presentations for quarternary or quinternary phase diagrams are even greater and reference must be made to specialist descriptions if phase diagrams are needed for such complex systems.

FURTHER READING

Gibbs, J. W. (1961) *The Scientific Papers, Vol. 1*, Dover Publications Inc., New York.

Kubaschewski, O. and Evans, E. L. L. (1967) *Metallurgical Thermochemistry*, Pergamon Press, Oxford.

West, D. R. F. (1982) *Ternary Equilibrium Diagrams*, Chapman & Hall, London.

Appendix C Preparation of metallographic samples

Metallography is the study of the structure of materials by examining the appearance of polished and frequently etched cross-sections. As its name suggests this type of study was first applied to metals, but it is now applied with equal utility to ceramics. The preparation of a sample for metallographic study reveals its constitution as defined by the proportions, shapes and sizes of different phases that may be present, and of the agglomerations of crystals and grains, that can constitute an extensive phase. Such information can be of critical importance in understanding many physical, chemical and mechanical characteristics. Additionally, interactions between different materials can change the microstructure and hence properties of the interfaces. Some of this information can be derived from an examination of fracture surfaces, but in practice it is much more often obtained by examination of specially prepared cross-sections through components and bonded assemblies.

Such studies can be macroscopic, using magnifications of perhaps one to ten times to reveal the presence of coarse porosity and, sometimes, the flow of metal produced by mechanical working. The most common form of metallography, however, is the study of microscopic features. This was developed initially to exploit the resolving power of optical microscopes that permitted magnifications of up to 2000 times but has since progressed to include the use of electron microscopes and other beam instruments such as the electron probe microanalyser. The use of sophisticated instruments can greatly facilitate the interpretation of microstructures, but only if the cross-sections have been prepared with care.

Optimization of the preparation techniques for samples prior to metallographic study has been the subject of much study. Many excellent procedures have been devised for different metals and alloys and readers are referred to the detailed descriptions in the further reading list given at the end of this appendix. For now, all that will be attempted is a general introduction that is sufficient to appreciate the comments on microstructures given within the main text.

The preparation of a sample for metallographic study usually involves five sequential stages: cross-sectioning, mounting, grinding, polishing and etching. The purpose of sectioning is to reveal a representative plane through the sample that is free from possible near surface effects such as penetration of grain boundaries by oxides or, conversely, to reveal the depth and nature of such effects. Sectioning of metal samples can be accomplished by sawing but it is often preferable, and in the case of ceramic samples usually necessary, to slit with an abrasive wheel. When used correctly, such slitting has the advantage of producing a planar cross-section. The normal practice is then to encase the sample in a mount of organic material such as a phenolic resin (Bakelite), an epoxy resin (Araldite), or polymethylmethacrylate (Perspex). When the sample is of a very hard or wear material, this practice may be modified by using one of the normal compounds as the matrix of a composite containing wear-resistant particles.

Any cutting action distorts the material near the cut surface and this distortion can be as deep as 100 μm when a metal sample has been coarsely sawn or slit. Such distorted layer will make interpretation of the microstructure difficult and it needs to be removed. A common and useful first step towards removal is lightly to machine the cut surface using a sharp tool. This itself will create another distorted layer, albeit considerably thinner, and the purpose of the grinding stage is to create a flat surface with a minimal distorted layer by using paper pads coated with particles of Al_2O_3, SiC or other hard material. Removal of the distorted layer is achieved by using long strokes to rub the sample on a static paper pad or by firmly holding it on a similar rotating pad. Grinding is completed by using a series of pads coated with progressively finer grades of abrasive that create flat surfaces with thinner and thinner distorted layers. Effective use of any one pad removes all evidence of the cutting produced by the previous coarser abrasive in a minute or two and often after only a few seconds.

The papers used in the grinding process are lubricated by flowing water or paraffin to promote the cutting action and minimize pick up of the abrasive. It is good practice to wash and rinse or ultrasonically agitate the sample after each stage of grinding to remove loosely retained particles of abrasive. The particles themselves are angular to maximize their cutting action and pads are routinely discarded after a short time of use before their particles become rounded and the rate of metal removal slows. If this is not done the surface of the sample can become burnished, that is the abrasive loaded paper will deform the metal surface rather than remove it to produce a smooth surface, but of very distorted material.

Successful completion of the grinding stage produces a flat and relatively smooth and undistorted surface for a sample comprising only one phase. Similar satisfactory results of grinding can be achieved with samples containing more than one phase if particular care is taken to use very hard abrasives and firm pads that are well lubricated. Failure to do so can result in slight height differences at the phase interfaces, generally not as steps but as gradients. This effect is particularly severe when the phases are of markedly different hardnesses, as when a hard intermetallic reaction product layer is formed at a metal–metal interface or when the sample includes a ceramic–metal interface. The presence of such gradients then makes it difficult to bring the interface into focus when using an optical microscope at high magnifications. Sharp images of interfaces are generally easier to achieve using a scanning electron microscope because of its greater depth of focus, but the tilting of the surface near an interface can nonetheless complicate the interpretation of the backscattered electron images and the information derived from analysis of X-ray spectra, EDAX.

The final polish of a satisfactory ground surface to remove surface striations caused by the cutting action of the abrasive particles can be achieved using several techniques. The most commonly employed is mechanical polishing using a low nap cloth pad whose surface has been coated with a paste containing fine particles of Al_2O_3, MgO, Cr_2O_3 or diamond. Once more, progressively finer grades of abrasive are used so that the final pad may be loaded with diamond particles of no more than a micrometre in size that produce cutting tracks only a fraction of a micrometre deep.

For some metals and alloys, an alternative approach to the finishing of a sample is electrolytic polishing in which the sample is made the anode of a cell so that material is removed, and particularly so from high spots and ridges. A DC power unit is used to drive the process, supplying a current of a few amperes at a potential that can vary from less than 5 volts to more than 50 volts for different sample–electrolyte combinations. Polishing is usually effected within a minute or few minutes by close control of the voltage. The electrolyte solutions used depend on the chemistry of the sample metal or alloy but three that are commonly used are a 2:1 mixture of methanol and nitric acid, for polishing stainless steel, a 7:3 mixture of phosphoric acid and water for Cu alloys and a 3:7 mixture of perchloric acid and acetic anhydride for Ni alloys. While effective, extreme care has to be taken when using such electrolytes because the mixtures can be explosive, and that of perchloric acid and acetic anhydride can detonate spontaneously. To minimize the chances of this happening, the mixture must be kept cool and away from Bi and organic materials such as cellulose. More rarely, the final polishing stage can be achieved by exposure to acidic chemical mixtures without the imposition of a potential to induce electrolysis. The times taken to achieve a satisfactory polish vary from system to system and with process conditions such as the temperature but they generally fall within the range of 10 seconds to 10 minutes. Additionally, for some alloys it is convenient to complete the polishing process by applying a low potential to produce etching.

Examination of polished surfaces can provide valuable information about microstructure and microchemistry, but it is usually beneficial to include an additional stage of etching before examining metal samples. Etching employs aggressive chemicals, usually mixtures containing an acid, preferentially to attack certain phases or the boundaries between phases or grains. Thus etching facilitates identification of phase boundaries and sometimes of phase chemistries. The etches that can be used are many and varied since their compositions are tailored to achieve the optimum delineation and definition of phases and grains in specific alloy systems; examples of etchants used with metals and alloys of frequent interest are given in Table C1.

The preparation of ceramic samples is generally similar to that of metals but there can be some significant differences. Once more it is essential to

Table C1 Commonly used etchants for metal samples

Metal or alloy	*Etch*	*Comment*
Al alloys	Dilute $HF/HCl/HNO_3$ 15 seconds	Colours some phases
	Dilute NaOH swabbing	Colours some phases
Cu	$NH_4OH/H_2O_2/H_2O$	Reveals grain boundaries
Cu alloys	$FeCl_3/HCl/H_2O$	Colours some phases
	$CrO_3/HNO_3/H_2O$	Colours some phases
Steels	Ethanol/HNO_3	Reveals grain boundaries in ferritic steels
	Oxalic acid/H_2O	Reveals carbides and grain boundaries in stainless steels
	$KMnO_4/NaOH/H_2O$	Reveals carbides in tool steels
Ni and Ni alloys	HCl/HNO_3/glycerol	Reveals grain boundaries in Ni and Ni-Cr alloys
	HNO_3/acetic acid/acetone	For Ni and Ni-Cu alloys
	HCl-$FeCl_3$-H_2O	For pure Ni

ensure that the slitting and grinding stage produces a flat and relatively smooth surface and this is generally achieved using firm or even rigid pads loaded with very hard abrasive particles of, usually, diamond. Distortion of the sub-surface layer is not such a problem because of the hardness of ceramics, but their brittleness can result in chipping that may be mistaken for porosity. The polishing stage can result in steps or gradients being produced at phase and grain boundaries, and this can be desirable because of the difficulty in identifying etchants that can be used to delineate the grain microstructures of ceramics. Finally, the surfaces of ceramic cross-sections to be examined using scanning electron microscopy are usually coated with a very thin layer of an electrically conductive material such as Au or C to prevent charging and distortion of the images achieved by collection of the back scattered electrons.

FURTHER READING

American Society for Metals (1973) Metallography, structures and phase diagrams, *Metals Handbook*, **8**.

Greaves, R. H. and Wrighton, H. (1967) *Practical Microscopical Metallography*, 4th edn, Chapman & Hall, London.

Jacquet, P. A. (1956) Electrolytic and chemical polishing, *Metallurgical Reviews*, **1**, 157.

Tegart, W. J. McG. (1959) *The Electrolytic and Chemical Polishing of Metals in Research and Industry*, Pergamon Press, Oxford.

Appendix D Mechanical testing

Mechanical testing is a fundamental part of most material development and acceptance programmes and is written into many acceptance and quality control specifications. Comparable standardized testing procedures have been adopted in most industrialized countries so that values reported by assessors from different parts of the world can be used with a high level of confidence. Details of a range of tests can be found in national standards, but of particular relevance to the evaluation of the mechanical properties of brazed and diffusion bonded joints are the tensile, impact, fatigue, hardness and toughness tests. References to some relevant British standards are given at the end of this appendix.

D.1 TENSILE TESTS

The tensile test is in principle simple to conduct and to interpret. It involves extending the length of a rod or strip with a narrow central section, or gauge, until it fractures. The strip or rod held by the chucks of a tensile test machine, one of which is attached to the fixed plate of the machine and the other to a beam that is moved away at a steady rate to extend the test piece. An automatic record is made of the load required to achieve this movement and hence to extend the test piece. The extension rates used in these tests can vary but often lie within the range 0.1 to 1 mm/minute.

A sketch of a tensile test piece is shown in Figure D.1 along with some typical dimensions for both circular and rectangular samples. The particular tabulated summary is that of the British Standards Institution but other national standards, including the American, are very similar and the small differences have no major effect on the reported values of the mechanical properties. These property values are derived from the records of the tensile load and the test piece extension such as that shown in

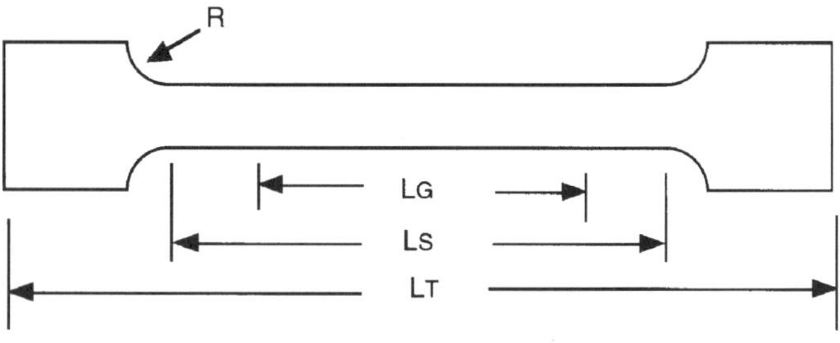

Typical test piece dimensions
Circular test pieces:
Gauge diameter 5.64 mm, LG 28 mm, LS 31 mm, R 6 mm
Rectagular test pieces:
Gauge width 6 mm, LG 25 mm, LS 28 mm, R 12 mm, LT 100 mm

Figure D1 The recommended configurations of tensile test pieces to be used for assessing the strength of a metal. The dimensions of the test pieces with circular or rectangular cross-sections are taken from BS 18:Part 2 (1987)

Figure D.2 which is schematically typical of those produced by ductile metals such as Al or Cu. The plot in the figure shows three distinct regions: initially the load has to be increased linearly to maintain a steady rate of extension, a middle region in which there is a progressive diminishment in the rate at which the load has to be increased to maintain a steady

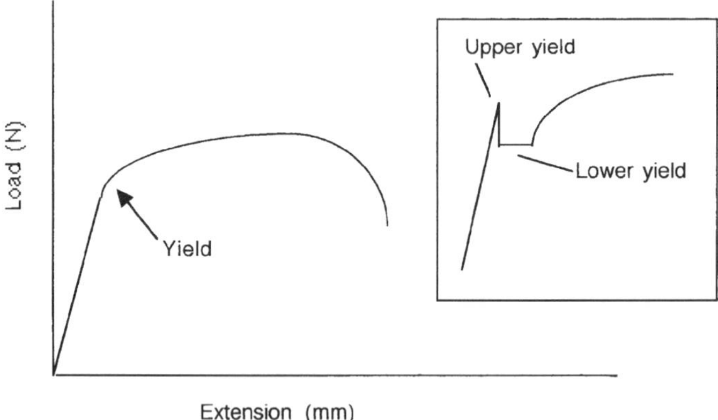

Figure D2 Sketch of a typical load-extension tensile test curve for a pure ductile metal. The insert illustrates the yielding behaviour of alloys such as mild steel

extension, and a final region in which the load decreases as the test piece continues to extend.

The first region of Figure D2 involves elastic extension, so that the test piece regains its original shape and size if the load is removed. The slope of this portion of the plot is the Elastic or Young's Modulus. This is derived from the ratio (Load/cross-sectional area)/(Length of the gauge section at that load/Original length of the gauge section) and has the dimensions of MPa. Permanent or plastic, deformation of the test piece occurs in the second region when the decrease in the slope of the load-extension plot in part reflects a significant decrease in the cross-sectional area of the gauge section of the test piece. The onset of this second stage is called yielding and occurs when the (Load/Cross-sectional area) exceeds a certain stress level, the yield stress, of so many MPa. In practice it can be difficult to identify precisely the load and extension at which yielding occurs and hence it is a common practice to define the yield, or proof, stress as that at which the extension was a certain small value such as 0.2%.

The maximum load that the sample withstands during a test divided by the original cross-sectional area of the gauge section is used to derive the Ultimate Tensile Strength, UTS, of so many MPa and this parameter is the most commonly quoted tensile test value. Applying such a maximum load causes the onset of 'necking', or localization of deformation and hence the reduction of the cross-sectional area of the test piece in a small portion of the gauge section. In fact although the load needed to continue extension of the test piece diminishes once necking commences, the stress or load divided by the actual cross-sectional area continues to increase in the necked portion.

Other parameters frequently derived from tensile tests include:

- the percentage extension of the gauge length at the moment that failure occurs, called the elongation to failure;
- the reduction in the cross-sectional area of the gauge length caused by both the generalized extension and the localized necking before the test piece failed, the reduction in area.

The yielding behaviour of some metal alloys such as mild steel is more complex, as shown by the insert in Figure D2, due to the influence of the C solute, but that of ceramics is generally simpler in that yielding is usually immediately followed by failure so the yield strength and UTS values for ceramics are effectively identical.

D.2 CREEP TESTS

It is necessary to consider mechanical property values at elevated temperatures to select suitable materials for some applications. Tensile tests can be used but the extension rates they use generally result in failure of the test

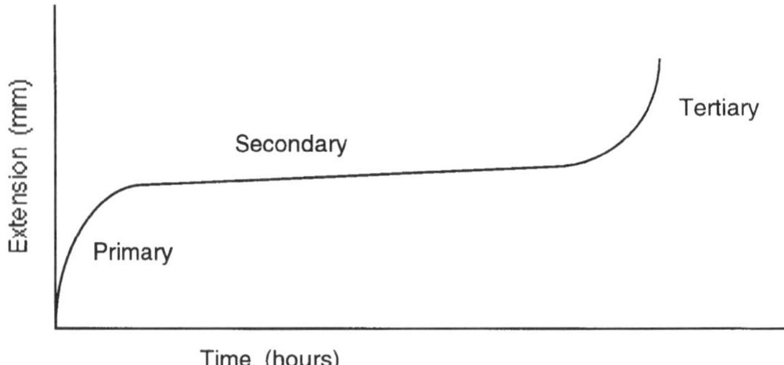

Figure D3 An extension vs. time curve produced using a constant load applied at a high temperature that illustrates the main stages of a creep test

piece within a few minutes and in practice quite low stresses can induce slow extension of components at high temperatures that can continue for very long times. This 'creep' behaviour can be characterized using samples similar to but generally smaller than those employed in tensile tests. Typically a creep test applies a constant load to extend the sample and lasts several thousands of hours. The resultant extension vs. time curve displays several stages as illustrated in Figure D3. The first stage of primary creep is essentially similar to the extension that occurs during a tensile test and often accounts for most of the extension that occurs during a test. The second stage displays markedly different behaviour since extension of the test piece progresses at a steady or linear rate and can do so for many thousands of hours. It is this linear rate that is usually quoted in compilations of creep data. The final, tertiary, stage of accelerating creep extension is usually of short duration and accompanied by the formation of cavities and rapid sliding of grains over one another. To gain an understanding of the creep resistance of a material it is necessary to consider the behaviour observed when different loads or stresses are applied at a range of temperatures.

The results of creep tests are usually reported in terms of the effect of temperature and stress on the linear extension rate displayed during the secondary stage, but other frequently quoted properties are the stress needed to cause rupture after a time such as 10 000 or 100 000 hours or to cause an extension of typically 1%.

D.3 FATIGUE TESTS

Components are often subjected to fluctuating stress levels when in service and the ability of materials to resist such conditions is measured by a fatigue test. In this test, a sample similar to those used in tensile

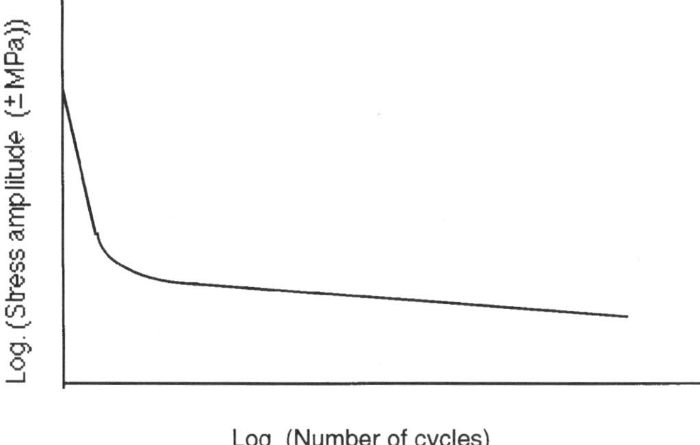

Figure D4 The effect of the stress amplitude on the number of cycles needed to cause failure during fatigue testing

tests is caused to fail while subjected to rapidly cycled stress levels that may have a positive or zero average value. The test sample frequently takes the form of a cantilever beam, one end of which is fitted in a rotating chuck while a weight is attached to the other. Thus the upper surface of the sample is in tension and each part of the surface experiences a complete stress cycle with every rotation. This repeated stressing can cause a microscopically small surface crack to form that grows slightly with each stress cycle until eventually it is large enough to cause failure. The number of stress cycles needed to induce such failure increases as their amplitude diminishes and hence fatigue life lifetimes can be established for differing stress levels as illustrated schematically in Figure D4. If the stress level used in a test is too low to nucleate a crack, the lifetime of the sample is infinite and the maximum stress at which this occurs is called the endurance limit.

D.4 IMPACT TESTS

The Izod and Charpy tests assess the energy required to break a notched sample subjected to a sharp blow. An impact sample has a circular or square cross-section that contains a U-, keyhole- or V-shaped notch and the energy required to cause fracture is derived from the decreased swing of a pendulum to which a striker is attached. In the Izod test, the sample is gripped so that the notch is in the same plane as the top surface of the grips and the striker hits the notched face at a fixed height of typically 22 mm above the notch. Standard test pieces either have a diameter of 11.4 mm or a side length of 10 mm and the Izod value is reported as so many J, with the energy of the striker at the moment of impact usually being 164 J. In the Charpy test, a square cross-section sample is gripped above and below the notch. The standard side length is 10 mm as for the Izod test but smaller samples are used where material is in short supply.

D.5 HARDNESS TESTS

Hardness tests employ machines that impress a rigid indenter into the surface of a sample for a fixed time and then use a microscope to measure the indentation size after the indenter has been removed. The indenters may be of a hardened steel or of a diamond and take the form of a hemisphere, cone, pyramid or rhomboid. Of particular relevance to this book are the Vickers and micro-hardness tests that use angular diamond-tipped indenters.

The Vickers test uses a right pyramid with a square base and the diagonal lengths of the impression produced by applying a load of 1 to 100 kgf to the indenter for typically 15 seconds is used to derive a Vickers Hardness Number, VHN, by substitution in the formula

$$\mathrm{VHN} = 1.854\,\mathrm{F_V}/\mathrm{D_{vi}^2} \tag{D1}$$

where F_V is the load in kilogramme force (1 kgf = 9.81 N) and D_{vi} is the diagonal length in mm. To characterize the hardness of a particular sample requires half a dozen or so well-spaced indentations be made to permit derivation of a statistically meaningful average value. When the material being tested is brittle, such as a ceramic, problems may be caused by chipping or cracking and hence the shapes as well as the sizes of indentations need to be examined before they can be regarded as of use in deriving hardness values.

Microhardness tests permit much more detailed assessment of hardness variations within a sample by employing loads of 1 to 1000 gf to produce small impressions which require careful measurement at high magnifications. Forming even a small indentation causes deformation of the adjacent material and hence changes its hardness, so care should be taken to separate indentations by a distance equal to at least five times their diagonal length when, for example, using the technique to derive a hardness gradient that may have been caused by interdiffusion.

The indenter may be a right pyramid as in the Vickers hardness test or an elongated rhombic pyramid as in the Knoop test. Hardness values are derived either from

$$\mathrm{VHN} = 1854.4\,\mathrm{f_v}/\mathrm{d_v^2} \tag{D2}$$

when using a Vickers indenter where f_v is the indenting force in gf and d_v is the average length of the diagonals in μm, or from

$$KHN = 14229\, f_k/d_k^2 \qquad \{D3\}$$

when using a Knoop indenter where f_k is the indenting force in gf and d_k is the length of the long diagonal of the indentation in μm.

The actual values derived from these micro-hardness tests vary for a given material with the load that is employed because of the differing importance of the elastic recovery in changing the shape of the permanent impression, so that in general hardnesses derived using the smaller loads are higher.

D.6 FRACTURE TOUGHNESS AND BEND STRENGTH TESTS

A material can be inherently strong but a component still fails at a low external stress because it contains cracks that propagate readily. It is not tough.

A measurement of toughness was for many years an elusive target but there now exist standard tests that provide reproducible measures by determining the load needed to fracture a notched and previously cracked sample. This load can be applied to a notched beam either to bend it or extend it in tension. The use of only the bend test will be outlined here but both types are described in standards specifications.

The beam sample is relatively small and has a rectangular cross-section, of typically 3 mm × 6 mm, that is notched in the centre of its length and has been fatigued to create a short crack at the foot of the notch, as illustrated in Figure D5. The beam is supported on the notched face at two points symmetrically located about its centre and a load is applied to the opposite face. This load is applied at the centre of the beam or at two locations closely and symmetrically located about the centre. These two variants are described as the 3-point and the 4-point bend test.

The load is increased progressively to a value of F_L at which failure occurs and the toughness of the sample, specifically its plane strain fracture toughness, K_{Ic} can be calculated from the expression

$$K_{Ic} = Y.F_L.e.a^{1/2}/(b.w'^2) \qquad \{D4\}$$

where Y is a dimensionless correction factor that depends on (a/w′), b is the breadth of the sample, e is the distance between the load application points or point and the support beams and K_{Ic} is in $MPa.m^{1/2}$. Inherent in the use of equation {D4} to derive a K_{Ic} value in accord with standards specification are the assumptions that both a and b are greater than $2.5(K_{Ic}/\sigma_Y)^2$ where σ_Y is the 0.2% yield strength of the material.

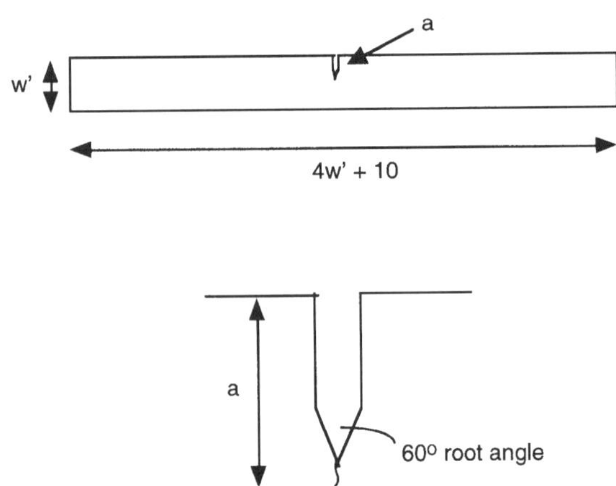

Sample thickness = b = 0.5w'
Effective crack length = a =0.45 to 0.55w'
Dimensions in mm.

Figure D5 Standard bend test piece showing the relationship between the dimensions

The same 3-point or 4-point tests can be used to derive a bend, or more correctly flexure, strength value for a notched rectangular sample by substitution into the expression

$$BS = 3F_L.e/(b.w'^2)$$

where BS is the bend strength and is normally expressed as so many MPa.

FURTHER READING

BS 18:Part 2 (1987) Tensile testing of metallic materials.
BS 131 :Part 1 (1961) Notched bar test: Izod impact tests for metals.
BS 427:Part 1 (1961) Hardness testing of metallic materials.
BS 5447 (1977) Fracture mechanics toughness tests.

Index